INNOVATIVE METHODS OF MARINE ECOSYSTEM RESTORATION

Edited by
THOMAS J. GOREAU
ROBERT KENT TRENCH

CRC Press
Taylor & Francis Group
Boca Raton London New York

CRC Press is an imprint of the
Taylor & Francis Group, an **informa** business

Front Cover: Biorock coral reef at Karang Lestari coral reef restoration project, Pemuteran, Bali. Project designed and installed by Wolf Hilbertz, 2005. Note how the fish populations organize themselves at various levels. Photo copyright EunJae Im http://ej.uwimg.net/gallery-list. Used with permission. All rights reserved.

CRC Press
Taylor & Francis Group
6000 Broken Sound Parkway NW, Suite 300
Boca Raton, FL 33487-2742

© 2013 by Taylor & Francis Group, LLC
CRC Press is an imprint of Taylor & Francis Group, an Informa business

No claim to original U.S. Government works

Printed in the United States of America on acid-free paper
Version Date: 20121015

International Standard Book Number: 978-1-4665-5773-4 (Hardback)

This book contains information obtained from authentic and highly regarded sources. Reasonable efforts have been made to publish reliable data and information, but the author and publisher cannot assume responsibility for the validity of all materials or the consequences of their use. The authors and publishers have attempted to trace the copyright holders of all material reproduced in this publication and apologize to copyright holders if permission to publish in this form has not been obtained. If any copyright material has not been acknowledged please write and let us know so we may rectify in any future reprint.

Except as permitted under U.S. Copyright Law, no part of this book may be reprinted, reproduced, transmitted, or utilized in any form by any electronic, mechanical, or other means, now known or hereafter invented, including photocopying, microfilming, and recording, or in any information storage or retrieval system, without written permission from the publishers.

For permission to photocopy or use material electronically from this work, please access www.copyright.com (http://www.copyright.com/) or contact the Copyright Clearance Center, Inc. (CCC), 222 Rosewood Drive, Danvers, MA 01923, 978-750-8400. CCC is a not-for-profit organization that provides licenses and registration for a variety of users. For organizations that have been granted a photocopy license by the CCC, a separate system of payment has been arranged.

Trademark Notice: Product or corporate names may be trademarks or registered trademarks, and are used only for identification and explanation without intent to infringe.

Library of Congress Cataloging-in-Publication Data

Innovative methods of marine ecosystem restoration / editors, Thomas J. Goreau and Robert Kent Trench.
   p. cm.
Includes bibliographical references and index.
ISBN 978-1-4665-5773-4 (hardback)
1. Marine ecosystem management. 2. Restoration ecology. I. Goreau, Thomas J. II. Trench, Robert Kent.

QH541.5.S3I537 2012
333.95'6--dc23
                                       2012023188

Visit the Taylor & Francis Web site at
http://www.taylorandfrancis.com

and the CRC Press Web site at
http://www.crcpress.com

Contents

Foreword vii
Acknowledgments ix
CD Contents xi
Contributors xiii

Chapter 1
Dedication to Wolf Hilbertz 1

Thomas J. Goreau

Chapter 2
Innovative Methods of Marine Ecosystem Restoration: An Introduction 5

Thomas J. Goreau

Chapter 3
Restoring Reefs to Grow Back Beaches and Protect Coasts from
Erosion and Global Sea-Level Rise 11

Thomas J. Goreau, Wolf Hilbertz, Abdul Azeez Abdul Hakeem, Thomas Sarkisian, Frank Gutzeit, and Ari Spenhoff

Chapter 4
Reef Restoration Using Seawater Electrolysis in Jamaica 35

Thomas J. Goreau and Wolf Hilbertz

Chapter 5
Electrically Stimulated Corals in Indonesia Reef Restoration Projects
Show Greatly Accelerated Growth Rates 47

Jamaludin Jompa, Suharto, Eka Marlina Anpusyahnur, Putra Nyoman Dwjja, Jobnico Subagio, Ilham Alimin, Rosihan Anwar, Syarif Syamsuddin, Thri Heni Utami Radiman, Heri Triyono, R. Ahmad Sue, and Nyoman Soeyasa

Chapter 6
Biorock Reef Restoration in Gili Trawangan, North Lombok,
Indonesia 59

Lalu Arifin Aria Bakti, Arben Virgota, Luh Putu Ayu Damayanti, Thri Heni Utami Radiman, Ambar Retnowulan, Hernawati, Abdus Sabil, and Delphine Robbe

Chapter 7
Electrical Current Stimulates Coral Branching and Growth
in Jakarta Bay 81

Neviaty P. Zamani, Khalid I. Abdallah, and Beginer Subhan

Chapter 8
Electricity Protects Coral from Overgrowth by an Encrusting Sponge in Indonesia 91

Jens Nitzsche

Chapter 9
Gorgonian Soft Corals Have Higher Growth and Survival in Electrical Fields 105

Diannisa Fitri and M. Aspari Rachman

Chapter 10
Suitability of Mineral Accretion as a Rehabilitation Method for Cold-Water Coral Reefs 113

Susanna M. Strömberg, Tomas Lundälv, and Thomas J. Goreau

Chapter 11
Utilization of Low-Voltage Electricity to Stimulate Cultivation of Pearl Oysters *Pinctada maxima* (Jameson) 131

Prawita Tasya Karissa, Sukardi, Susilo Budi Priyono, N. Gustaf F. Mamangkey, and Joseph James Uel Taylor

Chapter 12
Increased Oyster Growth and Survival Using Biorock Technology 141

Nikola Berger, Mara G. Haseltine, Joel T. Boehm, and Thomas J. Goreau

Chapter 13
Electrical Stimulation Increases Oyster Growth and Survival in Restoration Projects 151

Jason Shorr, James Cervino, Carmen Lin, Rand Weeks, and Thomas J. Goreau

Chapter 14
Restoration of Seagrass Mats (*Posidonia oceanica*) with Electrical Stimulation 161

Raffaele Vaccarella and Thomas J. Goreau

Chapter 15
Electrical Fields Increase Salt Marsh Survival and Growth and Speed Restoration in Adverse Conditions 169

James Cervino, Dajana Gjoza, Carmen Lin, Rand Weeks, and Thomas J. Goreau

Chapter 16
Postlarval Fish Capture and Culture for Restoring Fisheries 179

Gilles Lecaillon

CONTENTS

Chapter 17
Mariculture Potential of *Gracilaria* Species [Rhodophyta] in Jamaican Nitrate-Enriched Back-Reef Habitats: Growth, Nutrient Uptake, and Elemental Composition 189

Arlen Havenner Macfarlane

Chapter 18
Sustainable Reef Design to Optimize Habitat Restoration 245

Mara G. Haseltine

Chapter 19
Marine Ecosystem Electrotherapy: Practice and Theory 263

Thomas J. Goreau

Index 291

Foreword

INNOVATIVE METHODS OF MARINE ECOSYSTEM RESTORATION

It's such a simple idea. All life generates electricity. Can electricity help regenerate life? The answer is simple, too. Yes.

Chapters in this book provide experimental proof that mild electrical stimulation results in increased settlement, increased growth, and reduced mortality for a wide variety of marine organisms, including corals, oysters, sponges, seagrasses, and salt marsh grasses. These benefits are dramatic and they are important. The world is a dynamic place. Wind and waves, earthquakes and tsunamis mean that all structures require maintenance. In the natural world, maintenance is done by the inhabitants of the structure. Inhabitants on the reef, build the reef. Form and function are seamless in the natural world. By contrast, humans struggle to maintain their "crumbling infrastructure." Natural edifices however are self-replicating, self-repairing, and self-sustaining. The natural world glories in its self-reliance.

As studies in this book show, the secondary benefits of electrical stimulation are also impressive. These include reducing harmful bacteria in seawater, improving water quality, and reducing the number of beach closures. There is also an indication that these techniques can enhance local biodiversity and improve the survival of endangered species. Results also include enhancing ecosystem resilience, improving ecosystem function, and improving restoration outcomes. In sum, these benefits improve ecosystem services on which we and our livelihoods depend.

At present, more than half of the world's population lives within 60 km of the ocean. This is expected to rise to 75% by 2020. Food, drugs, and coastal protection from the sea are just a few of the ecosystem services provided by marine organisms. As human-induced climate change raises temperatures and sea levels across the globe, these ecosystem services are in decline. We need to get these services back. We need to get them back now. This book identifies simple, cost-effective ways to achieve this goal.

The specifics of coastal decline are daunting. Ninety percent of beaches around the continental United States are receding. Louisiana's coastline is being lost at an average rate of 50 feet per year. Some of America's best-known and most-visited beaches may have to be closed permanently due to high rates of erosion. This fate is being discussed for Folly Beach, South Carolina; Ocean Beach, California; and Robert Moses State Park, New York. Along the world-famous Atlantic City Boardwalk, the beach is being "fortified" with 1.3 million cubic yards of sand, spread along just 5 miles of coastline, and at a cost to taxpayers of $18 million in 2012 alone. It is widely accepted that without these expensive and ongoing "beach renourishment" efforts, there would be no recreational beaches in front of Florida's Miami Beach, South Beach, or Fort Lauderdale. These massive investments of time, money, and technology buy us few lasting solutions, and certainly no permanence. It is little more than spitting in the wind. This book shows that electrical currents, applied in systematic and appropriate ways, can enhance the growth of the same plant and animal species that provide lasting shoreline protection for the humans who live behind salt marshes and reefs just offshore. These solutions accrue by enhancing natural processes, not by resisting them.

Three aspects of this book will strike the reader. The diversity of ecosystems covered, the diversity of geographic regions represented, and the diversity of international contributors to this book make it the first truly global study of marine environmental restoration. This book covers coral reefs, oyster reefs, deep reefs, salt marshes, and seagrass meadows. The ecosystems studied are both temperate and tropical, and from the New World and the Old World. Befitting this diversity of subject matter and locale, the book's contributors likewise come from across the globe, from Albania, Australia, Belize, Canada, China, France, Germany, Indonesia, Italy, Jamaica, the Maldives, Panama, Sudan, Sweden, Thailand, and the United States.

This book highlights the innate restorative power of nature. It deflects the discussion away from debates over dubious "geoengineering" proposals, and instead focuses on learning how to enhance natural processes rather than to impose unnatural controls. It shows that survival of the natural world, and the ecosystem services that nature provides us, will follow only if we learn to work with nature to enhance its preexisting assets. It makes the point that most marine engineering, whether it is on levees in New Orleans before the arrival of Hurricane Katrina or on beach renourishment before sea-level rise, is nothing but hubris and human folly.

Humankind has passed its environmental Rubicon. We, and the rest of life with whom we share this planet, will have to survive in the future with an ocean that is hotter, higher, and more acidic than at any time in the recent evolutionary past. The ideas presented in this book will buy us time. They highlight the possible, and empower us with simple, cost-effective, and sustainable solutions for universal problems. They encourage us to reach into the only toolkit we have for broad-scale, long-lasting solutions. They exhort us to harness the restorative power of nature. A central tenet of life is growth. This book teaches us to promote it. When (if) humans look back in 1000 years at this disastrous period of human environmental modification (a geological epoch now being called the Anthropocene), it is likely that one of the few good ideas to emerge will be to tap into and promote natural processes. This is the unabashed thesis of this book.

This book promotes unconventional solutions to the largest challenges facing humankind. It is also interesting to highlight the fact that these unconventional ideas are coming from unconventional sources. Rather than emerging from tradition-laden technical schools and engineering colleges, these solutions emerge from classically trained biologists who seek solutions by observing the inner workings of the organisms and ecosystems affected. The beauty and power of this approach are that it exploits more than a billion years of evolutionary success. A billion years of natural selection and adaptation are more instructive on how to survive on planet Earth than a lifetime of study on marine engineering or coastal zone management.

The survival of both humankind and the natural world are now inexorably intertwined. *Innovative Methods of Marine Ecosystem Restoration* presents an elegant "how-to" manual dedicated to meeting the needs of both. The stakes are high. This book proposes how we can make the natural world our natural ally.

James W. Porter, PhD
Josiah Meigs Professor of Ecology and Marine Sciences
University of Georgia, Athens, Georgia

Acknowledgments

We would like to thank John Sulzycki, Dennis Troutman (diacriTech), Joette Lynch, Kathryn Everett and the entire production team at CRC Press for their exhaustive checking and constructive suggestions. Their eagle eyes and dedicated work resulted in many clarifications and improved this book in countless small ways for which we are very grateful.

CD Contents

1. Color versions of all photographs and graphs in the book.
2. PowerPoint presentations from 2011 World Conference on Ecological Restoration.
    a. Goreau et al. Shore Protection
    b. Jompa et al. Coral Restoration in Indonesia
    c. Alimin et al. Coral Restoration at Gili Trawangan
    d. Arifin et al. Coral Reef Restoration at Gili Trawangan
    e. Abdallah et al. Coral Branching
    f. Beddoe et al. Coral Restoration in Tobago
    g. Stromberg et al. Deep Sea Coral Restoration
    h. Nitzsche Coral-Sponge Interactions
    i. Karissa et al. Pearl Oyster Restoration
    j. Berger et al. Oyster Restoration in Tanks
    k. Shorr et al. Oyster Restoration in the Field
    l. Vaccarella et al. and Cervino et al. Seagrass and Salt Marsh Restoration
    m. Lecaillon Fisheries Restoration
    n. Goreau Electrotherapy
3. Yayasan Karang Lestari (Protected Coral Foundation in Indonesian) Presentation for the 2012 United Nations Development Programme Equator Award for Community-Based Development for Coral Reef and Fisheries Restoration Projects in Pemuteran, Bali, Indonesia, Awarded in Rio de Janeiro Brazil, June 2012.
4. Video of 2-year-old and 6-month-old Biorock Reef Restoration Projects at Pulau Gangga, Sulawesi, Indonesia, by Tom Goreau, June 9, 2008.

Contributors

Khalid I. Abdallah
Red Sea University
Port Sudan, Sudan

and

Bogor Agricultural University
Bogor, Java, Indonesia

Ilham Alimin
Jakarta Fisheries University
Ministry of Marine Affairs and Fisheries
Jakarta, Java, Indonesia

Eka Marlina Anpusyahnur
Hasanuddin University
Makassar, Sulawesi, Indonesia

Rosihan Anwar
Jakarta Fisheries University
Ministry of Marine Affairs and Fisheries
Jakarta, Java, Indonesia

Lalu Arifin Aria Bakti
Mataram University
Mataram, Lombok, Indonesia

Nikola Berger
City University of New York
New York, New York

Joel T. Boehm
The River Project
and
Queens College
City University of New York
New York, New York

James Cervino
Coastal Preservation Network
Queens, New York

and

Woods Hole Oceanographic Institution
Woods Hole, Massachusetts

Luh Putu Ayu Damayanti
Mataram University
Mataram, Lombok, Indonesia

Putra Nyoman Dwjja
Udayana University
Bukit, Bali, Indonesia

Diannisa Fitri
Hasanuddin University
Makassar, Sulawesi, Indonesia

Dajana Gjoza
Francis Lewis High School
Queens, New York

Thomas J. Goreau
Global Coral Reef Alliance
Cambridge, Massachusetts

and

Yayasan Karang Lestari
Pemuteran, Bali, Indonesia

and

Gili Eco Trust
Gili Trawangan, Lombok, Indonesia

and

Discovery Bay Marine Laboratory
Discovery Bay, Jamaica

Frank Gutzeit
Sun & Sea e.V.
Hamburg, Germany

Abdul Azeez Abdul Hakeem
Former Maldives Director of Agriculture
Malé, the Maldives

Mara G. Haseltine
New School for Social Research
New York, New York

and

G.A.I.A.: Artist Studio
Brooklyn, New York

Hernawati
Mataram University
Mataram, Lombok, Indonesia

Wolf Hilbertz (deceased)
Sun & Sea e.V.
Hamburg, Germany

Jamaludin Jompa
Hasanuddin University
Makassar, Sulawesi, Indonesia

Prawita Tasya Karissa
Gadjah Mada University
Jogjakarta, Java, Indonesia

Gilles Lecaillon
Ecocean
Montpellier, France

Carmen Lin
Francis Lewis High School
Queens, New York

Tomas Lundälv
University of Gothenburg
Strömstad, Sweden

Arlen Havenner Macfarlane
Global Coral Reef Alliance
Sherbrooke, Quebec, Canada

N. Gustaf F. Mamangkey
Sam Ratulangi University
Manado, Sulawesi, Indonesia

Jens Nitzsche
Eberswalde University of Applied Science for Sustainable Development
Eberswalde, Germany

Susilo Budi Priyono
Gadjah Mada University
Jogjakarta, Java, Indonesia

M. Aspari Rachman
Hasanuddin University
Makassar, Sulawesi, Indonesia

Thri Heni Utami Radiman
Jakarta Fisheries University
Ministry of Marine Affairs and Fisheries
Jakarta, Java, Indonesia

Ambar Retnowulan
Mataram University
Mataram, Lombok, Indonesia

Delphine Robbe
Gili Eco Trust
Gili Trawangan, Lombok, Indonesia

Abdus Sabil
Mataram University
Mataram, Lombok, Indonesia

Thomas Sarkisian
Biorock Thailand
Bangkok, Thailand

Jason Shorr
Pace University
New York, New York

Nyoman Soeyasa
Jakarta Fisheries University
Ministry of Marine Affairs and Fisheries
Jakarta, Java, Indonesia

Ari Spenhoff
Sun & Sea e.V.
Hamburg, Germany

Susanna M. Strömberg
University of Gothenburg
Strömstad, Sweden

Jobnico Subagio
Udayana University
Bukit, Bali, Indonesia

Beginer Subhan
Bogor Agricultural University
Bogor, Java, Indonesia

R. Ahmad Sue
Jakarta Fisheries University
Ministry of Marine Affairs and Fisheries
Jakarta, Java, Indonesia

Suharto
Hasanuddin University
Makassar, Sulawesi, Indonesia

Sukardi
Gadjah Mada University
Jogjakarta, Java, Indonesia

CONTRIBUTORS

Syarif Syamsuddin
Jakarta Fisheries University
Ministry of Marine Affairs and Fisheries
Jakarta, Java, Indonesia

Joseph James Uel Taylor
PT. Cendana Indo Pearl
Denpasar, Bali, Indonesia

Robert Kent Trench
University of California at Santa Barbara
Santa Barbara, California

Heri Triyono
Jakarta Fisheries University
Ministry of Marine Affairs and Fisheries
Jakarta, Java, Indonesia

Raffaele Vaccarella
Marine Biology Laboratory
Province of Bari, Puglia, Italy

Arben Virgota
Mataram University
Mataram, Lombok, Indonesia

Rand Weeks
Alternative Energy Design Associates
Brooklyn, New York

Neviaty P. Zamani
Bogor Agricultural University
Bogor, Java, Indonesia

CHAPTER 1

Dedication to Wolf Hilbertz*

Thomas J. Goreau

CONTENTS

Dedication 1
For More Information 4

DEDICATION

Wolf Hilbertz, a pioneer in building with unusual natural materials that grow themselves (what he termed "cybertecture"), tirelessly visited islands all around the world, seeking ways to save them from the impacts of global climate change. Wolf was born in 1938, and, after a childhood as a refugee, grew up in Detmold, Germany. Leaving high school before graduating to work as a laborer, he quickly realized he was wasting his time and enrolled in architecture at the Hochschule fur Kunste (College of Arts) in Berlin. Interested in novel materials, he submitted a thesis on building with plastic, which was immediately rejected on grounds that no one had heard of such an outrageous thing.

In 1965, he moved to New York, where an uncle, Max Urbahn, was an architect. He presented his rejected Berlin thesis to the Admissions Office at the University of Michigan, which promptly accepted him to pursue his graduate degree in architecture. A special influence was the laser group at the university, who were developing the first holograms, and Wolf quickly developed applications for laser-controlled design and machining of objects, which later became integrated into computer-controlled fabrication methods. Wolf said later that if they had thought to patent their work at the time, they would have been fabulously rich.

Wolf then moved to Baton Rouge, Louisiana, where he founded the architecture department at a small Black college, Southern University. His students could not stay with him in motels on field trips in the racially segregated South, so Wolf traveled with his students by bus and stayed in the colored sections of town. When funding ran out for the architecture program at Southern University, Wolf was offered a faculty position at the University of Texas in Austin, where he became professor of architecture. This led to the remarkable Ice City Project in Fargo, North Dakota, during January 1973 using computer-controlled nozzles to spray water in subfreezing temperature to build dome-like houses grown from ice. The concept was flawless in principle, working well at first, but the

---

* A previous version of this memoriam was published in "The Green Disc: New Technologies for a New Future," 2009, Gibby Media Group, Inc., and is adapted here with the permission of the publisher.

weather did not cooperate, and the warmest winter on record (to that time) melted down the Ice City as fast as his team of students could build it, working nonstop.

Inspired by the fact that corals and snails make their skeletons and shells from chemicals dissolved in seawater, Wolf saw that if limestone could be grown out of seawater in precisely controlled forms, construction material could be made in any size or shape, and whole islands could be grown in the sea. Wolf and his students began using low-voltage direct current to cause electrolysis of seawater to grow limestone structures at sites along the Texas and Louisiana coasts.

The growth rate of the material they produced in arches, blocks, cylinders, and other forms depends on the salt content of the water. It grew fast in salty lagoons of southern Texas and more slowly in muddy Louisiana mixed with Mississippi River water. But it grew very hard and was rapidly overgrown by oysters, which quickly covered the structures with layer upon layer of shells, a technique now being used for restoring oyster reefs. The first applications were protecting deteriorating wood or concrete structures by replacing them with stronger minerals. These were not only growing but also actually self-repairing, with physically damaged areas growing right back.

Wolf quickly found his way to the tropical islands (Figure 1.1), where his ideas were to achieve their greatest value. Solar- and wind-powered projects were set up in St. Croix in the Virgin Islands in 1976. These worked remarkably well for a while, with nearby coral growing along and over them, but were soon destroyed by a hurricane. The same happened to a large solar- and wind-powered project in Louisiana. With available funding exhausted, and realizing that no one would believe that large, strong structures could be economically grown in seawater unless they could see it themselves, he launched pilot projects in an astonishing variety of locations, anyplace people would let him, including Santa Catalina Island off the coast of California, the Cayman Islands, the Turks and Caicos Islands, Cartagena in Colombia, Isla Margarita in Venezuela, Majorca, Ireland, and many others, even thousands of feet down in the Cayman Trench, and north of the Arctic Circle in Alaska and Norway.

Figure 1.1 Photo of Wolf preparing a new design for a Biorock structure at Helen Reef, Palau. (Photograph by Caspar Henderson, 2004.)

Success followed success in terms of technical results, but money failed to follow because potential funders failed to see the possibilities or were not willing to wait for years to grow the limestone, which was strongest when grown at no more than 1–2 cm per year. His first patents, on what he called "Seacrete" or "Seament," made no money because he was cheated by his financial partner, who, unknown to him until later, turned out to be a disbarred lawyer convicted of cattle rustling in Oklahoma. In the 1970s, By chance invitation, Wolf sailed on a small yacht from Texas to the southernmost Caribbean, the San Blas Islands of Panama. There, he saw the Kuna Indians living on islands that they had painstakingly built on top of sandy reef flats, mining the corals from the reef to build their islands one coral head at a time. Wolf was astonished to see people who had preserved their ancient traditions untouched and was inspired by their mastery at building islands in the sea from living material. He also realized that the impacts on the reefs had been small when there were few people and many corals, but this had changed when the people became many, the corals became few, and the lobster and crabs and fish were disappearing the way the manatee and turtles already had. Wolf also saw that they could grow their own islands, using solar energy to produce the material that by now he had begun to call "Mineral Accretion," but he could not find the funding to get back until chance again intervened decades later.

In 1987, the author of this memoriam heard of Wolf's work and, after a trial experiment at the Discovery Bay Marine Laboratory, invited Wolf to come to Jamaica and collaborate on applications toward protecting coral reefs. Within months, corals were found to grow up to three to five times faster than normal, along with large amounts of white limestone sand-producing "calcareous" algae. Wolf and I then worked together as closest colleagues for 20 years, until his death. We again renamed the mineral material "Biorock," or living stone, because this better conveyed what the process was all about.

In the last 20 years of his life, Wolf, myself, and a constantly growing group of students and colleagues proceeded to build coral-reef restoration projects at many locations in Jamaica, Panama, Turks and Caicos Islands, Mexico, St. Maarten, Tobago, Cuba, Thailand, Seychelles, Maldives, Saya de Malha, India, Sri Lanka, Papua New Guinea, Dubai, Qatar, Corsica, Indonesia, Philippines, Palau, Marshall Islands, Fiji, French Polynesia, Bahamas, and the United States. The range of projects and habitats are far too extensive to be described here, except for an exceptional few. The project in Pemuteran, Bali, is the world's largest coral-reef restoration project, with more than 80 Biorock reefs totaling more than half a kilometer in length. Similar projects in other parts of Indonesia, Philippines, and Thailand are working with community-based fisheries management programs to restore the nearly destroyed fisheries habitat and restocking them with fast-growing corals, fish, and shellfish. These projects have been winners of many Indonesian and international ecotourism awards because they have taken formerly devastated areas and rapidly turned them into flourishing coral reefs packed with fish. In the Maldives, Biorock corals had a 16–50 times higher survival rate than nearby reefs after the record heat of 1998 and kept corals and fish alive that vanished elsewhere. A Biorock reef turned a severely eroding Maldives beach into 15 m growth of sand in a few years. In the Saya de Malha Banks, a huge shallow bank in the middle of the Indian Ocean, a floating solar-powered raft powered coral-restoration projects on the sea floor below. In Helen Reef, Palau, they built a solar-powered coral reef to protect the tiny island from eroding. And Wolf was finally able to return to the San Blas Islands in Panama and work with Kuna Indians and Panamanians to grow solar-powered coral reefs and fish and lobster habitat.

Wolf was larger than life in his remarkable career, embracing the possibilities to the fullest. His pioneering work in Cybertecture brought fame and followers, but not fortune. He tirelessly taught and promoted the new field he had conceptually created, which he called "Seascaping," the science and art of marine environmental restoration, growing reefs, and creating new habitat where it had disappeared and even where it had never been before. His wide network of loyal friends valued his genius as an innovator of a new, living, growing architecture, using the sea, "the world's largest mine," as he called it, and he inspired students around the world. He was not simply a revolutionary

Figure 1.2 Underwater bust of Wolf in Pemuteran, Bali, Indonesia, characteristically smoking a cigar. (Photograph by Rani Morrow-Wuigk.)

designer using novel growing materials, but he was also a builder, artist, and sculptor who built with his bare hands and understood that craft was as important as vision.

It is said, "A genius is someone who shoots at a target no one else can see, and hits it." Wolf did this again and again, but too often the crowd around never even noticed that the target had been hit, and long-term funding proved impossible to find. Wolf never despaired; he knew he was doing the right thing, doing the most he possibly could in areas where he had to lead alone and hope others would follow. He always clung to "das Prinzip du Hoffnung," the principle of hope, that the importance of his innovations would be recognized in time to free him to explore their potentials to the fullest while he still had the time. Sadly, that failed to materialize, but Wolf kept leading and inspiring to the end, always saying, "Onward through the fog!" and "Hope dies last," and remembering Martin Luther's saying that "even if I knew the world would end tomorrow, I would still plant my apple seeds today."

Wolf left behind five remarkable children, Kai (architect and computer expert, Munich), Derrick (artist and musician, New York), Halona (artist and musician, New York), Navassa (artist and fashion designer, Los Angeles), and Alissa (student, Berlin), with three wonderful and inspiring wives, Regina Piper (publisher, editor, Munich), Frances Carvey (Galveston, Texas), and Ursula Rommerskirchen (diplomat, Berlin). In his last 20 years, Wolf's work was mainly supported by his wife, "Uschi." Her assignments as a diplomat for the German Ministry of Foreign Affairs led them from Texas to Boston, Montreal, Ireland, Bangkok, and finally Dubai. From these bases, Wolf traveled tirelessly to islands around the world, starting coral-reef and fisheries restoration projects wherever people really cared to bring their dying reefs back. Wolf left the world a dazzling array of Biorock projects and a global network of colleagues committed to seeing his work continued and to be greatly expanded as urgently as possible as a critical tool to preserve, restore, and enhance our environment against global warming and sea level rise in the future. His body failed him on August 11, 2007, but his legacy keeps growing (Figure 1.2).

FOR MORE INFORMATION

http://en.wikipedia.org/wiki/Wolf_Hilbertz
http://www.wolfhilbertz.com

CHAPTER 2

Innovative Methods of Marine Ecosystem Restoration
An Introduction

Thomas J. Goreau

CONTENTS

Crisis of the Oceans Can Be Reversed Only through Large-Scale Restoration 5
Louisiana: A Tragic History of Missed Opportunities 7
Synopsis of Chapters 9
Conclusions 10
References 10

CRISIS OF THE OCEANS CAN BE REVERSED ONLY THROUGH LARGE-SCALE RESTORATION

This book, the first publication of innovative new methods to reverse the major threats now destroying our ocean planet's biodiversity and productivity, offers a ray of hope in an increasingly gloomy crisis. Here you will find new methods to greatly increase the settlement, growth, survival, and resistance to stress of marine ecosystems, fisheries, and eroding shorelines; to maintain biodiversity and productivity where it would be lost; to rapidly restore them in devastated places where there has been no natural recovery; and to create new methods for sustainable and biodiverse mariculture. If these proven technologies were applied on the scale that was needed, our major marine resource problems now spiraling downward could be reversed for a better future.

Reversing these negative trends is the critical issue for this century if we are to maintain and nurture our renewable marine resources. The accelerating crisis in the oceans is now widely known: impending mass extinction of coral-reef ecosystems from global warming and pollution; the crash of one fishery after another from overharvesting; disappearing species; the conversion of oceans from sinks of CO_2 to sources; the impending billion or so people who will be flooded from coastal homes to become global sea-level-rise refugees, with whole nations disappearing beneath the waves; and more.

Reversing these looming catastrophes cannot be solved by the conventional solution to all marine management problems: marine-protected areas (MPAs) that exclude fishermen. The widely touted claims that these ecosystems are "resilient" and "will bounce back all by themselves," thanks to the sagacity of their managers, is in fact almost never observed in practice because most MPAs are intrinsically incapable of reversing the root causes of the major threats that are laying waste our habitats. Every coral reef MPA is full of dead or dying corals that no local management can prevent. But so strong is the lobby of governments, funding agencies, and big international nongovernmental organizations (BINGOs) for MPAs that their failure cannot be admitted. Active restoration solutions are rejected out of hand, because to admit their need would be an admission that money has been wasted and that existing policies are futile and will fail even more in the future as global warming, sea-level rise, and pollution escalate.

These man-made threats are based in our unwise overexploitation and disruption of the natural mechanisms that regulate our atmosphere, ecosystems, soils, water, and climate. While it is a wonderfully praiseworthy task to protect the few healthy marine ecosystems that still survive, if they cannot be protected from the real causes of mass mortality, they will die anyway, perhaps only a little later. And if we don't restore the vast majority of the ecosystems that we have already destroyed or severely damaged, where will future fisheries come from? We are often told that restoration is pointless, because we can't possibly restore it all. We answer that "we certainly can't restore it all, but if we don't restore all that we can, what else will we leave for future generations?"

Even more bizarre is the claim often thrown at restoration experts by the BINGOs, who say that it is very dangerous to say you can restore anything, because then you are encouraging people to destroy it! The perversity of this argument is staggering; in effect, they are accusing people who spend all their time planting trees of aiding, abetting, and rationalizing the destruction of the rain forests! In fact, they are doing the very opposite: restoring badly degraded areas that have not recovered by themselves, so that they can become a productive source of ecosystem services; growing them back by actively undoing the damage wrought by others, so they can be sustainably harvested.

There is a well-established lobby for continuing business as usual, and honest admission of the MPA strategy failures would threaten more money flowing to those whose only vision is to repeat yet again the traditional mistakes of the past and expect a different result. But one could eliminate all the subsistence fishermen, and the fish won't come back wherever the habitats that supported them are largely gone. They are simply not recovering by themselves, despite what the politically motivated dogma of "resilience" so popular with governments, funding agencies, and BINGOs dictates. The sooner self-deception is discarded, the sooner we can start growing back what we have lost.

New technologies are critically needed that directly address the critical limiting factors preventing the growth of sustainable marine resources, and that is precisely what this book is all about. The bulk of the chapters in this volume were presented at the "Symposium on Innovative Methods of Marine Ecosystem Restoration" at the World Conference on Ecological Restoration in Merida, Yucatan, Mexico, August 22–24, 2011. PowerPoint slides from those presentations are contained in full in the CD in the back of the book, which greatly complements the papers in this book and adds new photographs and information to the text. It was possible to present these papers only thanks to the kind support of the Breck Foundation to cover tickets and expenses.

These papers included the first presentations of data on electrical marine ecosystem restoration, from work going as far back to the 1970s, and with one exception, these results are all published here for the first time. In addition, several additional papers are included from authors who submitted abstracts but were not able to get to the symposium and from authors whose data became available only after the abstracts for the symposium had been submitted. These papers discuss innovative new methods for greatly increasing the recruitment, growth, survival, and economic value of restored marine ecosystems. They provide an essential tool kit for transformation from economic overexploitation based on ecosystem degradation toward a restoration ethos, replacing what has been lost using the best science to accelerate restoration by identifying and remediating the major specific imbalances that now prevent recovery—namely; lack of recruits, lack of shelter, lack of food, lack of energy, and suboptimal nutrient ratios.

This book, therefore, is the first presentation of revolutionary new methods for restoring damaged marine ecosystems that could be immediately used to manage the coastal zone to reverse catastrophic global declines in renewable marine resources. A brief synopsis of the chapters follows at the end of this introduction.

LOUISIANA: A TRAGIC HISTORY OF MISSED OPPORTUNITIES

Object lessons in the systematic failure of coastal zone managers to use modern methods of ecosystem restoration could be elucidated anywhere in the world, but perhaps no single place shows these failures more abjectly than Louisiana.

Louisiana has extraordinary coastal management challenges. The coastline is retreating at staggering rates, in places hundreds of meters per year, and the fisheries have suffered from the BP oil-spill catastrophe. The challenges facing Louisiana are multiple. First, Louisiana is naturally sinking, and humans have accelerated that process by pumping out oil, gas, and sulfur from beneath it, adding artificial subsidence to nature's own. Masses of sediment that used to pour out of the Mississippi River, building the coast outward as the long-shore currents moved the mud westward, were at first greatly accelerated when European conquerors arrived, cut down the forests, ploughed the grasslands, and vastly accelerated soil erosion and coastal sedimentation. But the sediment supply was later reversed and virtually cut off as dams on the rivers blocked their flow to the ocean, clogging reservoirs with mud instead of building shorelines. The waters have been poisoned by oil, pesticides, fertilizers, and sewage. So much nutrients flowed into the sea that they triggered massive algae blooms, and when the organic matter fell to the bottom and rotted, bacteria stripped all the oxygen out of coastal waters, killing all life more complex than bacteria in expanding dead zones. Salt marshes, which had been the only thing holding the mud together and protecting it from being washed away, retreated inland as the shoreline and salt water moved inland, and the huge oyster beds, which had protected the salt marsh from erosion by waves, were wiped out by overharvesting. In parts of Louisiana, the coastline is retreating by hundreds of meters a year as land is washed into the sea.

Louisiana was precisely where the late Wolf Hilbertz began his pioneering work that led directly to almost all the work described in this book. In 1976, he built the first projects to grow rock structures in the ocean for experiments in producing building materials in the sea. He quickly showed that he could produce solid structures of any size or shape that could be used as prefabricated structures for use on land. He discovered that this was the only marine construction material that gets stronger with age (while all the rest get weaker with time) and had the unique properties of being self-repairing: damaged areas grew right back (Hilbertz 1979, see photos in Goreau 2012). Even more astonishing was the completely unexpected discovery three months later that the pilot project had been completely overgrown by multiple layers of oysters that had settled on them and grown to adult sizes at what seemed to be record rates. The technology Wolf Hilbertz developed there provides the only possible long-term solution to the Louisiana's massive coastal-management failures but has been steadfastly ignored where it was born, despite the enormous environmental, ecological, and economic benefits it would provide there (Figures 2.1 and 2.2).

How could the methods described in this book help Louisiana?

1. Oyster reefs could be grown back all along the eroding coasts, reversing erosion and producing sustainable oyster harvests as well as juvenile fish habitat.
2. Salt marsh could be restored where it is retreating, and extended seaward, growing out the land.
3. Fish and shellfish populations could be greatly expanded by growing habitat for them, especially in the critical juvenile phases.
4. The four thousand abandoned oil and gas platforms could be turned into growing reefs for fisheries, mariculture, and sustainable energy production (Sammarco et al. 2004; Kolian et al. 2011).

Figure 2.1 Early solar-powered Biorock project in Louisiana salt marsh. The data in this book show that salt marsh grass can be restored and grown seaward using Biorock. (Photographed by Wolf Hilbertz, 1985.)

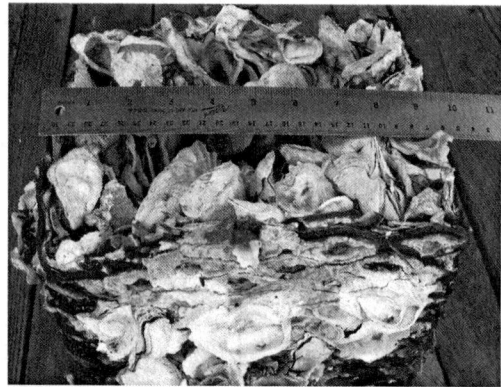

Figure 2.2 Oysters grown in Louisiana with Biorock in 1984. This small metal cage was wired to a solar panel. No oysters were placed in it, but they spontaneously settled on it and grew to large size in less than one year, completely filling it. The data in this book show that oyster reefs could easily be restored in Louisiana. The cage was in the water less than a year and then removed. It then sat outdoors for many years, which is when all the rusting took place. (Photographed by Eric Vanderzee, 2011.)

Louisiana is home to 4000 oil and gas rigs, many of which are no longer in production. They produce prolific ecosystems densely populated with oyster, corals, and fish. In fact, the growth of corals and other marine life, including oysters, is so prolific on them that a major task for oil company commercial divers is to break it off! Fishermen are lobbying to designate the structures to be officially recognized as Essential Fish Habitat (EFH) and be protected. With the application of the methods in this book, these platforms could become the basis for a vastly more productive fishery. All it would take would be a little support from Lousiana's policy makers, funding agencies, and private sector to restore what they have lost. Sadly, there has been little sign at all of this since 1976. With all of the money spent by BP and the federal government on lawyers to deny or minimize the oil-spill impacts, it is tragic that none of this funding has yet been used to restore Louisiana's fisheries, oyster reefs, salt marshes, and the very ecosystems that shape the coastline, produce the food and income, and protect the shores from erosion. Already most of the economically valuable red snapper harvest comes from oil platforms and artificial reefs, but the platforms are not recognized as EFH and are scheduled for demolition and removal instead of expansion!

The methods in this book could transform coastal management in Louisiana, where these methods were first pioneered. No place could benefit from them more, but none has resisted for longer all efforts to use the weapons of mass restoration to restore their own resources, or wasted more money watching them vanish. One only hopes that they come to their senses and use the best restoration technologies while there is still time. The first glimmer of hope came on April 19, 2012, when the Gulf of Mexico Fishery Management Council, the official national regulatory agency managing the resource, voted unanimously "to direct staff to develop a plan amendment to evaluate the appropriateness of designating fixed petroleum platforms and artificial reefs as EFH and develop alternatives that they be recognized as habitat areas of particular concern." Once they are designated EFH, there will be strong legal barriers to their removal and reasons to protect them permanently from corrosion and to accelerate growth, so that they can truly become "Rigs to Reefs."

SYNOPSIS OF CHAPTERS

This chapter sets the scene for the technical chapters and data.

Chapter 3 by Goreau et al. shows the physical results of shore-protection projects using Biorock technology to grow reefs in front of severely eroding beaches, resulting in dramatic new beach growth in a few years. These methods could be used to restore ecosystems, protect eroding coastlines from global sea-level rise, and even grow artificial islands.

Chapter 4 by Goreau and Hilbertz is the first study showing the dramatic increase in coral growth, settlement, and survival in electrical fields.

Chapters 5 and 6 by Jompa et al. and Arifin et al., respectively, summarize many studies made in Indonesia with different coral species at different locations and under different conditions, showing dramatic increases of coral growth and fish populations in electrical fields.

Chapter 7 by Zamani et al. shows that coral budding and branching is greater in the presence of electrical fields.

Chapter 8 by Nitzsche shows that the electrical field stimulates coral growth around structures as much as on them, but seems to have the opposite effect on sponges, which appear to have greatly increased recruitment, but not growth, on electrified structures.

Chapter 9 by Fitri and Aspari shows that electrical fields increase growth rates and survival of soft corals.

Chapter 10 by Strömberg et al. shows that deep-sea cold-water corals bud and branch better with low electrical fields, which might be used to restore deep reefs damaged by trawling and oil and gas exploration.

Chapter 11 by Karissa et al. shows that pearl oysters have faster growth and survival in electrical fields.

Chapter 12 by Berger et al. shows dramatically higher growth and survival of the North American oyster in electrical fields under controlled laboratory conditions.

Chapter 13 by Shorr et al. shows that oyster growth and survival are greatly increased by electrical fields under field conditions.

Chapter 14 by Vaccarella and Goreau shows that seagrass growth and survival are increased in electrical fields.

Chapter 15 by Cervino et al. shows that salt marsh growth and survival are greatly increased in electrical fields.

Chapter 16 by Lecaillon demonstrates that coastal fisheries can be greatly increased with new methods that eliminate almost all the mortality that affects juvenile fishes.

Chapter 17 by Macfarlane shows that careful attention to nutrient concentrations and ratios can optimize growth and value of algae in mariculture. Since algae form the base of the food chain, optimizing their growth has huge implications for the rest of the ecosystem.

Chapter 18 by Haseltine discusses how to optimize the physical designs of restored marine ecosystems to maximize biological function.

Chapter 19 by Goreau presents a summary of the biological effects of electrical fields and the fundamental biological mechanisms by which they benefit all organisms.

CONCLUSIONS

We can only hope that coastal managers start to apply all these innovative methods of marine ecosystem restoration as quickly as possible! It is no longer the case that our only option is to "protect" an area and hope that it recovers all by itself; the pace of degradation is so rapid that active restoration is needed. The good news from this book is that the techniques to do so have been developed and are very broadly applicable in almost all marine ecosystems, although further work will be needed to optimize the results for specific species in specific settings.

While the authors of this book come from Asia, Europe, North America, Central America, the Caribbean, Africa, and Australia, most of the authors of this volume are scientists from developing countries; in fact, the majority of the authors are from Indonesia, the country with the largest area of coral reefs and highest marine biodiversity of any country in the world. But other authors come from Jamaica, Belize, Maldives, and Sudan, although they are mostly residents abroad. It is important to realize that the best work on marine ecosystem restoration is now being done by researchers from developing countries working with little or no funds, rather than by jet-setting researchers from rich countries doing "tourist science" in the poor countries where most of the reefs are. The future of coral reefs will depend on building up the endogenous capacity of developing countries to do the work needed to save their ecosystems and the services they provide, which simply cannot be done by the traditional model based on short-term foreign visitors. We hope this book will be a contribution to help build up the capacity of developing countries to solve their own problems. But at the same time, the methods in this book are of universal application and can be applied, with adaptations, in all marine ecosystems.

To keep the price of the book as low as possible, we have opted for black-and-white figures, even though full color images best show the results. The CD in the back of this book greatly adds to the chapters in the book by adding full color images of the figures, additional images and video, and slide presentations made at the Symposium on Innovative Methods of Marine Ecosystem Restoration at the 2011 International Conference on Ecological Restoration in Merida, Yucatan, Mexico, where most of the papers in this book were first presented.

REFERENCES

Goreau, T. J. 2012 (in press). Marine electrolysis for building materials and environmental restoration. In V. Linkov (Ed.), *Electrolysis*, InTech Publishing, Rijeka, Croatia.

Hilbertz, W. H. 1979. Electrodeposition of minerals in sea water: Experiments and applications. *IEEE Journal on Oceanic Engineering* 4:1–19.

Kolian, S., S. Porter, and P. Sammarco. 2011. National Environmental Policy Act (NEPA) analysis of the removal of retired offshore oil and gas platforms. EcoRigs Platform Removal Brief No. 2. EcoRigs, Baton Rouge, LA.

Sammarco, P. W., A. D. Atchison, and G. S. Boland. 2004. Expansion of coral communities within the Northern Gulf of Mexico via offshore oil and gas platforms. *Marine Ecology Progress Series* 280:129–143.

CHAPTER 3

Restoring Reefs to Grow Back Beaches and Protect Coasts from Erosion and Global Sea-Level Rise

Thomas J. Goreau, Wolf Hilbertz, Abdul Azeez Abdul Hakeem,
Thomas Sarkisian, Frank Gutzeit, and Ari Spenhoff

CONTENTS

Global Shore-Erosion Crisis 11
Global Warming, Sea-Level Rise, and Erosion 12
Hard-Shore Protection 13
Soft-Shore Protection 20
Biorock: A Superior Marine Construction Material 22
Biorock Shore Protection in the Maldives 25
Biorock Shore Protection in Indonesia 27
References 33

GLOBAL SHORE-EROSION CRISIS

Worldwide, most beaches are suffering from serious and accelerating erosion (Pilkey and Young 2009). Major factors responsible for this erosion include rising sea levels, sand mining, regional subsidence from tectonic causes (amplified in many places by oil and gas extraction), increasing tropical storm intensity driven by global warming, use of groynes that block long-shore sand transport (stealing sand from the neighbors), and seawalls that are built too close to the shoreline (to protect buildings that should never have been permitted so near the water's edge) due to lax or irresponsible planning (Pilkey and Wright 1988; Pilkey and Dixon 1996). The only beaches that show clear net growth are a few yellow or brown quartz sand beaches, made up of mineral grains washed into the sea from erosion of rocks on land. These are growing in a few places where massive deforestation has greatly increased erosion of soils and rocks, causing increased transport of sand to beaches near river mouths.

In the tropics, the major factor is death of reef-building corals from global warming, new diseases, pollution, soil erosion, physical damage from human activities, or unsustainable fishing practices. In cooler waters, massive loss of oyster reefs from overharvesting is causing the same result; loss of biological reef structures that protect and build beaches. This results in loss of the organisms whose dead skeletons make up white beach sand grains, and concomitant loss of growing protective reefs shielding the shore from waves. Consequently, beaches suffer from two effects simultaneously: sand erodes faster due to higher wave energy, while less new sand is available to replace it. The massive worldwide destruction and degradation of coral reefs in the last two decades have resulted in most white limestone sand beaches suffering serious net erosion, shown clearly by trees, buildings, roads, and airport runways collapsing into the sea.

When reefs become degraded and need to be replaced by seawalls in order to keep the beach from washing away, the typical costs for concrete or rock seawalls are about $10–15 million per kilometer. A standard estimate of global coastline length is 350,000 km (although this surely ignores the fine-scale details), so protecting all of that could cost up to $5.25 trillion. These estimated costs are for a sea-level rise of about 1 m, but runaway greenhouse warming owing to unabated, increased CO_2 levels that melts the ice caps could result in ultimate sea-level rises of up to 100 m!

Naturally, not all coastlines are eroding, because a small proportion of coasts is actually growing as a result of massive inland soil erosion caused by deforestation and bad land management, and also not all shorelines have inhabited infrastructure that needs to be protected. But, to put that crude estimate into perspective, the (officially revealed) US military budget is less than one trillion dollars, and that makes up roughly half of global military expenses. So, protecting coasts against sea level rise of only 1 m might cost more than "national security" from the worst enemies.

GLOBAL WARMING, SEA-LEVEL RISE, AND EROSION

Global sea-level rise is accelerating faster than the highest model-based projections made by the Inter-Governmental Panel on Climate Change (Rahmstorf et al. 2007), and is certain to increase greatly in coming decades (Zhang et al. 2004; Vermeer and Rahmstorf 2009; Rahmstorf 2010; Goreau 2011). Our present course of uncontrolled greenhouse-gas emissions is pushing us inexorably toward greenhouse climate conditions similar to those 55 million years ago, when there were no glacial ice caps and sea levels were about 100 m higher than today.

Global sea-level rise will be one of the biggest costs of global warming, by any estimate. If our "leaders" cared about the future, adapting to sea-level rise and increasing intensity of storms are likely to be among the largest single costs of global warming, and the amounts will be staggering, even if all we do is make a billion or so people run away to higher ground! These problems will get far worse in coming years as sea-level rise accelerates and hurricane maximum wind speed increases, and coastal erosion costs will become immense when whole countries disappear and millions, eventually billions, of people are forced from their homes to become environmental refugees. Yet no serious efforts are being made to prepare for these imminent problems, not even by the countries that are most affected, with the sole exception of the Netherlands, which plans to spend billions of dollars to raise their dikes by a meter. They do not seem to realize that they will need to do so every decade for thousands of years if global warming is not reversed!

Despite this rapidly escalating global shore-erosion crisis, there remains little or no anticipatory planning for shore erosion. Instead of preparing for the future, we simply wait until catastrophic flooding and erosion strikes, and then appeal for short-term disaster aid to shovel more rocks into the sea and build concrete walls on the shoreline. This ensures that the sand in front of them will be washed away, and the inevitable and never-ending future repairs will be even more costly. Ultimately, these prove long-term failures, lucrative only to the contractors who must rebuild these structures over and over again every time they fail.

A few years ago, searching for information on how much the world spends every year on shore protection, we did a computer search on "shore protection." Every single hit we got was in fact for "offshore asset protection"; we learned the names of every money-laundering bank in the world, but absolutely nothing about global shore-protection expenditures. The world's policy makers appear to have their heads buried in the sand!

HARD SHORE PROTECTION

The fundamental strategy behind almost all shore protection projects is "hard" protection: building a solid wall to reflect waves backward and protect the land behind it (Wiegel 1992; Schiereck 2004). This hard protection suffers from an intrinsic and unavoidable physical flaw; all or most of the wave energy is concentrated onto the plane of the wall. In fact, the wave is reflected backward from a vertical wall with the same speed and energy with which it came in, so the force on the wall itself is twice the energy of the wave due to reversing the energy flow vector. Solid breakwaters protect land immediately behind them, but focus wave energy in one plane, eroding all sand in front of them and causing intense scour that undermines the breakwaters, ultimately causing settling, collapse, failure, and endless cycles of rebuilding. This causes higher-velocity currents to form in front of such walls, eroding sand in front and underneath the structures, with impacts on downstream near shore marine ecosystems (Airoldi et al. 2005).

Current methods of shore protection have largely proven to be very costly long-term failures. Solid breakwaters, made from cement, rocks, or geotextile rubber tubes filled with sand, typically cost in the range of $10,000–$15,000 per meter of shoreline, or $10–$15 million per kilometer. In the March 11, 2011 Japan tsunami, almost every tsunami barrier, constructed at the cost of billions of dollars, was ripped apart, collapsed, or fell over and proved not fit for the purpose (Cyranoski 2011).

This tremendous concentration of energy in a small area is very different from the way that a sloping beach gradually shoals and slows down waves, breaking the biggest waves farther offshore. Hard seawalls and breakwaters therefore cause all the sand in front of them to be washed away, speeding up erosion and resulting in inevitable scouring of sediment under the structure, causing it to be undermined, settle, crack, and collapse. As a result, all such structures must be constantly rebuilt every time they fail, as they all do sooner or later. People selling such hard "engineering solutions" know that they have a guaranteed repeat business, forever, solving their own financial problems, but not those of their customers, who in desperation are forced to endlessly repeat their expenditures or lose their shorefront investments!

A classic example of such failures is Galveston, Texas. After the worst loss of life in a hurricane in US history in 1900, a massive seawall was built. This was extended along the shoreline of Galveston Island later. The beach in front of it immediately washed away. Every year, vast sums of money is spent in dumping sand in front of the seawall so that there is a beach for tourism, and every year it disappears. Now there is no more sand in the region left to dredge, because so much has been swept out to deeper water and lost. As waves hitting the wall form an intense coastal long-shore current, they greatly increase the erosion downstream. Time series aerial images show the vast bulk of erosion down-current took place immediately after the wall was built (Figure 3.1). It was later increased when sand was excavated from large pits in downstream beaches to dump in front of the wall. These borrow pits were then breached in subsequent hurricanes.

Recently, Rice University issued a major report summarizing all of the geological data and past shoreline information and concluded that it was now so hopeless to protect west Galveston Island from sea-level rise that they should not even waste any money trying—they should just abandon the place immediately and move to higher ground (Hight et al. 2011). That same week, Texas governor Rick Perry ordered removal of the words "global sea level rise," and "global warming" from all official Texas State environmental assessment and planning documents, even though while Texas suffered from record high temperatures and drought.

(a)

(b)

(c)

Figure 3.1 (a–t) Galveston shore erosion time series, 1954–2010. This site is just west of the end of the seawall that was extended in 1974, and dramatically shows the greatly increased erosion caused by the strong longshore current created by wave reflection off the wall, as well as increased erosion caused by excavation of beach sand dug up to replace that eroded in front of the sea wall, as well as the impact of a major hurricane. The position of the cross marks in the first 1954 image should be compared to those in the final 2010 image to see the net beach erosion over this 56-year period. (Images from Google Earth.)

(d)

(e)

(f)

Figure 3.1 (*Continued*)

(g)

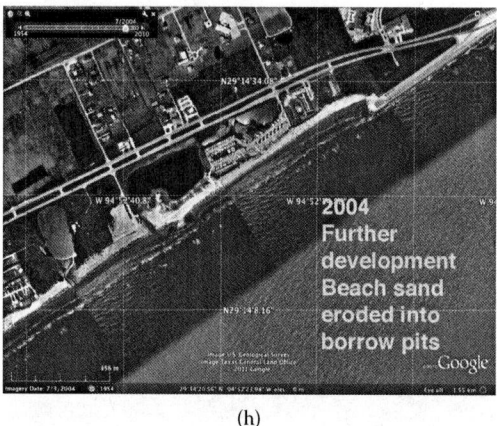

(h)

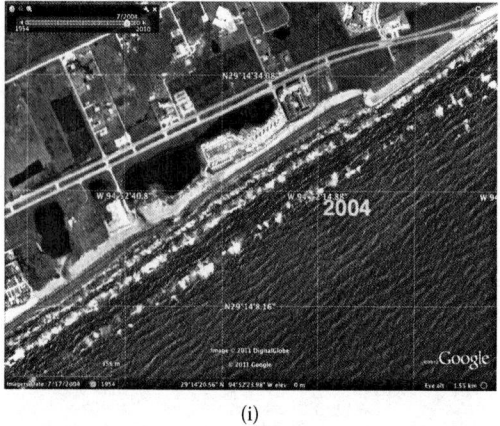

(i)

Figure 3.1 (*Continued*)

RESTORING REEFS TO GROW BACK BEACHES AND PROTECT COASTS

(j)

(k)

(l)

Figure 3.1 *(Continued)*

(m)

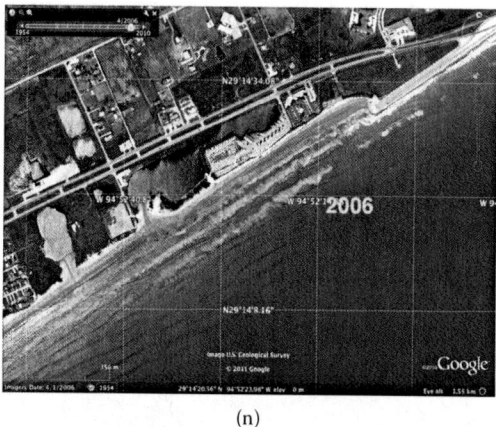

(n)

(o)

Figure 3.1 (*Continued*)

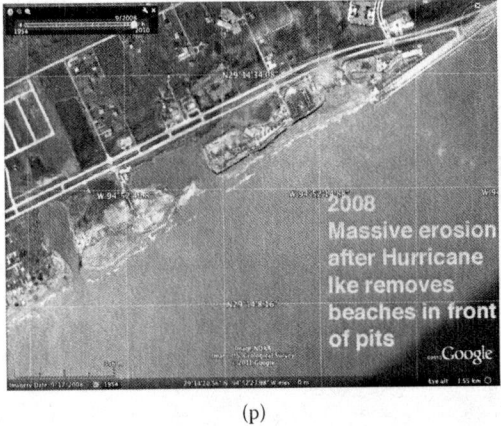
(p)

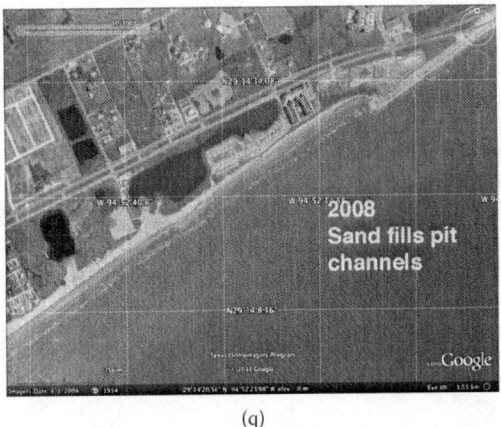
(q)

(r)

Figure 3.1 *(Continued)*

(s)

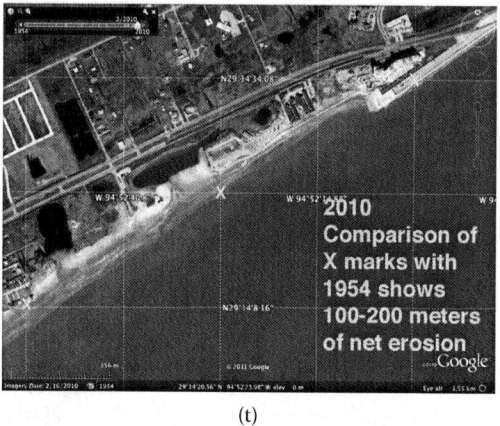

(t)

Figure 3.1 (Continued)

SOFT SHORE PROTECTION

"Soft" shore protection works in a very different way, the way that natural coral reefs and oyster reefs act to grow beaches. "Permeable" barriers are built which are open structures, rather than monolithically solid. They dissipate a portion of the wave energy through friction and allow some to pass through, gradually reducing wave energy. Instead of reflecting waves, they refract them, so very different physical principles are involved. When the waves reach the shoreline, they have much less force, depositing sand instead of eroding it (Figures 3.2 and 3.3).

Coral reefs form nature's best shore protection (Munk and Sargent 1948). Pacific atoll islanders stand on the outer edge of windward reef flats in waters that are calm and only ankle deep and toss fishing lines into huge pounding waves breaking only meters away. The energy of large ocean swells is attenuated and can be completely dissipated by the complex structures of the coral reef in front of them, though attenuation is a complex function of the wave field, depth, and tidal height (Wolanski 1994; Hardy and Young 1996; Gourlay and Colleter 2005). If corals are broken by exceptional wave events, they grow right back. Coral reefs have evolutionarily optimized the art of maximizing the flow and energy dissipation through them, bringing food to reef organisms and flushing out their wastes, with minimal damage to the structure itself. Intense roughness and complexity on all scales is the fundamental base of the extraordinary productivity of coral reefs. But their hydrodynamics remain poorly described or understood (Monismith 2007).

Figure 3.2 Bikini Atoll windward reef flat edge. Pounding wave surge energy at the reef flat edge is almost entirely dissipated by the coral reef in front of it before the water reaches the reef flat. (Photograph by Fritz Goro, 1946.)

Figure 3.3 Reef surge channel in Windward reef flat, Bikini Atoll. The complex channels serve to dissipate wave energy by high surface area and friction. (Photograph by Fritz Goro, 1946.)

As long as reef corals are alive and able to recover, they protect the shore for free, better than any other alternatives. Sadly, most reef corals worldwide are now dying or dead. During the December 26, 2004 Indian Ocean tsunami, wherever the coral reefs were intact, there was a minimal damage and loss of life, but in nearby places where reefs had been destroyed by mining or other causes, waves went much further inland, and there was a massive loss of life (Samarawickrama et al. 2008). There was very little physical damage to coral reefs from the tsunami; most damage took place in coastal fringing reefs, and that was not owing to the wave forces but to the backwash of water containing smashed buildings and human construction debris. The waves were less threat to the corals than trash was! Thomas Sarkisian took photographs of coral reefs on the offshore Similan Islands in the Andaman Sea the week before and the week after the tsunami, and most of the sites suffered little or no visible damage, even to brittle branching corals. There were small local areas with heavy physical damage, but these were all on promontories and headlands that focused wave energy onto them. It is often not appreciated that a tsunami has only a few very large waves, well separated in time, and causes far less damage to coral reefs than a hurricane, where big waves can strike every few seconds for days on end.

"Soft" or "permeable" shore protection structures are typically built of open frameworks or of many small solid units with spaces in between the waves, letting a fraction of the wave energy to pass through, rather than reflecting the waves (Goudas et al. 2003). As a result, these structures use far less material than do those for hard shore protection, and promote shoreline growth rather than erosion. They are often built as cubes, or "bumper blocks," in staggered rows. These, however, can cause scour in front of them if the wave hits a vertical plane, as is typical of most designs. A further disadvantage is deterioration of the structural material itself, as steel rusts and corrodes, and concrete deteriorates and crumbles, or as geotextile tubes are ripped apart by severe storms, dumping their sand and littering the (eroding) beach with shredded rubber.

BIOROCK: A SUPERIOR MARINE CONSTRUCTION MATERIAL

A superior material for construction would be one that does not rust, corrode, or crumble, that can be grown in place from seawater, costs less than conventional construction materials like steel, concrete, or reinforced rubber mats, and is self-repairing and self-attaching. The only construction material known that meets these criteria is Biorock material (Hilbertz 1992; Goreau 2012).

Biorock materials were invented by an architect, the late Professor Wolf Hilbertz, who called the ocean "the world's largest mine." Inspired by the fact that corals, snails, and other marine organisms could convert dissolved minerals in the ocean into precisely shaped skeletons and shells, he developed the methods to grow prefabricated construction materials from these natural materials in any form or dimension (Hilbertz 1979).

Biorock material is composed of natural calcium and magnesium minerals dissolved in seawater that crystallize out on top of conductive metal surfaces that are given a small electrical charge. By maintaining this charge, structures of any size or shape can be grown out of seawater. When the charging rate is kept sufficiently low, and the rock is grown at rates of less than about 2 cm per year, they are predominantly made up of calcium carbonate, or limestone. Limestone has been used as a construction material since the first pyramids of ancient Egypt.

Engineering tests of mature Biorock material show that they have compressive (or load-bearing) strength of about 80 N/mm^2 (MegaPascals), about three times that of concrete made from Portland cement, the most widely used building material in the world. The metal frameworks on which Biorock is grown are completely protected from all rusting and corrosion by the electrical charge, so they never deteriorate, unlike reinforcing bars in concrete, which rust, expand, and crack and stain concrete, limiting its lifetime.

Furthermore, Biorock materials in the ocean are uniquely "self-repairing" (Figures 3.4 through 3.7): if the Biorock material is broken off, the damaged areas grow back preferentially. No other marine construction material has these properties, making them ideal for breakwaters and submerged structures. Also, the affordable cost of the electricity needed to grow them from seawater makes them considerably more economical than concrete structures of the same dimensions, although the extent depends on local electricity, labor, and cement transport costs (Goreau and Hilbertz 2005).

Biorock reefs promote amazing growth of corals and other marine organisms, typically two to eight times faster than normal, depending on the species and the conditions, so they rapidly create lush coral reefs, or oyster reefs in colder waters, that are swarming with fish. Furthermore, the corals growing on Biorock reefs are found to have 16–50 times higher survival from severe high temperatures than corals in surrounding reefs. This means that coral reefs and oyster reefs, nature's best shore protection, can be kept alive where they would die, and restored in a few years in places where no natural recovery is taking place, making them not only shore-protection devices but also providing highly valued fisheries habitat and ecotourism resources in front of the beaches they protect (Hilbertz and Goreau 1996; Goreau and Hilbertz 2005). Although growth of corals and oysters make Biorock reefs much more effective more quickly, Biorock reefs can also provide shore protection without them in places where corals or oysters will not grow due to excessive physical or chemical stresses.

Hard structures from steel, rock, or concrete cause water flow to diverge around them, speeding it up and causing erosive scour of sediments around them. Even those concrete structures that have holes in them, like reef balls, pyramids, and cubes, cause severe scour of sediments around them to a horizontal distance away from them roughly equal to the height of the structure, and to a depth of

Figure 3.4 This Biorock reef at Pemuteran, Bali, was damaged by the impact of a large boat, which cracked off the hard limestone coating and exposed the original rebar frame. Photograph was taken on April 19, 2011 shortly after the collision. (Photograph by Rani Morrow-Wuigk.)

Figure 3.5 Closeup of impact, April 19, 2011 shows clean white limestone material break, with a very thin surface coating of algae. Notice that there is no rusting on the steel, which has been in the sea for 11 years. (Photograph by Rani Morrow-Wuigk.)

Figure 3.6 Same area on March 25, 2012, nearly a year later. The damaged area has grown a new coating of limestone rock. (Photograph by Rani Morrow-Wuigk.)

Figure 3.7 Closeup of damaged area, March 25, 2012. (Photograph by Rani Morrow-Wuigk.)

half the height (Shyue and Yang 2002). In sharp contrast, open frame Biorock structures show no scour around their bases; instead, the sediment builds up around and under them as water velocity slows due to surface friction interactions. It makes far more sense to design seawalls to mimic the shape of coral reefs than of buildings.

Biorock structures are able to survive severe hurricane wave forces that would destroy or move solid structures. For example, structures in 5–7 m of water in Grand Turk, Turks and Caicos Islands, survived the two most severe hurricanes in their history, which hit a few days apart and damaged or destroyed 80% of all the other buildings on the island. Even though the Biorock structures were sitting unattached on bare sand and were weakly built, they suffered only minor structural damage, only the electrical cables had to be replaced as the insulation was sandblasted off. Thousands of corals had been transplanted onto them a few weeks before from an area where corals were dying from sedimentation caused by dredging for a cruise ship terminal, and although many of these corals were only loosely attached, the vast majority survived the two hurricanes unharmed (Wells et al. 2010). About one-fourth to one-third of a meter of sand built up around the base, but the structures did not sink into the sand. In dramatic contrast, concrete reef balls nearby caused scour and undermining that made them dig their way down through the sediment, and many of them sank down and were almost completely buried.

The forces on a structure can be calculated from the Universal Drag Equation:

$$F_d = 0.5 C_d D A V^2$$

where F_d is the drag force in the direction of the flow, C_d is the drag coefficient, D is the density of the liquid, A is the cross-sectional area perpendicular to the direction of flow, and V is the velocity. To compute the forces on the entire structure, one needs to compute the forces on each element and integrate them. This can be done with numerical finite element mathematical analysis. When the drag force exceeds the shear strength of the structure, it breaks or bends.

The key parameter is the drag coefficient, which depends strongly on the shape and orientation of each structural element. The drag coefficient for a flat sheet or plate parallel to the flow is 0.001 for laminar flow and 0.005 for turbulent flow, but for one perpendicular to the flow it is 1.28, roughly a thousand times higher. For example, the drag coefficient of the Empire State Building and the Eiffel Tower are in the range of 1.3–2 (http://en.wikipedia.org/wiki/Drag_coefficient). On land, the stresses are predominantly the vertical force of gravity, and horizontal wind forces are small because the density of air is about 90 times less than water. In the ocean, where gravitational forces are reduced by buoyancy, and horizontal forces of breaking waves are orders of magnitude higher, most buildings would be knocked flat. During Hurricane Andrew, every conventional artificial reef made from sunken ships in southern Florida was destroyed, even in deep water, and the fragments scattered over huge distances. In contrast, the volume of the Biorock reefs were over 99% water, and the drag forces were minor as the wave surge passed through the structures without destroying them. If the Biorock reef had been a solid structure, it would have been ripped apart.

BIOROCK SHORE PROTECTION IN THE MALDIVES

The Maldives is one of the lowest-lying countries in the world, and will be one of the first to be lost in case global sea level rises. Every one of the 1200 islands has erosion problem. The capital island, Male, is surrounded by a rock wall, since all the beaches vanished after the nearby reefs were mined for construction material. Every resort island is surrounded by a rock wall made from a living reef that was mined and killed, or by sandbags. The rock walls rapidly fail and need constant rebuilding, while the sandbags are ripped apart by waves, leaving shredded plastic on the beach, and must be constantly replaced with new ones (Figures 3.8 through 3.11).

A 50 m long Biorock shore protection project was constructed in the Maldives in front of a severely eroding beach with a 1 m (3–4 ft.) high erosion scarp, with trees falling into the sea and buildings and decks threatened with imminent collapse (Figures 3.12 and 3.13). The Biorock reef was an open mesh framework made of welded steel bars, with corals transplanted on top, but almost entirely empty inside. When it was built, waves passed through it as if it were not there, but as the minerals and corals grew, the waves interacted visibly with the structure, slowing down as they passed through it and depositing sand around it and on the formerly eroding beach. Within three years, the beach grew by 15 m (50 ft.). The cost of the project was around one-tenth of what a concrete wall the same dimensions would have cost. Photographs showing the evolution of this project can be found at Goreau et al. (2004).

Figure 3.8 A day at the beach. The shoreline of Male, the capital island of the Maldives, is entirely lined by concrete costing 13 million US dollars per kilometer. (Photograph by Wolf Hilbertz, 1997.)

Figure 3.9 Sandbags to protect the beach from erosion at Ihuru Island, North Male Atoll, Maldives, now the Banyan Tree Angsana Maldives Ihuru Resort. As the wind reverses direction every six months, the sandbags have to be rebuilt twice a year at opposite ends of the island. (Photograph by Wolf Hilbertz, 1997.)

Figure 3.10 Breakwaters that surround Furanafushi Island, North Male Atoll, Maldives, now the Sheraton Full Moon Resort, are made up of corals mined from the reef. An entire reef was killed to make these walls. (Photograph by Wolf Hilbertz, 1997.)

Figure 3.11 Closeup of the Furanfushi breakwater. Expensive plastic-coated galvanized steel mesh was used for the installation less than a year before, and already the steel in starting to rust. Such structures rarely last more than a year or two before collapsing and needing to be rebuilt. (Photograph by Wolf Hibertz, 1997.)

The Biorock reef had 50 times higher survival of corals than the surrounding reef after the catastrophic heat stroke mortality of 1998, attracting dense fish populations from surrounding newly dead reefs, and making it a huge tourism attraction, which a rock wall would not have done. The Biorock reef, and the beach behind it, suffered no loss of corals or sand from the December 26, 2004 tsunami, which passed right over the island (Figures 3.14 through 3.18).

Figure 3.12 The beach on the south side of Ihuru Island before the start of the project in 1997. Sandbags have been piled in front of a meter and half high erosion scarp because trees were falling into the sea, and buildings about to collapse. (Photograph by Wolf Hilbertz.)

Figure 3.13 Necklace shore protection reef at Ihuru Island is located just in front of the eroding beach in the previous figure. Here, it is shown shortly after construction in 1997 after corals had been transplanted onto it. These corals were found loose on nearby reefs and attached to the Biorock structure. At this point, there is no limestone coating on the rebar. (Photograph by Wolf Hilbertz. For more photos of the development of this project see http://globalcoral.org/MALDIVES%20SHORELINES.%20GROWING%20A%20BEACH.htm.)

BIOROCK SHORE PROTECTION IN INDONESIA

The Biorock Anti-Wave (BAW) is the latest and most rapidly effective form of Biorock shore protection yet developed. The first 12 units were deployed at Gili Trawangan, Lombok, Indonesia, at the end of December 2008 (Figures 3.19 and 3.20). The name comes from their action in breaking up wave energy, and from their shape like an upside-down wave, or anti-wave. Waves were observed to slow down as they passed through them immediately after installation, and this effect was greatly increased when they were filled with rocks.

The BAW is an extremely stable structure that is rapidly and cheaply constructed in any size. It cements itself to bedrock and acts very effectively to slow waves. Units can be packed with rocks for immediate effect, or left empty, as in the Maldives, or internal steel mesh compartments can be made, some of which are rock-filled and others left empty to provide habitat for fishes. The units can be deployed side by side in a row, but are most cost effective when deployed as staggered rows. They can be subtidal or intertidal (when they will grow only while submerged, and from the bottom up).

Two days after the units were installed in the intertidal, the rust had already largely disappeared, and white limestone rock was starting to grow on the structures from the bottom up, cementing

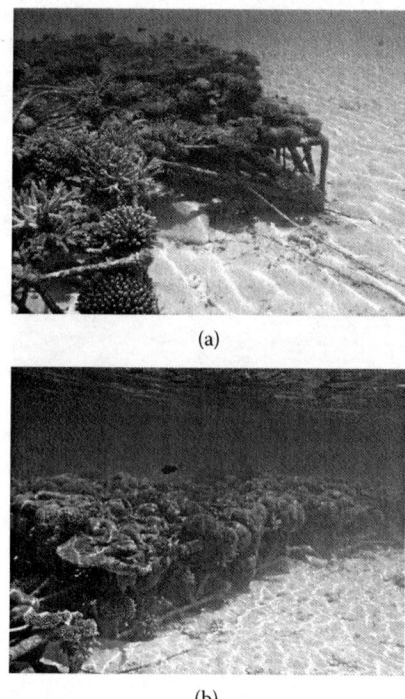

Figure 3.14 (a, b) After 15 years, there is now a well-developed reef, slowing down the waves. (Photograph by Azeez Hakeem on April 20, 2012.)

Figure 3.15 (a, b) After 15 years, the bright corals and fish on the mature reef is one of the major snorkeling attractions for Maldivian ecotourism. (Photograph by Azeez Hakeem on April 20, 2012.)

Figure 3.16 Within a couple of years, 15 m (50 ft.) of new beach grew and stabilized behind the necklace. Here the width of the beach is being measured. When the project began, the deck and the beach bar building at right were about to collapse into the sea, and the hotel had frantically piled sandbags in front of it, but were certain that there was no way they could save it and that they would have to build a new structure further from the shore. The Asian Tsunami passed over the island but did not damage the Biorock reef or the beach. (Photograph by Azeez Hakeem, 2003.)

Figure 3.17 The necklace Biorock reef is the dark line in front of the new beach. The sandbags in Figure 3.12 were located around the middle of the shore. (Photograph by Azeez Hakeem, 2003.)

them to the base rock and rubble. Within two weeks of installation, sand was already accumulating behind them, even though it was the height of the monsoon shore-erosion season at the site. After four months, there was heavy growth of oysters and mussels on the structures, and coral rubble and sand was piling up in front of and behind the BAWs. This was despite the fact that this period was the monsoon season with very heavy wave activity that had caused severe erosion of beaches on the island. The accumulation of sand and reef rubble in front and behind these structures increased greatly during the following calm season. They first reduced erosion of the beach, followed by beach growth as the BAWs became progressively more effective.

Figure 3.18 Biorock grown reef remains stable 13 years after the Biorock reef was built. (Photograph by Norman Quinn, 2010.)

Figure 3.19 Heavy erosion in the monsoon season was causing trees to collapse into the sea and resort owners to pile sandbags along the shore at Gili Trawangan Island, Lombok, Indonesia. (Photograph by Delphine Robbe, 2008.)

Figure 3.20 Biorock Anti-Wave structures placed in front of this beach, shown at low tide. High wave energy and currents prevent any accumulation of sand on top of the reef flat, which consists of dead coral rubble. Sand produced on the reef flat either is washed into deeper water or onto the beach. (Photograph by Tom Goreau, 2009.)

RESTORING REEFS TO GROW BACK BEACHES AND PROTECT COASTS

Within eight months after installation, the growth of new beach behind each of these projects was clearly visible on Google Earth (Figures 3.21 and 3.22). Trees that had their roots exposed and were about to fall down were now protected by a new broader beach. A seawall at one resort that was being undermined by erosion at the start of the project had sand pile up in front of it (Figure 3.23). In contrast, nearby properties that did not have Biorock projects continued to erode (Figures 3.24 and 3.25). A seawall built on an adjacent property had completely collapsed and fallen into the water a year later. It should be noted that the Biorock reef did not interfere with long-shore drift of sand along the beach the way that groynes would have done. Instead they acted by reducing the wave energy at the shoreline by absorbing energy offshore.

Figure 3.21 Google Earth image of Biorock Anti-Wave (BAW) sites 1 and 2, eight months after installation, shows beach already starting to grow. The BAW structures are clearly visible. The right-hand site of growth is where the trees had been falling into the sea eight months earlier in Figure 3.17.

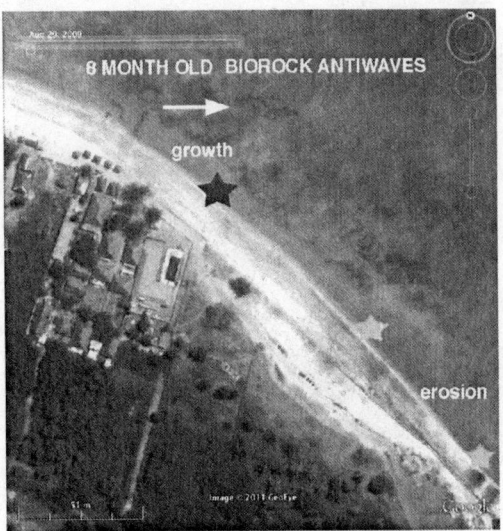

Figure 3.22 Google Earth image of Biorock Anti-Wave site 3, eight months after installation. Beach growth has already taken place behind them at the left star. The two right-hand star sites continue to erode.

Figure 3.23 Site marked by left star in Figure 3.20, one year after installation. At start of the project, the seawall was being undermined and about to collapse. After one year, half a meter of sand had built up in front of it. Biorock Anti-Wave structures out of image, to the right. (Photograph by Tom Goreau, 2009.)

Figure 3.24 Site of middle star in Figure 3.20, one year after installation. This area, which was not protected, is continuing to erode, trees are falling into the sea, and sandbags have been placed to prevent the coastal road from being washed away. (Photograph by Tom Goreau, 2009.)

Figure 3.25 Site of right-hand star in Figure 3.20. A seawall built a year and a half before has already been undermined and collapsed into the sea. (Photograph by Tom Goreau, 2009.)

Figure 3.26 Design of part of a 400 m long Biorock Anti-Wave shore protection project to be built in 2012 to protect a severely eroding beach at the largest resort in Jamaica's major tourism area. (Design by Thomas Sarkisian.)

Despite these immediate results from the demonstration pilot project, the project was too small to do more than create a very localized effect, and to protect the entire area it needed to be expanded. However, the shore resorts failed to expand it or even keep the projects under power, preventing rusting and increasing strength. As a result, the benefits were purely mechanical, and will be completely lost when the structures rust and collapse. By early 2012, an extreme storm caused massive erosion along the shore that could have been prevented had they maintained the power and expanded it on the scale needed.

We are confident that Biorock reefs will prove to be the most cost-effective form of shore protection known. Plans are already under way to apply them in many more places around the world (Figure 3.26). Wherever coral or oyster reefs once protected the shores are now gone, they can be grown back with Biorock Technology, using energy from the sun, winds, waves, and currents. Biorock reefs not only protect the shore against erosion, they are also sand factories from the rapid growth of corals, oysters, and sand-producing calcareous algae that grow on them, so they can produce new sand to replace that lost from erosion. Tropical islands now vanishing from global sea-level rise can be protected with Biorock reefs that can keep up with sea-level rise. Artificial islands can readily be grown on shallow seafloor.

REFERENCES

Airoldi, L., M. Abbiati, M. W. Beck, S. J. Hawkins, P. R. Jonsson, D. Martin, P. S. Moschella, A. Sundelof, R. C. Thompson, and P. Aberg. 2005. An ecological perspective on the deployment and design of low-crested and other hard coastal defense structures. *Coastal Engineering* 52:1073–1087.

Cyranoski, D. 2011. Japan faces up to failure of its earthquake preparations. *Nature* 471:556–557.

Goreau, T. J. 2011. Fools' paradises and castles on the sand: Coastal erosion and global sea level rise. http://www.globalcoral.org/Fools%20paradises.1.pdf (accessed August 25, 2012).

Goreau, T. J. 2012 (in press). Marine electrolysis for building materials and environmental restoration. In V. Linkov (Ed.), *Electrolysis*, Zagreb: InTech Publishing.

Goreau, T. J., and W. H. Hilbertz. 2005. Marine ecosystem restoration: Costs and benefits for coral reefs. *World Resource Review* 17: 375–409.

Goreau, T. J., W. H. Hilbertz, A. Azeez, and A. Hakeem. 2004. Maldives shorelines: Growing a beach. http://globalcoral.org/MALDIVES%20SHORELINES.%20GROWING%20A%20BEACH.htm (accessed August 25, 2012).

Goudas, C., G. Katsiaris, V. May, and T. Karambas (Eds.). 2003. *Soft Shore Protection: An Environmental Innovation in Coastal Engineering*, Kluwer, Dordrecht, The Netherlands.

Gourlay, M. R., and G. Colleter. 2005. Wave generated flow on coral reefs: An analysis for two-dimensional horizontal reef-tops with steep faces. *Coastal Engineering* 52:353–387.

Hardy, T. A., and I. R. Young. 1996. Field study of wave attenuation on an offshore coral reef. *Journal of Geophysical Research* 101:14311–14326.

Hight, C., J. Anderson, M. Robinson, and D. Wallace. 2011. *Atlas of Sustainable Strategies for Galveston Island*, Rice University School of Architecture, Houston, TX.

Hilbertz, W. H. 1979. Electrodeposition of minerals in sea water: Experiments and applications. *IEEE Journal on Oceanic Engineering* 4:1–19.

Hilbertz, W. H. 1992. Solar-generated building material from seawater as a sink for carbon. *Ambio* 21:126–129.

Hilbertz, W. H., and T. J. Goreau. 1996. Method of enhancing the growth of aquatic organisms, and structures created thereby. Patent 5,543,034. United States Patent Office.

Monismith, S. G. 2007. Hydrodynamics of coral reefs. *Annual Reviews of Fluid Mechanics* 39:37–55.

Munk, W. H., and M. C. Sargent. 1948. Adjustment of Bikini Atoll to ocean waves. *Transactions of the American Geophysical Union* 29:855–860.

Pilkey, O. H., and K. L. Dixon. 1996. *The Corps and the Shore*. Island Press, Washington, DC.

Pilkey, O. H., and H. L. Wright. 1988. Seawalls versus beaches. *Journal of Coastal Research* 4:41–64.

Pilkey, O. H., and R. Young. 2009. *The Rising Sea*. Island Press, Washington, DC.

Rahmstorf, S. 2010. A new view on sea level rise: Has the IPCC underestimated the risk of sea level rise? *Nature Climate Change* 4:44–45.

Rahmstorf, S., A. Cazenave, J. A. Church, J. E. Hansen, R. F. Keeling, D. E. Parker, and R. J. Somerville. 2007. Recent climate observations compared to projections. *Science* 316:709.

Samarawickrama, S. P., S. Balusubramanian, H. J. S. Fernando, and S. S. L. Hetiarachchi. 2008. Investigating the hydraulic performance of coral reefs by simulated porous structures. *Proceedings of COPEDEC VII* 113:1–9.

Schiereck, G. J. 2004. *Introduction to Bed, Bank, and Shore Protection*. Taylor & Francis, London.

Shyue, S. W., and K. S. Yang. 2002. Investigating terrain changes around artificial reefs by using a multi-beam echosounder. *ICES Journal of Marine Science* 59:S338–S342.

Vermeer, M., and S. Rahmstorf. 2009. Global sea level linked to global temperature. *Proceedings of the National Academy of Sciences* 106:21527–21532.

Wells, L., F. Perez, M. Hibbert, L. Clervaux, J. Johnson, and T. J. Goreau. 2010. Effect of severe hurricanes on Biorock coral reef restoration projects in Grand Turk, Turks and Caicos Islands. *Revista Biologia Tropical* 58:141–149.

Wiegel, R. L. 1992. *Oceanographical Engineering*. Dover, Mineola, NY.

Wolanski, E. 1994. *Physical Oceanographic Processes of the Great Barrier Reef*. CRC Press, Boca Raton, FL.

Zhang, K., B. C. Douglas, and S. P. Leatherman. 2004. Global warming and coastal erosion. *Climatic Change* 64:41–58.

CHAPTER 4

Reef Restoration Using Seawater Electrolysis in Jamaica*

Thomas J. Goreau and Wolf Hilbertz

CONTENTS

Introduction 35
Materials and Methods 36
Results 38
Discussion 43
Acknowledgments 44
References 45

INTRODUCTION

Jamaican coral reefs are undergoing accelerating deterioration wherever human activity physically disturbs reefs, degrades water quality, or overharvests key species (Goreau 1992). Many reefs no longer function as vital ecosystems: the coral-dominated wave-resistant upward-growing structures are turning into benthic ecosystems with a minor component of isolated corals. These ecosystems are coral communities rather than coral reefs, because biodiversity is severely degraded and the reef structure, being bioeroded faster than it grows, is less able to protect shorelines, keep up with rising sea level, or provide shelter and food for the many other organisms that live between corals. Degraded reefs have fleshy algae dominant over calcareous algae and can no longer provide beach sand to replenish that lost to erosion after damaged reef crests allow increased wave energy to reach the shore.

Several types of "artificial reefs" have been built as wave-resistant barriers and hiding places for fish (Goodwin and Cambers 1983). They have a poor record, because structures built from steel, poured concrete, stone, concrete blocks, gabions (wire baskets containing rocks), sandbags, sunken ships, wrecked airplanes, or old automobiles unavoidably rust, corrode, and are broken by waves. Their fate is ultimate destruction by storms, requiring expensive and inevitably futile replacement. They turn into dangerous projectiles in hurricanes. After hurricane Andrew hit southern Florida,

---

* This chapter, the first to describe the results of the first eight years of work on reef restoration using mineral accretion, was presented at the Eighth International Coral Reef Symposium in Panama in July 1996 and is published here for the first time.

Figure 4.1 Top of artificial reef after two years of mineral accretion growth. This structure reaches within 0.5 m of the surface. (Photograph by Dr. Peter D. Goreau.)

a survey of "artificial reefs" found that all had moved. Some had one or several fragments found, but many had vanished entirely. Although fish will hide behind any underwater obstacle, hard corals will not colonize them for a very long time, if ever, and they are mainly colonized by soft corals and sponges rather than reef-building corals. "Artificial reefs" made of automobiles (a popular excuse for creating marine junkyards) rust and break apart before corals will settle on them (Goodwin and Cambers 1983). The failure of exotic materials to instigate natural hard coral reefs is caused by unsuitable surface chemistry and leaching of toxic hydrocarbons and metals from engines, paints, plastic fillers, concrete, and steel.

A novel technology, developed by architect W. Hilbertz in the 1970s, uses electrolysis of seawater to precipitate calcium and magnesium minerals to "grow" a crystalline coating over artificial structures to make construction materials (Hilbertz 1975, 1979). The mineral accretion, largely aragonite ($CaCO_3$) and brucite ($Mg(OH)_2$), is very similar in chemical and physical properties to reef limestones (Hilbertz 1992), which are primarily the remains of the aragonite skeletons of corals and green calcareous algae. This chapter describes the results of work done in Jamaica since 1988 building and growing mineral-accretion artificial reefs for enhanced coral growth and reef restoration (Figure 4.1).

MATERIALS AND METHODS

Electrolysis of seawater results in mineral deposition at the cathode. The physical properties of the material depend on mineralogy and crystal size, functions of deposition rate, and electrical current parameters. Higher current densities result in faster growth but weaker material dominated by brucite, while lower current densities produce slower deposition dominated by harder aragonite (Hilbertz 1992). Mineral accretion materials have a mechanical strength comparable to, and often greater than, concrete (Hilbertz 1979).

Deposition of minerals results from alkaline conditions created at the cathode by the reduction reaction:

$$2H_2O + 2e^- = H_2 + 2OH^-$$

which precipitates calcium and magnesium minerals from seawater:

$$OH^- + HCO_3^- + Ca^{++} = CaCO_3 + H_2O$$

$$2OH^- + Mg^{++} = Mg(OH)_2$$

In contrast, the anode becomes acidic due to

$$2H_2O = 4H^+ + O_2 + 4e^-$$

and highly oxidizing conditions result in

$$2Cl^- = Cl_2 + 2e^-$$

The sum of the net reactions at both electrodes should be neutral with regard to hydrogen ion production, and hence with regard to CO_2 generation through acid–base equilibrium and carbonic acid hydrolysis:

$$2HCO_3^- = CO_3^{-2} + CO_2 + H_2O$$

Cathodes and anodes can be made in any size and shape, with current flow dependent on their spacing and surface area. Typically, the cathode is built out of expanded steel mesh constructed as simple geometric forms such as cylinders, sheets, triangular prisms, or pyramids, but we have also molded complex forms using square mesh or chicken-wire mesh. An experimental reverse catenary was even built supported by floating spheres—that is, a buoyant metal chain structure fixed to a cathodic base plate. This structure was initially flexible and became rigid with progressing mineral accretion. Other new applications include molding shapes out of powdered sand or other materials, containing a cathode to enhance cementation by mineral accretion.

Pilot artificial reefs have ranged up to 3 m high and 10 m across, but there is no theoretical limit on their size, provided sufficient current density is applied. Anodes are typically much smaller than cathodes and shaped as sheets, rods, or mesh, depending on the materials used. Cathode materials are entirely protected from rusting by reducing conditions, whereas anodes are subjected to rapid oxidation unless resistant material is used. We have used a wide variety of anode materials, including lead, graphite, and steel, but had best results with specially coated titanium (Figure 4.2).

Although any direct-current source will work, our preference is to use solar- and wind-generated power rather than alternating current generated from renewable fossil fuels that pollute the atmosphere with CO_2 and acid rain. Current is applied across the terminals from a variety of power sources. We have empirically found it best to use lower voltages and higher currents. We have used transformers and battery chargers at both 12 and 6 V, photovoltaic panels in a direct-charge mode at a range of voltages between 3.8 and 17 V, and have plans to use windmill-generated current as well. Electricity consumption of each structure is equivalent to a single light bulb. These current levels are entirely safe to swimmers and divers, and it is possible to feel only a slight tingle when one directly short-circuits the current by touching both anode and cathode simultaneously with bare hands. Electrical currents are transmitted using insulated copper cables, either mono- or multistrand. Anode cable connectors are protected by clear silicone to detect the green color formed if salt water corrodes the electrical contact.

Small pieces of corals were transplanted onto the structures and attached with plastic ties, iron wire, or monofilament line, or simply allowed to sit on them. These corals largely consisted of fragments, which had been naturally broken by storms and damaged by anchors, divers, or spearfishermen, or corals whose bases were so bioeroded that they would be broken by storms, as well as

Figure 4.2 Artificial reef seen from above. This structure is 8 ft. (2.2 m) tall. The prolifically branching *Porites porites* facing the camera grew from 10 to 30 cm in diameter in two years following transplantation onto the structure. The sides of the structure are largely covered with the red calcareous alga *Jania*. The base of the structure is solidly cemented to the limestone bottom by mineral accretion. The three hemicylindrical ballast chambers around the base were filled with rocks to stabilize the structure before mineral accretion attached it to the bottom. The chamber at the right contained an anode, and mineral accretion is seen to be much greater than on the chamber at left, due to the higher electrical current density. (Photograph by Dr. Peter D. Goreau.)

small pieces of branching corals from nearby "control" colonies. Most species of Caribbean corals have been tried (see Table 4.1 later in the chapter). Only corals were attached, but these included some epifaunal sponges, calcareous algae, and other organisms on their undersides.

Artificial reef structures have been built in depths ranging from 0.5 to 7 m, in locations ranging from extremely protected back reef sites, open sites on the leeward western end of the island, to open shores fully exposed to the direct impact of winter northers. One structure, located in a depth of 1.5 m, continues to work despite being exposed to breaking waves that can reach up to 7 m high. They have been built on seagrass beds, limestone hard ground, white sand, and mud bottoms. We also built control structures receiving different current levels or no current at all and structures that were allowed to accrete for a period of time and were then turned off. In addition, we have connected corals growing *in situ* directly to current sources via wires leading to artificial reefs.

RESULTS

Crystal growth and hydrogen gas bubbling began as soon as rust on the steel had been reduced to iron. The surface changes from red to black to gray and then white as minerals grow on it. Minerals accreted to a thickness of up to 20 cm over three years. Iron and steel remain bright and shiny as long as sufficient electrical current flows to maintain cathodic protection. They are protected from corrosion by overlying mineral layers after the current is turned off, unless this coating is broken.

Structures on limestone hard ground became solidly cemented onto it, while those on sand and mud remain loosely attached and are vulnerable to being toppled in severe storms.

In almost all cases, transplanted corals healed quickly and were cemented solidly onto the mineral accretion within weeks (Figures 4.3 through 4.7). They showed bright, healthy tissue pigmentation and prolific polyp feeding-tentacle extension. However, some *Acropora cervicornis* have been killed by bristle worm (*Hermodice carunculata*) and gastropod (*Coralliophila*) attack, and some were broken by severe storm waves. Transplanted corals grew skeletons at rates faster than the highest values measured in the field (Gladfelter et al. 1978), even though all sites had suboptimal water quality. Growth rates were determined by periodically measuring the diameter of colonies with a ruler or by measurements from sequential photographs or video images. Fragments of *Porites porites* grew from 5 to 30 cm across in two years. *Acropora cervicornis* branched prolifically and grew 5–8 cm in just 10 weeks. The tissue of corals attached to the structures via wires soon began to grow over the mineral accretion. Such corals were visibly brighter than adjacent corals of similar species, but became less colorful when the current was turned off for periods of up to two months and then regained bright pigmentation within days when the current was restored. Young corals colonized and grew on the mineral accretion. We found juvenile coral colonies up to 1 mm in diameter at densities of around 0.7 cm^{-2} on 3-year-old artificial reef substrate. One artificial reef was colonized by around a hundred young *Agaricia agaricites* and *Favia fragum* in two years, and these grew to a size of several centimeters across, in a polluted lagoon where little or no natural recruitment was observed.

Figure 4.3 Close-up of *Porites porites* growing on the artificial reef, showing prolific polyp extension. (Photograph by Dr. Peter D. Goreau.)

Figure 4.4 Close-up of *Porites astreoides* growing on the artificial reef. (Photograph by Dr. Peter D. Goreau.)

Figure 4.5 Close-up of *Diploria strigosa* growing on the artificial reef. (Photograph by Dr. Peter D. Goreau.)

Figure 4.6 Close-up of the artificial reef showing young colonies of *Acropora palmata* and *Porites divaricata* that spontaneously settled and grew on the structure. No *Acropora palmata* was observed growing in the surrounding back reef. (Photograph by Dr. Peter D. Goreau.)

Figure 4.7 *Acropora cervicornis* growing on an artificial reef in Negril, Jamaica. (Photograph by Wolf Hilbertz.)

Except for transplanted corals and a few small organisms encrusting their bases, all other species found on the artificial reef spontaneously settled on it or migrated to it. A highly diverse coral reef community (Table 4.1) established itself on the mineral accretion structures, including foraminifera, cyanobacteria, chlorophytes, rhodophytes, phaeophytes, porifera, bryozoans, cerianthids, coralliomorpharia, gorgonaceans, sabellid, serpulid, and nereid polychaetes, oysters, gastropods, octopods, squids, echinoids, holothurians, ophiuroids, crinoids, cleaning shrimp, crabs, hermit crabs, and spiny lobsters. A large variety of adult and juvenile fish became permanent or temporary residents, including morays, trumpetfish, squirrelfish, sea bass, fairy basslets, cardinalfish, grunts, drums, butterfly fish, angelfish, damselfish, wrasses, parrot fish, blennies, gobies, surgeonfish, filefish, and porcupine fish. The geometry of the structure appears to strongly affect the type of species recruited. Dolphins were observed swimming near the structures. No organism was observed to show aversive behavior.

The main difference between our artificial reefs and nearby natural reefs was the preponderance of fleshy algae, which were overgrowing corals on nearby reefs, while the artificial reefs had balanced coral and algal growth and the algae were predominantly sand-producing calcareous reds and greens with a much lower density of weedy algae than adjacent natural reefs. Large masses of calcareous *Jania*, *Amphiroa*, and *Halimeda* grew on the sides of the structures, generating sand. Mineral-accretion structures whose power was turned off subsequently had their calcareous algae

Table 4.1 Partial List of Species Seen on Artificial Reefs

These are species that were regularly found on the artificial reefs, including corals that had been transplanted, attached organisms that had spontaneously settled and grown on them, and mobile organisms that were repeatedly found on the structures and appeared to have taken up permanent residence on them. Many other organisms were also seen on the structures once or twice, but these are not listed, as they may have been only temporarily passing by.

Scleractinian Hexacorallia

Stephanocoenia michelinii, Madracis decactis, Madracis mirabilis, Acropora palmata, Acropora cervicornis, Agaricia agaricites, Agaricia tenuifolia, Agaricia fragilis, Helioseris cucullata, Siderastrea radians, Siderastrea siderea, Porites porites, Porites furcata, Porites astreoides, Porites branneri, Favia fragum, Diploria clivosa, Diploria strigosa, Diploria labyrinthiformis, Manicina areolata, Colpophyllia natans, Montastrea annularis, Montastrea cavernosa, Meandrina meandrites, Dichocoenia stokesii, Dendrogyra cylindrus, Mussa angulosa, Isophyllia sinuosa, Isophyllastrea rigida, Mycetophyllia ferox, Mycetophyllia lamarckiana, Eusmilia fastigiata

Other Invertebrates

Porifera	*Iotrochota birotulata*
Hydrozoans	*Millepora alcicornis, Stylaster roseus*
Cerianthids	*Ceriantharia* sp.
Actiniaria	*Lebrunia danae*
Coralliomorpharia	*Ricordea florida*
Gorgonaceans	*Briareum asbestinum*
Sabellid	*Sabellastarte magnifica*
Serpulid	*Spirobranchus giganteus*
Nereid	*Hermodice carunculata*
Terebellid	*Eupolymnia nebulosa*
Oysters	*Pteria colymbus, Isognomon* sp.
Gastropods	*Coralliophila* sp.
Octopods	*Octopus briareus*
Squid	*Sepioteuthis sepiodea*
Echinoida	*Diadema antillarum, Eucidaris tribuloides*
Holothuria	*Holothuria mexicana*
Ophiuroida	Various species
Crinoida	*Nemaster rubiginosa*
Shrimp	*Stenopus hispidus*
Isopods	*Anilocra* sp.
Crabs	*Percnon gibbesi, Mithrax* sp.
Hermit crabs	Various species
Lobster	*Panulirus argus*

Prokaryota

Cyanobacteria	*Schizothrix* sp.

Protozoa

Foraminifera	*Homotrema rubrum, Gypsina* sp.

Algae

Chlorophyta	*Halimeda opuntia, Caulerpa racemosa*
Rhodophyta	*Jania rubens, Galaxaura oblongata*
Phaeophyta	*Lobophora variegata*

Table 4.1 Partial List of Species Seen on Artificial Reefs (*Continued*)

Fish

Morays	*Gymnothorax funebris, Gymnothorax moringa, Muraena miliaris*
Trumpetfish	*Aulostomus maculatus*
Squirrelfish	*Holocentruus rufus, Myriprlstes jacobus*
Sea bass	*Epinephelus cruentatus*
Fairy basslets	*Gramma loreto*
Cardinalfish	*Apogon maculatus*
Grunts	*Haemulon flavolineatum*
Drums	*Equetus punctatus*
Butterfly fish	*Chaetodon ocellatus, Chaetodon striatus*
Angelfish	*Holacanthus tricolor, Pomacanthus paru*
Damselfish	*Eupomacentrus partitus, Eupomacentrus dorsopunicans, Eupomacentrus leucostictus, Abudefduf saxatalis*
Wrasses	*Thalassoma bifasciatum, Halichoeres maculipinna*
Parrot fish	Various species
Blennies	Various species
Gobies	Various species
Surgeonfish	*Acanthurus coelerus, Acanthurus chirurgus, Acanthurus bahiana*
Filefish	*Cantherhines macroceros*
Porcupine fish	*Diodon holocanthus*

and corals overgrown by fleshy algae. In sharp contrast to electrified structures, the control structures that received no current rusted and fell apart within months, and the rusting fragments were not colonized by corals or other organisms.

DISCUSSION

Rapid coral growth and recruitment even in areas of known poor water quality (Goreau 1992) show that our method is able to partly counteract coral-reef eutrophication due to coastal zone nutrient fertilization, and so can contribute to restoring damaged reefs and creating new ones in even degraded areas. As the structures become stronger with age, they are also able to contribute more and more to shore protection from waves and to keep pace with rising sea level. Unlike "artificial reefs" made of exotic materials, Biorock reefs get constantly stronger with time. As long as current is applied, they are self-repairing, since any cracks and breaks of mineral accretion are rapidly and preferentially filled in by new material. While some structures have been damaged by storm waves or impacting objects, such damage is easily repaired by itself as long as the electrical current is applied.

The stimulation of calcareous organisms of all types on the artificial reef, and the relative paucity of noncalcifying organisms, is probably largely due to the boost the former receive from locally alkaline conditions, which allow them to grow their skeletons at lower energetic cost because they do not have to use metabolic energy to pump protons away from calcification sites to maintain internal pH homeostasis (Goreau 1977). The bright colors of the colonies and their high degree of tentacle expansion may be due to the extra biochemical energy freed as a result. An alternative explanation could be due to the high density of electrons on the cathode, some of which may be trapped and used to generate adenosine triphosphate, but if this were the major factor, non-calcareous organisms would also be stimulated. This was later discovered to be the case. The general stimulation of growth of marine organisms by both processes on mineral-accretion substrates is covered under pending US patent 08/374993, issued to Hilbertz and Goreau (1995).

The view that mineral accretion is the most suitable substitute substrate for coral recruitment compared to natural limestone is supported by marine archaeology. Where only iron metal is found in shipwrecks, it rusts away and is not colonized by hard corals unless first covered by encrusting calcareous red algae. Where several dissimilar metals are found in wrecks as well, such as brass, bronze, copper, magnesium, or aluminum, the differing electromotive potentials of the metals results in electrolytic current flows that cause deposition of mineral accretion over the cathodic metals until the anodes are consumed, ending the reactions. Natural electrolysis is responsible for the preservation of most metal artifacts in shipwrecks dating as far back as the Bronze Age, which are found under thick concretions of limestone minerals. We have observed old iron anchors and chains completely covered with hard mineral accretion, allowing corals to settle and grow prolifically on them. This would probably not have happened without electrolytic mineral coatings and concurrent cathodic protection.

We believe that apart from protection of living reefs, mineral accretion is the best substitute for enhancing coral growth and restoring natural coral-reef ecosystems even under stressed conditions. Since the method is able to rely entirely on nonpolluting and renewable energy, it is suitable for remote areas. Laboratory experiments showed that 1.07 kg of mineral accretion was precipitated per kilowatt-hour of electricity. At Jamaican residential customer rates for imported fossil-fuel-generated electricity (US$ 0.10 per kilowatt-hour), resulting materials are nearly an order of magnitude cheaper than the equivalent weight in concrete blocks.

Typical costs for seawalls and breakwaters using conventional techniques run around US$13,000 per meter, the amount it cost the Maldives to replace mined-out reefs with stacked precast concrete tetrapod breakwater structures to protect the shore from erosion and the aquifers from saltwater intrusion. Unlike concrete blocks, mineral-accretion structures can be built in any size and shape, contain internal steel reinforcement, and get stronger with age rather than weaker. Submerged seawalls could therefore be built that would eventually become much stronger than concrete structures, at a fraction of the cost.

We expect that mineral-accretion technology will eventually become the preferred form of reef restoration and shore protection, where reefs have been degraded due to anthropogenic or natural causes, especially if sea level continues to rise more rapidly than coral reefs grow upward. The global average sea-level rise measured by the TOPEX/Poseidon radar satellite has been 3–4 millimeters per year, as fast as most healthy reef structures are accumulating but faster than degraded reefs or bleached corals can grow (Goreau and Macfarlane 1990).

ACKNOWLEDGMENTS

We are grateful for support from the European Union for the latest phase of artificial-reef construction under a grant to the Negril Coral Reef Preservation Society for the establishment of the Negril Marine Park. We thank Ursula Hilbertz-Rommerskirchen, Maya Goreau, Bill Wilson, Katy Thacker, Karen McCarthy, Martin Brinn, and Bert Bentley for assistance during artificial-reef construction and deployment. This work would not have been possible without permission from Richard Murray and David Cunninghame to use Tensing Pen property to house the power supply and their support for electricity bills. Anodes were donated by Heraeus Elektrochemie GmbH, Germany.

REFERENCES

Gladfelter, E., R. Monahan, and W. Gladfelter. 1978. Growth rates of five reef-building corals in the northeastern Caribbean. *Bulletin of Marine Science* 28: 728–734.

Goodwin, M. R., and G. Cambers. 1983. *Artificial reefs: A handbook for the Eastern Caribbean.* Caribbean Conservation Association. Barbados.

Goreau, T. J. 1977. Coral skeletal chemistry: Physiological and environmental regulation of stable isotopes and trace metals in *Montastrea annularis*. *Proceedings of the Royal Society of London—Series B: Biological Sciences* 193: 291–315.

Goreau, T. J. 1992. Bleaching and reef community change in Jamaica. *American Zoologist* 32: 683–695.

Goreau, T. J., and A. H. Macfarlane. 1990. Reduced growth rate of *Montastrea annularis* following the 1987–1988 coral bleaching event. *Coral Reefs* 8: 211–215.

Hilbertz, W. 1975. Towards self-growing structures. *Industrialization Forum* 6: 53–56.

Hilbertz, W. 1979. Electrodeposition of minerals in seawater: Experiments and applications. *Oceanic Engineering* 4: 94–113.

Hilbertz, W. 1992. Solar-generated building material from seawater as a sink for carbon. *Ambio* 21: 126–129.

Hilbertz, W., and T. J. Goreau. 1995. A method for enhancing the growth of aquatic organisms and structures created thereby. US Patent 5,543,034.

CHAPTER 5

Electrically Stimulated Corals in Indonesia Reef Restoration Projects Show Greatly Accelerated Growth Rates

Jamaludin Jompa, Suharto, Eka Marlina Anpusyahnur, Putra Nyoman Dwjja, Jobnico Subagio, Ilham Alimin, Rosihan Anwar, Syarif Syamsuddin, Thri Heni Utami Radiman, Heri Triyono, R. Ahmad Sue, and Nyoman Soeyasa

CONTENTS

Introduction 48
Methods 48
    Study Sites 48
        Barrang Lompo Island 49
        Samalona Island 49
        Pemuteran Village 49
        Gili Trawangan 49
Materials 49
Results 52
    Barrang Lompo Experiment 52
        Survival Rates 52
        Coral Growth Rates 52
    Samalona Experiments 53
        Water Quality 54
    Pemuteran Experiment 54
    Gili Trawangan Experiments 55
        Coral Growth 55
        Fish Populations 55
Discussion 57
Conclusion and Recommendation 57
Acknowledgments 58
References 58

INTRODUCTION

Coral reefs are an invaluable coastal ecosystem in tropical areas, because they provide not only significant amount of food for humans but also offer important physical protection for coastal areas. The hard structure of coral reefs is mainly formed or constructed by scleractinian corals (Class Anthozoa), by precipitating calcium carbonate into their skeleton (Veron 1993).

Indonesia has the largest area and highest biodiversity of coral reefs of any country in the world. In spite of their high economic value, Indonesian coral reefs face serious problems from heavy reef degradation mainly due to anthropogenic factors, such as destructive coral reef fishing (dynamite, cyanide, and overfishing), coral mining, pollution, and sedimentation. In addition, phenomena such as bleaching, *Acanthaster* outbreaks, and diseases also have been additionally threatening the future of coral reefs. These multiple stressors may cause serious effects on the sustainability of coral-reef fisheries as the sources of income for small-scale fishermen all over Indonesia, who provide most of Indonesia's food protein. In addition, many other ecosystem functions of coral reefs disappear or get worse, such as for coastal protection from erosion, habitat for spawning and nursery grounds of many valuable marine organisms, and source of medicine (Supriharyono 2000), as well as their use for marine ecotourism.

In response to these degraded coral-reef ecosystems, there have been significant efforts to introduce many different ways of rehabilitating the ecosystem by means of coral transplantation, artificial reefs, and Biorock. Many sunk reef structures made of concrete, shipwreck, old cars, and tires can sometimes function as artificial reefs, but become broken due to deterioration and wave action. Those structures that remain are usually dominated by algae, sponges, hydroids, soft corals, tunicates, and so on, while hard corals are rarely found to dominate (Goreau and Hilbertz 1996, 1998).

One method that has been developed for more advanced coral-reef rehabilitation is the mineral accretion (Biorock) method. This technique works by electrolysis using low-voltage electric current Hilbertz 1979; (Hilbertz and Goreau 1999; Lee 1997; Van Treeck and Schuhmacher 1997). Experiments by Hilbertz (1979) and Goreau and Hilbertz (Chapter 4) showed that minerals (mainly calcium carbonate or magnesium hydroxide) deposited on the substrate can grow up to 20 cm in two years, and corals transplanted onto them also grew faster; for example, *Acropora cervicornis* grew 5–8 cm in 10 weeks (Goreau and Hilbertz, Chapter 4). An additional advantage of the Biorock approach is that transplanted corals can better tolerate relatively poor water quality.

The amount of mineral deposited onto the substrate depends on the current and voltage. High voltage and current result in faster mineral precipitation, but it is less hard because the deposited mineral is dominated by brucite ($Mg(OH)_2$). With lower voltage and current, the electrolysis process is slower, resulting in slower mineral precipitation on the cathodes, but it is harder because the dominant mineral is aragonite limestone ($CaCO_3$). These mineral deposits are similar to those being made by coral skeleton (Hilbertz and Goreau 1998).

This research quantified the effect of different electric voltages on the growth and survival of various species of acroporid corals including *Acropora nobilis*, *Acropora formosa*, and *Acropora valenciennesi* (Wallace and Wolstenholme 1998) transplanted onto the Biorock structures at various places under different conditions. It is expected that this result will contribute in improving coral restoration, especially in specific areas where this technique is visible and strategic, such as close to resorts, hotels, or tourism sites.

METHODS

Study Sites

Studies to compare the growth rate of corals on the Biorock structures with genetically and environmentally identical controls not receiving electricity were conducted from 2002–2008 at several sites in Indonesia, including Pemuteran, Bali, the Spermonde Archipelago near Makassar,

Sulawesi, and Gili Trawangan, Lombok. The experiments were conducted at various depths on the reef flat or the reef slope. The following are the short descriptions of the main study sites:

Barrang Lompo Island

Barrang Lompo Island is a small, inhabited island around 15 km from Makassar city. The source of power was from a battery charger placed in a small house next to the beach. The Biorock structures were placed around 10 m from the house down to 4 m depth. The sea floor around this area is mainly soft sand and some seagrass and macroalgae. Coral reef patches develop only around 20 m from this site. The water transparency was relatively good in the study sites during this experiment.

Samalona Island

Samalona Island is a small, uninhabited island around eight km from Makassar city. The island has been a famous recreation destination, especially for local visitors. The Biorock structures were placed at around 5–9 m depth. The reefs around this island were in moderate condition with live coral cover around 30%. The water transparency was relatively good during dry season but becomes a bit turbid during rainy season, and it is affected by green water due to the proximity to Makassar, the largest city in Sulawesi.

Pemuteran Village

The location of this research site is in the Karang Lestari Reef Restoration Project in the Pemuteran Village Marine Protected Area, Gerokgak subdistrict, Buleleng Regency, Bali. The experimental units were placed around 4–6 m depth with relatively low visibility during the rainy season due to sediment runoff from the land.

Gili Trawangan

Details on the Gili Trawangan sites are provided in Arifin et al. (Chapter 6).

MATERIALS

There were two main components of this experiment at Samalona and at Barrang Lompo: the power supply and the Biorock transplantation structure. Electricity on this island is supplied by gasoline power generator, which runs only during nighttime. Therefore, we used a charger with battery unit to supply power to the electrodes (Figure 5.1) during the day, whereas during the night, we used power adaptors to supply the electrodes with lower-voltage electricity from charged batteries. To test the effects of different voltages, we used 6 and 12 V batteries during nighttime. These power supplies were connected to the electrodes using cables. On Bali and on Gili Trawangan, the projects were powered for 24 h a day from shore-based chargers running at 12 V at the source, although the actual delivered voltage at the end of the cables is less.

The second main part of the experiment is the transplantation unit, which was made of a standard steel frame for coral (*A. nobilis*) transplantation (Figure 5.2). The frame was made of reinforcing bar steel rods and wire mesh (cathode) on which the transplanted corals were attached, and the anode was positioned below the cathode (Figure 5.3). Both the cathode and anode were connected to the power supplies placed on the land (Figure 5.4). The coral fragments (around 10 cm) were attached to the cathode substrate using cable ties.

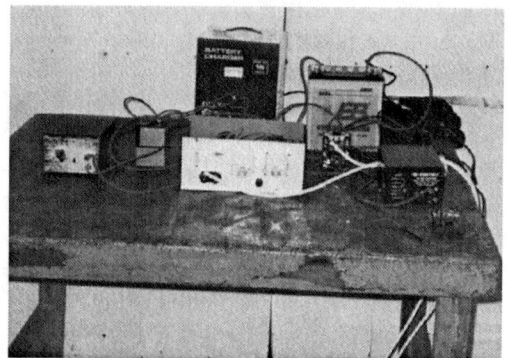

Figure 5.1 Power supply units using battery, battery charger, adaptor, and NYM cables at Barrang Lompo Experiment.

Figure 5.2 (a) Branching hard coral mother colony *A. nobilis* from which tips were fragmented (indicated by numbers). (b) Fragments transplanted onto the cathode frame.

Figure 5.3 Transplanted corals and experimental structure at Samalona Island.

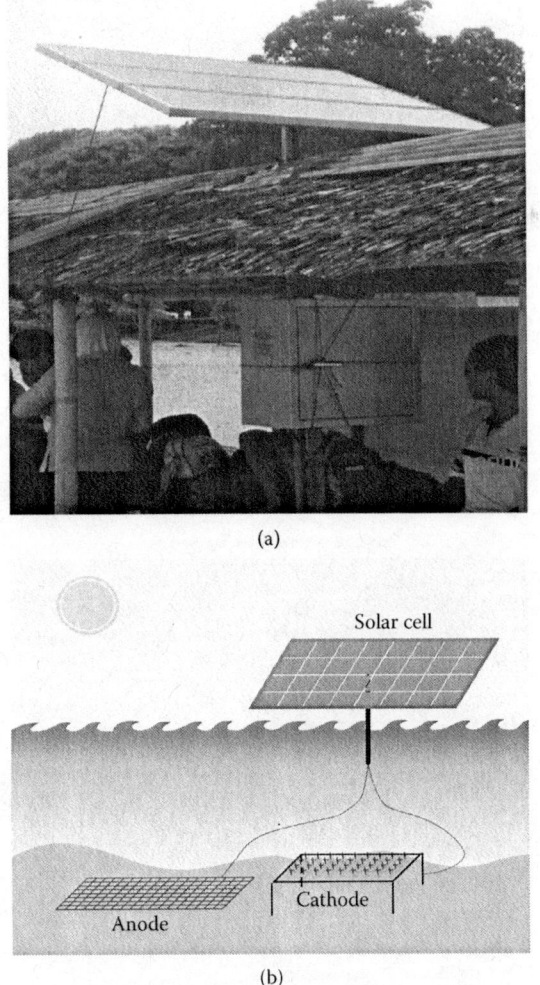

Figure 5.4 (a) Source of power (solar cell) used and (b) the experimental setup at Samalona Island.

RESULTS

Barrang Lompo Experiment

Survival Rates

In general, both transplanted corals and those living in the vicinity (control corals) survived throughout the period of experiment (3 months). However, there was a relatively low mortality (11%) on the 6 V treatment, due to external factors such as being eaten or removed by parrot fishes or entangled and removed by fish hooks. Therefore, this method at both voltage levels resulted in very good coral survival rates. The transplanted corals using this mineral-accretion method appear to survive better from algal overgrowth that usually invades transplanted corals, especially in a relatively low grazing level (Goreau and Hilbertz 1998; Jompa and McCook 2002a).

Coral Growth Rates

The growth rates of minerals at Barrang Lompo are presented in Figure 5.5, and the coral growth rates of each treatment are presented in Figures 5.6 and 5.7. About three times faster growth of minerals was found on the higher voltage structure, which is due to the higher electrical current caused by Ohm's law. The graphs show that all corals grew continuously. The total increase of coral length was measured every two weeks. Although there was a reduction of coral growth in the 6 V treatment, it was not due to the treatment but because some corals had their tips chewed off, possibly by parrot fishes. The average coral growth rates around 8 mm/2 weeks for 12 V treatment, 4 mm/2 weeks for 6 V, and 2 mm/2 weeks for control.

Coral transplants at this site without electrolysis had highest growth rates of 2.56 mm/2 weeks and lowest rate of 1.04 mm/2 weeks (Yuliantri et al. 2006). This result is lower than transplants using the electrolysis method. It was interesting that the coral fragments bitten off by the fish recovered very quickly and started to grow faster again in the following weeks. This is in line with the experience of Goreau and Hilbertz (1996) that corals transplanted using the electrolysis method tend to develop faster with better reproduction and more resistance to the natural disturbances.

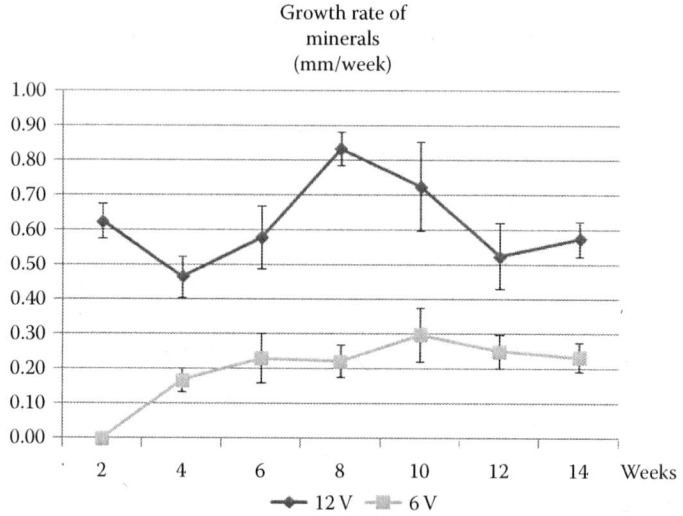

Figure 5.5 Growth of mineral coatings on structure (in millimeters) as a function of voltage.

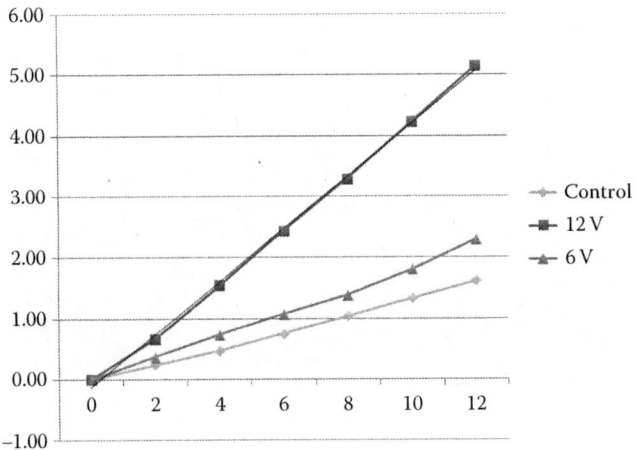

Figure 5.6 Changes in coral length (in centimeters) for each treatment.

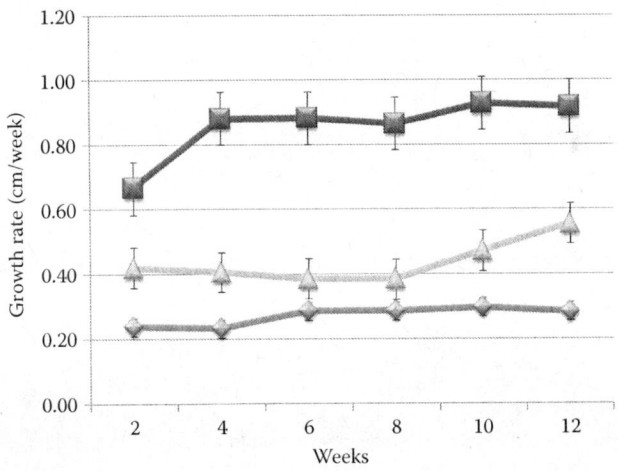

Figure 5.7 Coral growth rates for each treatment (control, 6 V, and 12 V). Squares are the 12 V treatment, triangles are the 6 V treatment, and diamonds are the controls.

Statistical analyses showed that the treatment (different voltages) resulted in significant differences ($P < .05$) in coral growth rates. The t-test demonstrates that the 12 V treatment gave the highest coral growth rate, whereas coral growth between the 6 V was intermediate and the controls were lowest. The 6 V treatment was affected by mortality from fish biting during the eighth week. Therefore, the effect of this mineral accretion method at both voltages resulted in a much better performance in terms of better growth rates.

Samalona Experiments

The results of this experiment show that coral growth rate at 5 m depth with electric power is significantly higher compared to other treatments (Figures 5.8 and 5.9). The electrified corals at

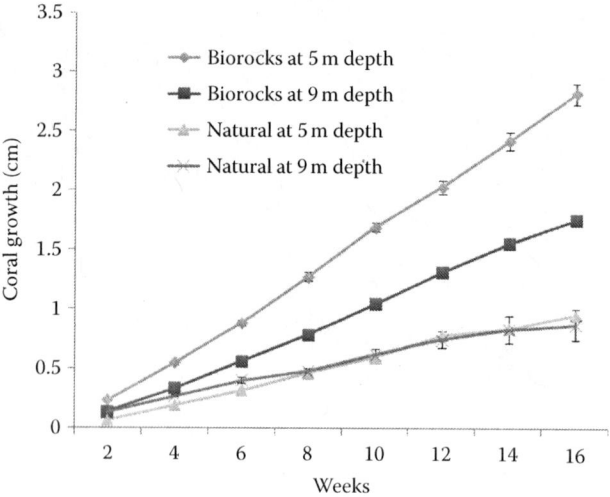

Figure 5.8 Changes in coral length for each treatment.

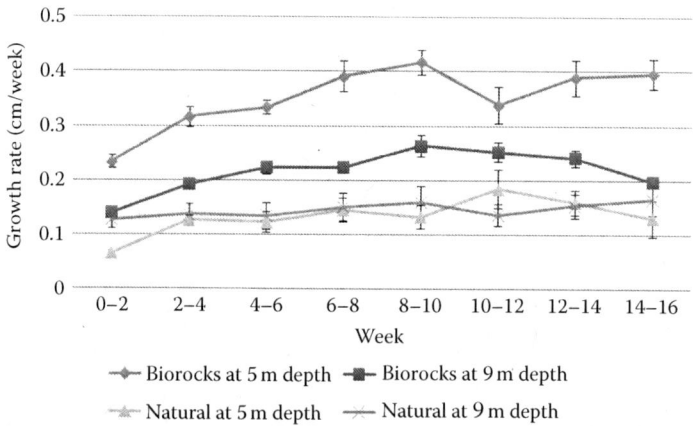

Figure 5.9 Coral growth rate (cm/week) for all treatments during 16 weeks experiment.

9 m depth were also significantly higher compared to those at nonelectrified corals (controls). These data indicate that all electrified corals (5 and 9 m depth) grew significantly faster than the controls (Figure 5.10).

Water Quality

The results of water quality measurements at all study sites showed that most of the parameters including temperature, pH, salinity, and water transparency were within a relatively normal range for seawater on coral reefs. Moreover, there were no significant fluctuations detected during study periods. Therefore, it can be assumed that the changes on coral growth rates observed here were mostly attributed to the effects of Biorock mechanism (Goreau and Hilbertz 1996).

Pemuteran Experiment

The results from Pemuteran experiments, (Figure 5.11) show that both at 4 and 6 m depth, corals on Biorock grew four times faster than controls. At this site, no difference with depth was

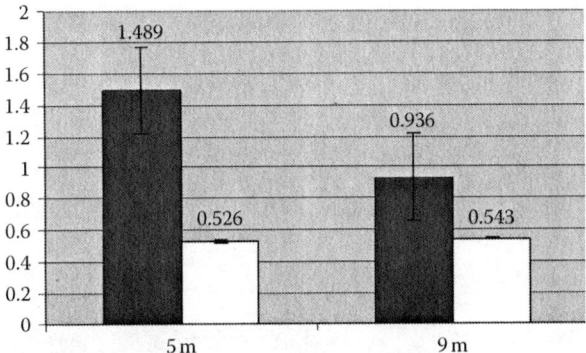

Figure 5.10 Comparison of groups. Black bars represent electrified corals and white bars represent controls.

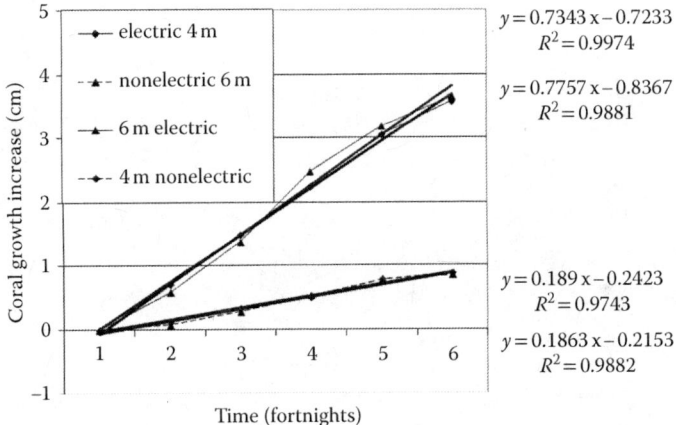

Figure 5.11 Coral growth (cm) measured every two weeks during experiment (12 weeks).

observed for either electrified or control corals. We also observed that corals on electrical substrate had darker color, and new branches emerged more rapidly than on the controls. This indicated clearly that the electrified corals significantly benefited from the electrolytic process.

Gili Trawangan Experiments

Coral Growth

Coral growth was measured for different corals, staghorn and table Acropora species, at different sites and at different depths, and compared to growth of undisturbed natural reef colonies as well as transplants to substrates that were not electrified. All electrified corals grew much faster than controls, up to around eight times faster, and both electrified and control corals grew faster in shallower depths (Figures 5.12 through 5.14).

Fish Populations

Fish populations were 6.25 times denser, 1.84 times more diverse, and 15.75 times more evenly distributed by species on Biorock structures than on surrounding natural reef areas (Figure 5.15). The control area had 3.6 times greater dominance of the most abundant species than the Biorock reef.

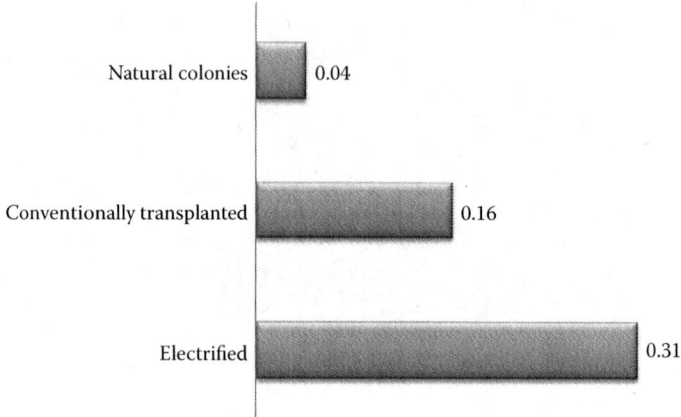

Figure 5.12 Growth rate (cm/week) of the staghorn coral *Acropora formosa* in three different treatments.

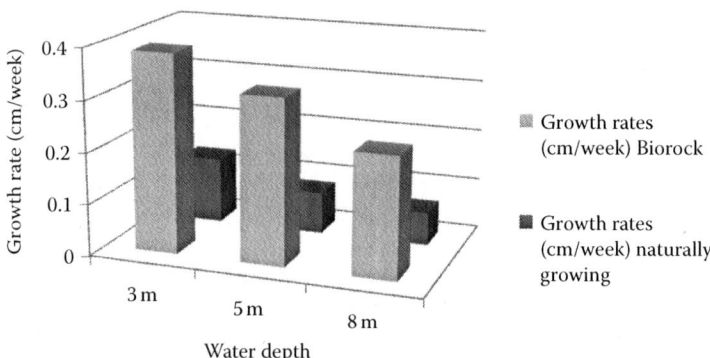

Figure 5.13 Comparison of *Acropora formosa* (cm/week) grown at three different depths versus local controls.

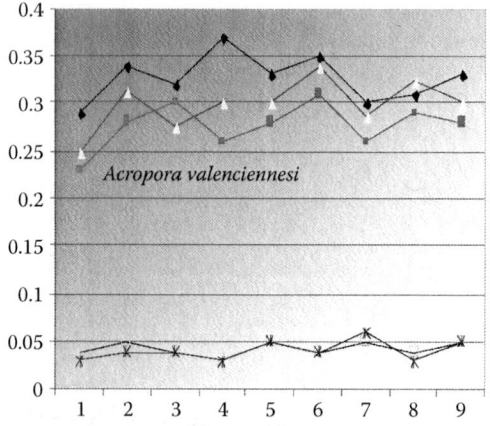

Figure 5.14 Comparison of growth rates (cm/week) over a nine-week period of the table coral *Acropora valenciennesi* grown at three different locations versus controls. Diamonds, Blue Marlin structure 7 m. Square, Big Bubbles 12 m. Triangle, Beach House 7 m. X, control 7 m. Box, control 12 m.

Parameter	Inside Biorock	Outside Biorock
Fish abundance (individual/m^2)	3.25	0.52
Biodiversity index		
Diversity	2.15	1.17
Evenness	0.63	0.04
Simpson's dominance	0.18	0.65

Figure 5.15 Fish population abundance, diversity, evenness, and Simpson's dominance on Biorock and on nearby natural reef.

DISCUSSION

The consistent results of all these different experiments conducted at different places and times under different conditions with different species strongly indicated that coral growth or coral skeleton deposition was significantly stimulated by low-voltage electricity using the Biorock method. These support the arguments and results previously described by Goreau and Hilbertz 1998. They suggested that increased pH around electrified cathodic frameworks resulting from electrolysis of seawater caused faster calcification and skeleton growth of electrically stimulated corals.

Another aspect that shows the well-being of electrified corals was their relatively high survival rates. The electrified structures not only provided the transplanted corals with better substrate to be strongly cemented onto but also increased the pH, which probably reduced the probability that the corals would be overgrown by macroalgae. Many species of macroalgae are detrimental to coral growth and survival, especially in areas of low herbivore and high nutrient levels (Jompa and McCook 2002b).

It is interesting to notice that although Samalona and Pemuteran experiments used the same species (*A. nobilis*) and also similar electric voltage, the results showed different growth rates. The electrified corals at Samalona Islands grew at slower rate (0.2 cm/week) compared to those at Pemuteran (0.3 cm/week). This could possibly be attributed to the different methods where the electrified corals in Pemuteran used a battery charger, while the source of the electricity at Samalona experiment came from a solar cell, which could be sometimes unstable due to cloud cover.

The fact that many Indonesian reefs (e.g., in the Spermonde Archipelago) have been seriously degraded (Jompa 2007) makes it urgent to seek alternative solutions to restore coral reefs and fisheries habitat. The Biorock method appears to be especially relevant to the degraded reefs where natural recovery is unlikely to occur or very slow due to recruitment failure.

CONCLUSION AND RECOMMENDATION

Corals of different reef-building species were grown on electrically stimulated Biorock reefs at different locations in Southwest Sulawesi and Northwest Bali, Indonesia. All species of corals were found to grow significantly faster on Biorock at all sites than nearby controls. This increase ranged from (1) two to three times faster growth of *A. nobilis* at Biorock sites compared to controls in Southwest Sulawesi at different voltages, (2) two to four times faster growth of *A. formosa* on Biorock than controls at another Southwest Sulawesi site at depths of 5 and 9 m, (3) Up to 4.01 times faster growth for Biorock *A. nobilis* versus controls in Bali at 4 and 6 m depth, and (4) more than six times faster growth in Gili Trawangan. These results show clearly that coral growth rates can be greatly increased with electrical trickle currents for habitat restoration. The responses appear to depend on species, local environmental conditions, and the amount of power received.

The Biorock method shows clear utility for restoring coral reefs and fisheries habitat in degraded reef areas. Therefore, this method could be used especially at those degraded reefs close to tourist destinations where electricity can be readily available. Because the specific benefits differ between species, location, and charging conditions, further work is needed to optimize the method for different species.

ACKNOWLEDGMENTS

We would like to thank very much Dr. Tom Goreau and Prof. Hilbertz for helping us applying the methods and providing us the anodes. Thanks also to the Center for Coral Reef Research at Hasanuddin University for the funding support. We also thank Agung Prana and the people of Pemuteran for their support to the Karang Lestari Project and the Gili Eco Trust for support to the Gili Trawangan projects.

REFERENCES

Hilbertz, W. H. 1979. Electrodeposition of minerals in sea water: Experiments and applications. *IEEE Journal on Oceanic Engineering* 4:1–19.

Hilbertz, W., and T. J. Goreau. 1998. Third Generation Artificial Reefs. http://globalcoral.org/Ocean%20Realm-1.pdf

Jompa, J. 2007. *The Fate of Degraded Coral Reefs under Anthropogenic Pressures in Spermonde Archipelago, South Sulawesi.* Torani, Indonesia.

Jompa, J., and L. McCook. 2002a. The effect of herbivores on competition between a hard coral and a macroalga. *Journal of Experimental Marine Biology and Ecology* 271:25–39.

Jompa, J., and L. McCook. 2002b. The effect of nutrients and herbivores on competition between a hard coral, *Porites cylindrica* and a brown alga, *Lobophora variegata*. *Limnology and Oceanography* 47:527–534.

Lee, E. R. 1997. Seament Research—Electroaccretion Physics. http://stanford/~erlee/seament/smphys.htm

Supriharyono. 2000. *Management of Coral Reef Ecosystems*. Penerbit Djambatan, Jakarta, Indonesia.

Van Treeck, P., and H. Schuhmacher. 1997. Initial survival of coral nubbins transplanted by a new coral transplantation technology—Options for reef rehabilitation. *Marine Ecology Progress Series* 150:287–292.

Veron, J. E. N. 1993. *Corals of Australia and the Indo-Pacific*. University of Hawaii Press, Honolulu, HI.

Wallace, C. C., and J. Wolstenholme. 1998. Revision of the coral genus *Acropora* (Scleractinia: Astrocoeniina: Acroporidae) in Indonesia. *Zoological Journal of the Linnean Society* 123:199–384.

Yuliantri, A. R., W. Moka, J. Jompa, and M. Litaay. 2006. The Successful Transplantation of *Acropora microphthalma* at Barrang Lompo Reef Edge, South Sulawesi. *Marine Research in Indonesia* 30:21–25.

CHAPTER 6

Biorock Reef Restoration in Gili Trawangan, North Lombok, Indonesia

Lalu Arifin Aria Bakti, Arben Virgota, Luh Putu Ayu Damayanti, Thri Heni Utami Radiman, Ambar Retnowulan, Hernawati, Abdus Sabil, and Delphine Robbe

CONTENTS

Background 60
Gili Trawangan as a Laboratory for Biorock Reef Restoration 60
The System and Principle of the Biorock Method 61
Methods of Biorock Study in Gili Trawangan 63
    Selection of *Biorock* Structures 63
    Selection of Coral Samples 63
    Measurements of Coral Growth and Survival Rate 63
    Measurements of Limiting Factors of Coral Growth 65
    Measurement of Survival Rate 66
    Measurements of Growth and Survival of Sponge *Clathria* 66
    Observation of Coral Fishes 68
    Measurement of Density of Coral Fish 69
    Index of Diversity (H'), Index of Uniformity (E), and Index of Dominance (C) 69
Results 70
    Growth of *Acropora formosa* on Biorock Substrate 70
    Growth Rate of Coral *Acropora formosa* at Three Stations 71
    Average Growth Rate of *Acropora formosa* Groups at Three Different Stations 71
    Survival Rate 72
    Growth Rate of *Acropora formosa* and *Montipora digitata* 72
    Environmental Condition of the Study Area 74
    Growth of *Clathria* Sponge on Biorock Structure 75
    Growth of Transplanted Sponge Fragments 76
    Rate of Basal Growth of Transplanted Sponge Fragments 77
    Rate of Survival of Transplanted Sponge 77
    Density of *Pomacentridae* and *Labridae* Coral Fishes at Biorock Substrate 78
Conclusions 79
References 80

BACKGROUND

The Gili Islands in Eastern Indonesia lie in the Coral Triangle, the most biodiverse region of the oceans. Unfortunately, the Gili Islands coral reefs have experienced substantial deterioration and destruction within the last two decades, mostly due to human activities causing marine resources overexploitation, destructive fishing methods like bombs and poisons, coral "heatstroke" from global warming, land-based sewage, global sea-level rise, overfishing, and direct physical damage from boats, anchors, tourists, reef harvesting, and coral diseases, compounded by the absence of appropriate management, poor enforcement capacity, and a lack of environmentally sound alternative sources of livelihood.

As a result, renewable marine resources are declining and endangering local food supplies, shorelines, and tourism income. The Biorock coral rehabilitation method can increase the growth rates of coral. Robbe et al. (2011) report the Biorock reef restoration project on the Gili Islands has been regenerating coral reefs for seven years. They add that measurable success can be clearly seen with regard to fish populations, coral growth and survival rates, ecotourism, education, and the halting of beach erosion.

This chapter explains the results of a study conducted to determine the growth rate of colonies of *Acropora formosa* and sponge *Clathria* as well as to measure the density of fishes growing on Biorock structures. This study was conducted in Gili Trawangan, one of the three Gili Islands, in 2010 and 2011.

GILI TRAWANGAN AS A LABORATORY FOR BIOROCK REEF RESTORATION

Several studies to assess the effect of Biorock on coral and sponge growth have been carried out in Gili Trawangan, located in the village of Gili Indah, Pemenang subdistrict of North Lombok district, West Nusa Tenggara, Indonesia. This most developed of the Gili Islands covers 342 ha. To the north and west is the Bali Sea, to the south is Lombok strait, and on the east is Tanjung Sire. Gili Indah lies between 8° 20'–8° 23' South and 116° 00'–116° 08' East, and the islands are surrounded by white sand. The depth of the bottom is less than 10 m at 20 m from the shore and is more than 20 m depth at 40 m away from the coast.

Since March 15, 2001, until March 4, 2009, the management of Gili Islands has been under the Natural Resources Conservation Unit or Balai KSDA NTB, Department of Forestry (under Ministry of Forestry decree number: 99/Kpts-II/2001). On March 4, 2009, the authority of the Gili management was handed over to the Department of Ocean and Fisheries following Ministry decrees BA.01/Menhut-IV/2009 and number BA.108/MEN.KP/III/2009. Based on the Ocean and Fishery Ministry decree number 67/MEN/2009 for selection of national water conservation areas, management authority for Gili Air, Gili Meno, and Gili Trawangan in West Nusa Genggara (September 3, 2009) is under the Director General of Ocean, Marine, and Small Islands (BKKPN).

The Gili Islands are dependent on a healthy marine habitat for their fisheries, tourism, sand supply, shore protection, and marine biodiversity. Unfortunately, the Gili Islands' coral reefs have experienced a substantial deterioration and destruction within the last two decades, mostly due to human activities but also including severe mortality from high temperatures in 1998 and from storm damage.

As a result, renewable marine resources are declining and endangering local food supplies, tourism income, and shorelines. Without large-scale restoration of degraded habitats to make them capable of supporting larger fish and shellfish populations, there will be fewer fish in the future, and without healthy growing corals, there will be fewer beaches or tourism income, affecting all businesses and residents on the island.

The Biorock reef restoration project on the Gili Islands has been regenerating coral reefs for seven years. Regular training workshops on Biorock restoration are facilitated by Satgas Gili Eco

Trust in collaboration with Centre for Environmental Studies in Mataram at the University of Mataram, with genuine support from all stakeholders in the Gili Islands. Robbe et al. (2011) reports that a measurable success can be seen with regard to fish populations, coral growth and survival rates, ecotourism, education, and the halting of beach erosion. Therefore, Gili Trawangan can be seen as a laboratory of Biorock reef restoration.

THE SYSTEM AND PRINCIPLE OF THE BIOROCK METHOD

The Biorock method was invented, developed, and patented by the late Prof. Wolf Hilbertz and Dr. Thomas J. Goreau. Biorock technology uses low-voltage direct currents (above 1.2 V) passing through a steel structure (Figure 6.1). The Biorock electrolysis process occurs between two metals receiving electricity in seawater, causing the steel cathode structure to grow solid limestone minerals and the anode to slowly disintegrate. These currents are safe to humans and all marine organisms. There is no limit in principle to the size or shape of Biorock structures, and they could be grown hundreds of kilometers long if funding allowed. The limestone is the best substrate for hard coral. The Biorock process is used to regenerate coral reefs, repopulate damaged reefs with corals and fish, break wave action, and grow beaches.

The formation of mineral deposits is not a direct oxidation reaction such as electroplating, but is an indirect process, in which mineral deposition occurs as a byproduct of changes in pH around the cathode due to the electrolysis of seawater. Oxygen and chlorine are produced at the anode, while hydrogen forms at the anode and dissolved magnesium and calcium, which are abundant in seawater, precipitate on the cathode (Figure 6.2). This deposited material is composed largely of calcium carbonate, which is the mineral that forms coral reefs (Goreau and Hilbertz 2005). Biorock corals grow dense branches, have bright colors, can recover from physical damage up to 20 times faster than natural reefs, and also have a survival rate up to 50 times higher after severe high-temperature bleaching events (Goreau and Hilbertz 2005).

Hard and soft corals, sponges, tunicates, and bivalves are observed to grow on Biorock materials at extraordinary rates. Fish and lobster population growth in these structures is extraordinary, especially juvenile fish, and depends on the shape of Biorock structures, which can be made to provide

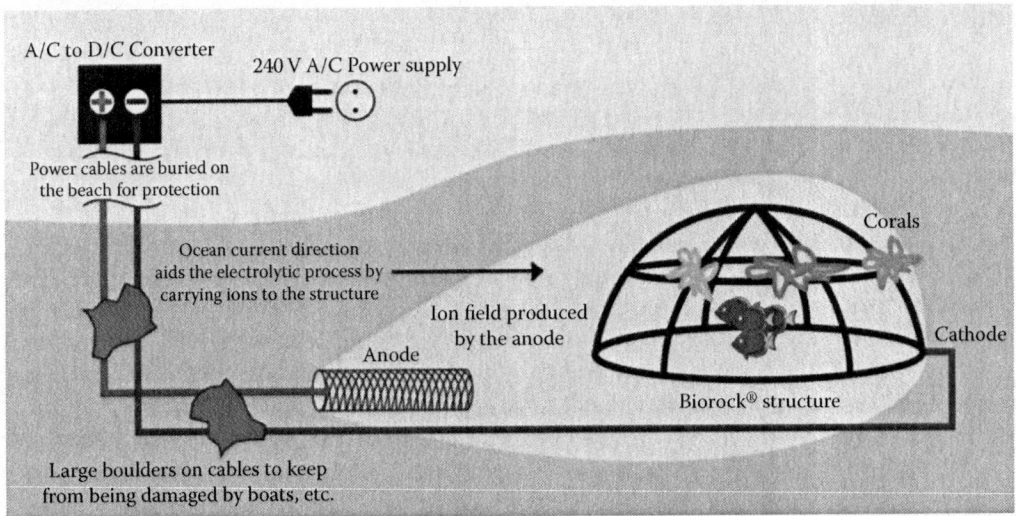

Figure 6.1 Scheme of the Biorock system.

Figure 6.2 Hydrogen bubbles produced by electrolysis on the surface of Biorock structures.

an extraordinary density of hiding places. Biorock reefs have turned severely eroding beaches into 15 m of beach growth in a few years by slowing waves, so that instead of eroding sand at the shore, they deposit it. They have been found stable in category 4 hurricanes and the Asian Tsunami because the open frameworks allow large waves to pass through (Goreau 2010; Goreau and Hilbertz 2005, 2007, 2008).

Biorock structures are made with metal bars, charged by a low-voltage current above 1.25 V. These structures are installed on the ocean floor, and pieces of corals are attached to them. These corals come from reefs in the neighborhood that had been naturally broken by various causes (unaware divers, strong waves, anchor damage, etc.).

The electric current, which is totally harmless for any organism, leads to electrolysis, causing a calcareous precipitation on the whole structure. This will not only prevent appearance of rust, which would weaken the structure, but since coral skeleton is made of limestone, the structure will, thanks to this reaction, become the best place for coral to grow.

Thus Biorock technology relies on a very simple principle: enhancement by electrolysis of the natural reactions occurring among coral, seawater, sun, and dissolved minerals.

Biorock technology acts to catalyze the natural reaction and enables coral growth two to six times faster than in usual conditions. Normally coral grows only around centimeters per year, so faster growth is a highly efficient way to restore damaged reefs. Moreover, corals on Biorock structures are more resistant to hazards they face.

Hard corals are not the only ones to grow on Biorock structures, but tunicates, bivalves, sponges, and soft corals also develop faster than average. On a Biorock structure, their survival and resistance rate is 20 to 50 times higher than in natural environment following severe high-temperature bleaching events (Goreau and Hilbertz 2005).

Finally, because Biorock technology relies on electrical fields, it benefits all corals and ecosystems around the metal structure in a perimeter up to about 10 m. These results by W. Hilbertz and T. Goreau are borne out since the 1980s by the efficiency of Biorock structures installed all around the world. These structures contributed to restoration of damaged coral reefs, enlargement of beaches affected by erosion, and repopulation of marine areas with many species of fishes and other sea organisms.

Since 2004, the Gili Eco Trust has launched the Biorock program around the Gili Islands. There are now nearly 75 Biorock structures around the Gili Islands, which fall into three categories of structure: (1) Biorock reefs to grow corals and provide new fish habitats, thus creating interesting new dive sites; and (2) Biorock antierosion reefs to grow corals and provide fish nurseries while causing sand to accumulate on the beach, (3) good shallow snorkeling sites, and (4) Biorock wavebreaker structures to stop and reverse the erosion.

Every Biorock reef or structure has been showing positive consequences on fish populations, coral biodiversity, and regenerating eroding beach. It is the best technology ever used to restore our corals reefs, in association with education to avoid further damage. Everyone on the Gili Islands now knows about the Biorock project, and it is a big step forward in ecotourism. We should continue to send out information and expand our existing reefs.

The antierosion reefs are working very well as fish nurseries and by accumulating sand. The wavebreaker reefs in Karma Kayak, Gili Eco Villas, and Kokita have shown incredible results very quickly by stopping the erosion process. The beach directly behind the structures has started to grow (Goreau et al. Chapter 3). The other Biorock reefs are growing coral faster and making divers happy through the biodiversity of fish and creatures they can observe on the structures and the surrounding area.

METHODS OF BIOROCK STUDY IN GILI TRAWANGAN

Selection of Biorock Structures

The research site for the Biorock study can be seen in Figure 6.3. The three stations for monitoring are located in front of "Tir Na Nog" and "Beach House" restaurants (Table 6.1).

Biorock structures were designed in various shapes and sizes following the seventh Biorock Reef Restoration Training and Workshop in Gili Trawangan on November 2010. They are made with metal bars, charged by a low-voltage current above 1.2 V. Three different structures were selected in this study, namely a Rack structure, which is located under 3 m depth; a Manta (Pari) structure, which is placed under 5 m depth; and a Dolphin (Lumba-lumba) structure, which is located under 8 m depth. These structures were installed on the ocean floor in the coordinate point as shown in Table 6.1, and pieces of corals are attached to them during the workshop. Control corals were grown at 3, 5, and 8 meters depth.

Selection of Coral Samples

The coral employed in this study was *Acropora formosa*. This coral is relatively fast-growing and is easy to find in the Gili Islands.

This coral comes from reefs in the neighborhood and was broken by various forces (unaware divers, strong waves). Various sized fragments were attached to the determined structures. *Acropora formosa* grown in nature was selected as controls.

Measurements of Coral Growth and Survival Rate

Growth of *Acropora formosa* and *Montipora digitata* were measured on the Biorock structures (Figure 6.4), which has mineral accretion treatment, and in natural condition on posts (Figure 6.5) as a control. Three coral samples were identified for observation at each different Biorock structures of 3, 5, and 8 m depth.

For the coral growth of *Acropora formosa* and *Montipora digitata*, the length of vertical and horizontal branching were measured using calipers at a given interval using the following formula:

$$\alpha = L_t - L_0$$

Figure 6.3 Map of research site for Biorock study.

Table 6.1 Coordinate Points of the Monitoring Stations

No	East	South	Structure of Biorock	Depth (m)
1	116 02' 33.55"	08 21' 27.49"	Rack (antierosion structure)	3
2	116 02' 33.87	08 21' 27.30"	Pari (Manta)	5
3	116 02' 33.83	08 21' 27.71"	Lumba-lumba (Dolphin)	8

where

α is the growth of the coral transplanted
L_1 is the height after 8 weeks
L_0 is the height at the first time transplanted

Figure 6.4 Biorock structure.

Figure 6.5 Block substrate away from Biorock.

while the coral growth rate was measured using the following formula:

$$\beta = \frac{L_{i+1} - L_i}{T_{i+1} - T_i}$$

where

β is the growth rate of coral fragments
L_{i+1} is the average of height colony on the measurement in $i + 1$
T_{i+1} is the time of measurement

Measurements of Limiting Factors of Coral Growth

In this study, some limiting factors of coral growth such as temperature, salinity, current velocity, water transparency, and pH were determined during the measurement of coral growth. The water transparency was measured using Secchi disc, while a dive computer was used to measure the temperature at the study area.

Water salinity was measured by hand refractometer. The current velocity was measured using current drogue. For pH of the surface water in the study area, a pH paper was employed. Good transparency is important for photosynthesis because penetration can reach deep water. Clear water is needed by coral for photosynthesis by *Symbiodinium*. The intensity of solar radiation at depth is related to water transparency (Supriharyono 2007). This is crucial for the growth of symbiotic or *hermatypic* reef-building corals. Without enough radiation, the rate of photosynthesis is reduced, thus the capacity of coral to form the reef from limestone precipitation ($CaCO3$) is also reduced (Dahuri 2003). To grow well coral reefs require temperatures ranging around 25°C–30°C (Nontji 2002; Soekarno et al. 1983). Fluctuation of temperatures will influence feeding behavior of corals. Their feeding behavior tends to decrease in extreme temperatures (Nybakken 1988). Salinity also influences the growth of corals (Thamrin 2006). Coral reefs grow well in salinities of 30‰–35‰ (Dahuri 2003). Corals tolerate water salinities of 27‰–40‰ (Nontji 2005). Current is important to bring food for the polyp of coral. Currents, water mass movement, and waves are needed for the transport of nutrients, larvae, suspended organic matter and sediment, oxygen, and plankton, and help keep the polyp clean (Nontji 2005). Higher water velocity is much better than still water for coral growth. Indonesian waters have relatively low salinity and pH due to high rainfall. Coral can grow well at low pH and high nutrients (Sadrun 1999).

Measurement of Survival Rate

The survival rate of transplanted corals on Biorock and in the controls was determined using the following formula (Sadarun 1999):

$$S = \left(\frac{N_t}{N_o}\right) \times 100\%$$

where

S is the rate of survival (%)
N_t is the number of survived coral at the last of observation (colony)
N_o is the number of survived coral at the first observation (colony)

Measurements of Growth and survival of Sponge *Clathria*

The sponge observed in this study is the genus *Clathria*, which is a member of the class Demospongia.

Concrete cement blocks were made as a substrate placed at a distance from Biorock structure just a week before transplanting the *Clathria* (Figure 6.6). The Biorock structure selected in this sponge growth study was designed and installed during the seventh Biorock workshop/training in 2010.

A *Clathria* fragment from the same parent colony was transplanted on Biorock structure and on concrete block substrates outside the Biorock (Figure 6.7 and 6.8). Three sponge fragments were attached in the structure and in concrete block substrate using cable ties with distance of 2, 4, 6, 8, and 10 m from the Biorock structure.

The sponge survival rate was determined by counting the fragments surviving from the start until the last observation around the Biorock and outside the Biorock fortnightly. The first reading was taken immediately after transplanting (T_0), which is followed by T_1, T_2, T_3, and T_4 reading at 2 week intervals.

Figure 6.6 *Clathria* sponge.

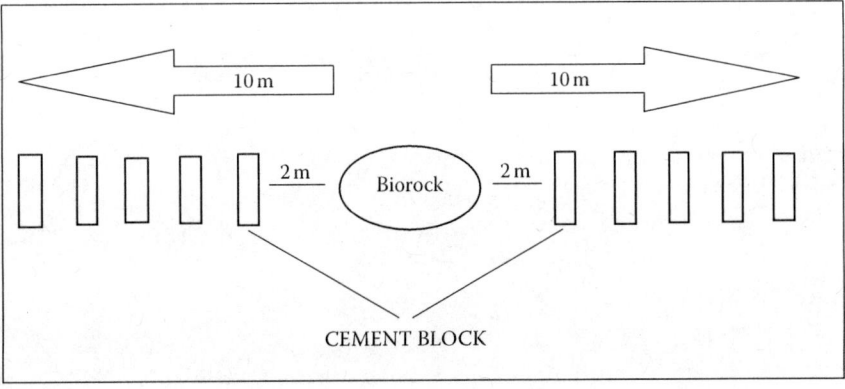

Figure 6.7 Design of the concrete cement block layout.

Measurement of vertical and basal growth rate of sponge was done using the following formula (Fauziyah 2006):

$$\beta = \frac{L_{i+1} - L_i}{T_{i+1} - T_i}$$

where

β is the growth rate of length and circle of sponge fragments transplanted
L_{i+1} is the average length and circle of sponge fragments at $i + 1$
T_{i+1} is the time of reading at $i + 1$

Figure 6.8 (a) Concrete block substrate. (b) Concrete block outside Biorock.

Rate of survival of sponges was measured using the following formula (Fauziyah 2006):

$$S = \left(\frac{N_t}{N_o}\right) \times 100\%$$

where

S is the rate of survival (%)
N_t is the number of surviving sponges at the last observation
N_o is the number of surviving sponges at the start of observation

Observation of Coral Fishes

Coral fish monitoring around Biorock structures was carried out by the underwater visual census method at 3 to 8 m depth. Population density of coral fish was monitored weekly at 08.00 to 11.00 a.m., along a 5 m distance between the Biorock structure and the non-Biorock structure. Observations were made at 5 m sight distance, with observed area of 2×5 m² for the Biorock Rack structure at 3 m water depth, 5×6 m² for the Manta structure at 5 m water depth, and 2.5×7 m² for Dolphin structure at 8 m water depth. For coral fish density observation outside the Biorock, a transect line, which has similar area with the observed Biorock structure, was made. Monitoring stations, which consist of soft coral, sand, and coral fragments, are located at 5 m distance from each Biorock structure.

Data recorded include all species and number of fish observed around Biorock and non-Biorock of 2 m distance from right and left of the transect line.

Measurement of Density of Coral Fish

A visual census technique was used to identify the populations of coral fishes living around Biorock habitat, for example, quantifying their density, diversity, uniformity, and dominance index. The population density of coral fishes was determined following Odum (1996):

$$X = \sum \frac{X_i}{n}$$

where

X is the density of coral fishes
X_i is the number of fishes at i-station
n is the observed transect area

Index of Diversity (H'), Index of Uniformity (E), and Index of Dominance (C)

1. Index of Diversity (H')
Index of diversity for coral fishes was determined by the following formula (Krebs 1989):

$$H' = \sum_{i=1}^{S} P_i \ln P_i$$

where
   H' is the index of diversity
   P_i is the proportional difference of fish species i
   S is the number of total fishes observed
   Criteria descriptions for the index of diversity are
      a. $H' < 2.30$: low diversity, very strong environmental pressure
      b. $2.29 < H' < 6.90$: moderate diversity, medium environmental pressure
      c. $H' > 6.90$: high diversity, balanced ecosystem
2. Index of Uniformity (E)
Index of uniformity of the observed fishes was calculated by the following formula (Krebs 1989):

$$E = \frac{H'}{H_{max}}$$

where
   E is the index of uniformity
   H_{max} is $\ln S$
   S is the number of observed coral fishes

Index value ranges from 0 to 1, which falls into three criteria (Krebs 1989):
 a. $0.00 < E \leq 0.05$ is low uniformity, community under pressure
 b. $0.05 < E \leq 0.75$ is medium uniformity, unstable community
 c. $0.75 < E \leq 1.00$ is high uniformity, stable community
3. Index of Dominance (C)
 Index of species dominance, which is also known as the Simpson index, was used to determine the value of dominance within the community following Odum (1996):

$$C = \sum_{i=1}^{S} (P_i)^2$$

where
 C is the index of dominance
 P_i is n_i/N (fish proportional difference at $-i$)
 Index of dominance ranges from 0 to 1, which can be classified as follows:
  a. $0.00 < C \leq 0.30$ is low dominance
  b. $0.30 < C \leq 0.60$ is medium dominance
  c. $0.60 < C \leq 1.00$ is high dominance

RESULTS

Growth of *Acropora formosa* on Biorock Substrate

A study was conducted to measure the growth of *Acropora formosa* on Biorock structures at different depths, to monitor the growth of coral under natural conditions around Biorock as controls, and to assess some environmental parameters that influence coral growth.

Based on the results, the average vertical growth of coral *Acropora formosa* was 0.293 cm/week at Biorock and 0.072 cm/week for controls (Figure 6.9). The average horizontal growth of *Acropora formosa* was 0.078 cm/week on Biorock and 0.027 cm/week for controls (Figure 6.10). Therefore, vertical and horizontal growth of *Acropora formosa* was four times and three times higher than controls, respectively.

The data collected at three different depths also showed a significantly accelerated growth of *Acropora formosa* transplanted onto Biorock structures compared to controls. At 3 m, the growth rate of *Acropora formosa* was significantly faster than at deeper structure position (0.315 cm/week). The corals grown at 5 m and 8 m depth reached about 0.291 cm/week and 0.071 cm/week vertical growth rate, respectively. The depth tends to reduce the growth rate of *Acropora formosa*. In this study, the coral survival rate with both Biorock and control treatments was 100%. Thus, the growth rate of coral *Acropora formosa* at Biorock was two to four times higher than the natural controls.

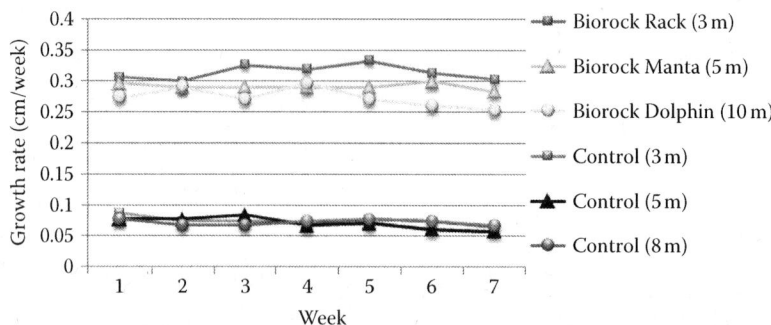

Figure 6.9 Vertical growth rate of *Acropora formosa* as a function of time.

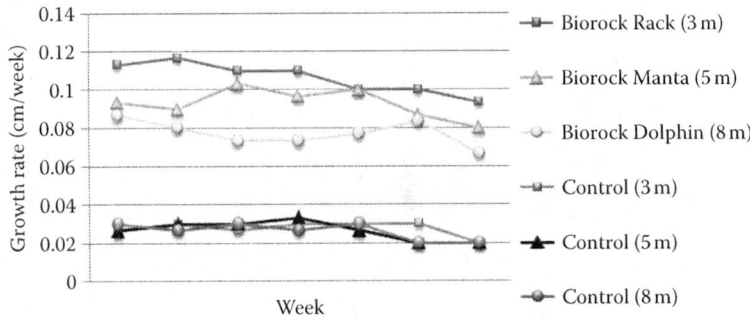

Figure 6.10 Horizontal growth rate of *Acropora formosa* as a function of time.

Growth Rate of Coral *Acropora formosa* at Three Stations

The growth of coral *Acropora formosa* vertically at Biorock was four times faster than the growth at control treatment.

This significant difference of coral growth is also supported by Soesilo and Budiman (2006), who stated that the growth of some corals may reach 10 times faster due to mineral accretion.

The average vertical and horizontal increment of *Acropora formosa* varied somewhat over time. This is possibly because of unstable electrical power supply that affects mineral accretion in the Biorock structures. It is normal in Gili Trawangan that the electrical power operated by the state electrical company (PSN) suffers frequent power outages. In turn this may cause varying growth of *Acropora formosa*.

Sudden changes in temperature may inhibit growth of corals and their feeding behavior (Nybakken 1992). Another factor influencing growth is turbidity, which is a function of suspended solid and dissolved organic matter concentrations in the water column. When particle content is high, penetration of solar radiation is reduced, lowering the compensation point for photosynthesis. According to Nontji (2005), solar radiation is needed for photosynthesis of symbiotic algae, which then provide energy to the host coral animal. Photosynthesis of stimulates growth of calcareous skeleton ($CaCO_3$) by corals.

Average Growth Rate of *Acropora formosa* Groups at Three Different Stations

Average vertical growth by *Acropora formosa* on the Biorock Rak was 0.315 cm/week, which is 4.257 times (425.7%) greater than the growth of control corals (0.074 cm/week). At Manta Biorock, the growth rate was 0.291 cm/week, which is 4.157 times faster (415.7%) than controls (0.070 cm/week). At Dolphin Biorock, the rate was 0.273 cm/week, which is 3.845 times faster (384.5%) faster than controls (0.071 cm/week). It is clear that the growth of coral is greater on Biorock (Figure 6.11 and 6.12).

The higher growth rate of coral on Biorock mineral accretion was achieved because coral metabolism is stimulated due to the formation of calcareous substances deposited on the steel structure by the electrical field. The Biorock influenced the growth of *Acropora formosa* significantly at different water depth. At 3 m depth, the coral transplanted on Biorock tend to grow faster than at 5 and 8 m depths, suggesting that depth factor contributes to the growth of coral on the Biorock. This is probably because of solar radiation intensity penetrating the water and influencing photosynthesis by the symbiotic algae. Solar radiation is needed by the symbiotic algae for photosynthesis (Nontji 2005; Suharsono 2004). However, unlike the Biorock condition, there was no significant difference in the control coral growth at three different water depths. This suggests the Biorock process may be increasing photosynthetic efficiency.

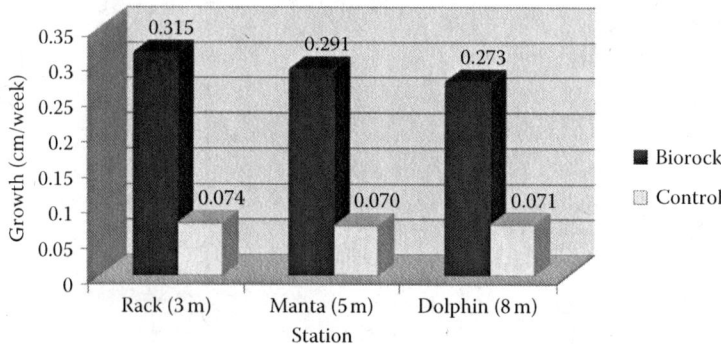

Figure 6.11 Average vertical growth rate of *Acropora formosa* at three sites.

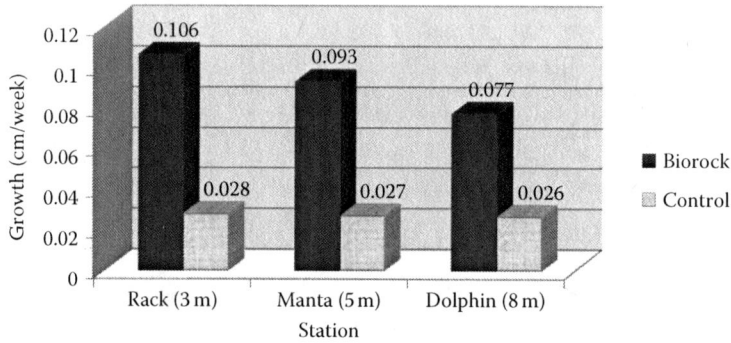

Figure 6.12 Average horizontal growth of *Acropora formosa* at three sites.

With 0.074 cm/week average increase of vertical coral growth in the controls, it is estimated that their height increase will be 3.848 cm per year. According to Romimohtarto and Sri (2007), branching corals (*Acropora*) may grow as much as 5–10 cm or even more per year. Based on this growth result, it can be concluded that the coral growth in Trawangan is slow due to turbidity influencing solar radiation penetration into seawater.

Survival Rate

The survival rate of all corals with mineral accretion and control corals was 100%. There was no case of coral death during the study. Thornton et al. (2000) state that coral rehabilitation is successful if the survival rate ranges from 50% to 100%.

Environmental conditions affect coral survival rates. Fluctuating currents and good water transparency are among factors influencing the development and growth of coral reefs (Bengen 2001).

Growth Rate of *Acropora formosa* and *Montipora digitata*

A comparison study was conducted to determine the growth rates of *Acropora formosa* and *Montipora digitata* growing on Biorock substrate and further away (up to 10 m) from a larger Biorock structure in Gili Trawangan. Data collection for the coral growth was taken by direct measurements (Supriharyono 2007) using a caliper. The reading was taken every two weeks on *Acropora formosa* and *Montipora digitata* attached on Biorock structure and placed on concrete blocks (Figure 6.2) away from the same Biorock structure (up to 10 m distance).

Acropora formosa on Biorock substrate showed the greatest increase in growth, 3.42 cm within 8 weeks, while *Acropora formosa* away from Biorock grew only 0.78 cm. *Montipora digitata* on the substrate Biorock grew 2.40 cm, while *Montipora digitata* away from Biorock grew only 0.72 cm (Figure 6.13). The result of *t*-test analysis showed significant differences between coral growth on the Biorock structure and away from it. The height of *Acropora formosa* colonies transplanted on Biorock structure (0.43 cm/week) is four times higher than the growth away from the Biorock structure (0.09 cm/week). Growth of *Montipora digitata* colonies on Biorock was 0.3009 cm/week, which is three times faster than the growth away from Biorock (0.009 cm/week).

The growth of *Acropora formosa* on Biorock was 4.39 times faster than the growth away from Biorock. This is consistent with research conducted earlier where the vertical growth of *Acropora formosa* on Biorock was up to four times faster than the growth of control corals. The growth rate of *Montipora digitata* on Biorock was 3.34 times faster than *Montipora digitata* controls away from the Biorock substrate.

The growth rate of each species is significantly different depending on species and distance from the Biorock structure, as well as varying over time (Table 6.2).

Growth of *Acropora formosa* and *Montipora digitata* on Biorock substrate is three to seven times faster than the growth on the blocks away from Biorock structure, with *Acropora formosa* growing faster than *Montipora digitata* on Biorock, although control rates were similar.

A similar study conducted by Murtawan in 2011 found a significantly higher growth rate of *Acropora formosa* on Biorock than nearby. The growth rate of *Acropora formosa* was measured by colony height and colony diameter. The height increase rate of *Acropora formosa* at the Biorock, north of Biorock, and south of Biorock was 0.57, 0.29, and 0.23 cm/week, respectively, whereas the diameter increase rate was 0.14, 0.48, and 0.24 cm/week. The correlation of colony height increase on the north and south near Biorock is −0.80 and −0.58, respectively, which means that the farther

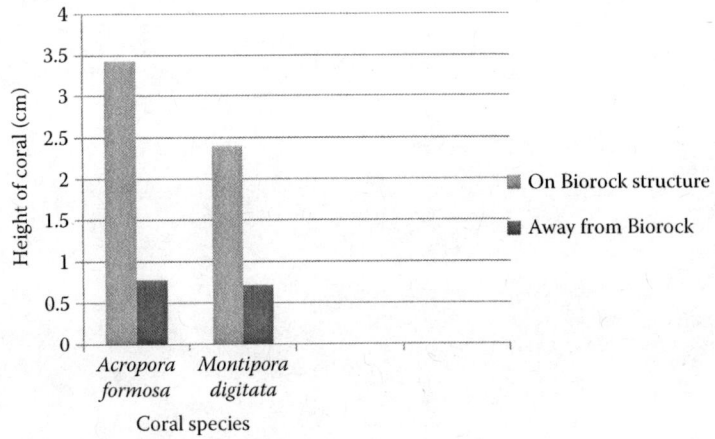

Figure 6.13 Growth of *Acropora and Montipora* on and away from Biorock.

Table 6.2 Coral Height Growth Rate

Time	A. formosa on Biorock (cm/2 week)	A. formosa away from Biorock (cm/2 week)	M. digitata on Biorock (cm/2 week)	M. digitata away from Biorock (cm/2 week)
T_0–T_1	0.68	0.12	0.44	0.13
T_1–T_2	0.766	0.2	0.62	0.21
T_2–T_3	0.98	0.18	0.68	0.18
T_3–T_4	1	0.28	0.65	0.2

corals are transplanted from Biorock, the slower they grow. On the diameter growth rate, the correlation degree of northern block is −0.47, while the southern block has −0.46. Based on a graphic analysis of *Acropora formosa* growth rate, Biorock structure can affect the growth of *Acropora formosa* that has been transplanted up to 10 m from Biorock.

Environmental Condition of the Study Area

Physical and chemical parameters assessed during this study were geomorphology, substrate condition, water transparency (Figure 6.14), temperature (Figure 6.15), salinity (Figure 6.16), current velocity (Figure 6.17), and pH (Figure 6.18). Geomorphologically, the study site is a slope, 8 m

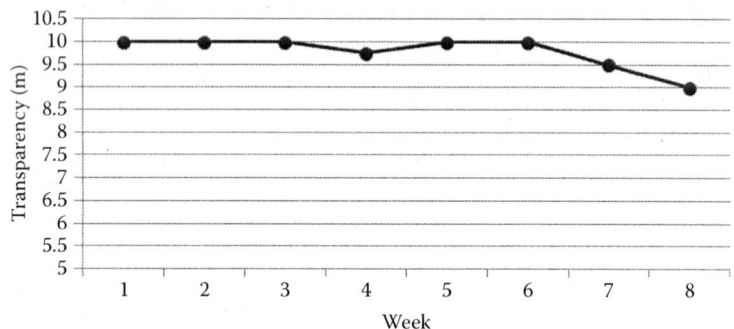

Figure 6.14 Transparency during the study (Secchi disk depth).

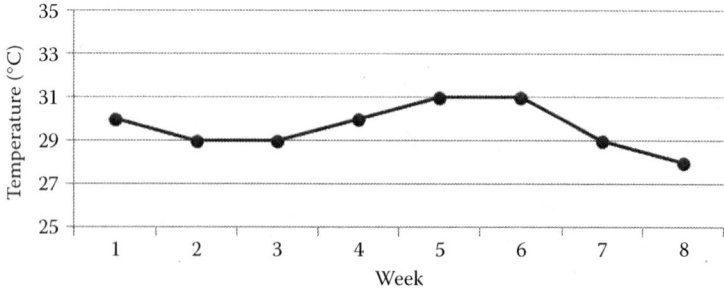

Figure 6.15 Temperature during study.

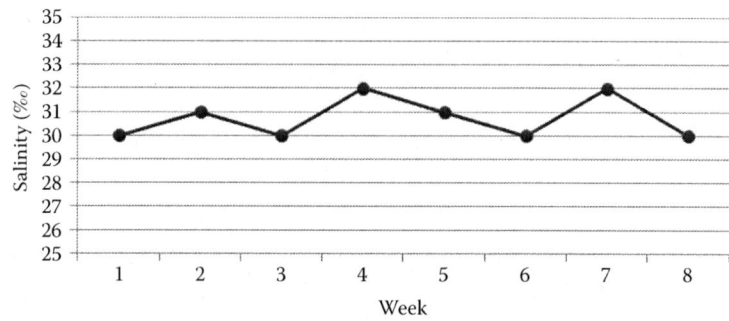

Figure 6.16 Salinity of the water during study.

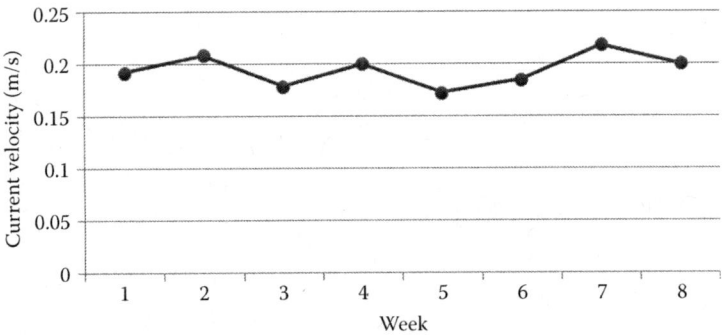

Figure 6.17 Current velocity during study.

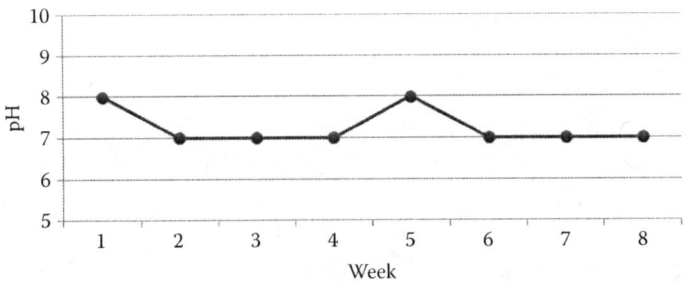

Figure 6.18 pH during study.

deep. The substrate is slightly sandy, which may influence coral and sponge growth. All parameters were considered as being in the healthy range for corals and sponges.

The location of this study has good transparency. The temperature fluctuated slightly at each reading, ranging from 27°C to 29°C. Change in temperature due to weather can be seen in Figure 6.19. The current velocity in the study area also varied at the time of measurements. Water current was between 0.061 and 0.224 m/s. It had a similar pattern with the temperature and rate of sponge growth: when the current velocity was high, the growth of sponge was also high. The pH of water during study is on the low side for growth of coral, but the precision of the measurement was only 1 pH unit.

The physiology of sponges is influenced by water flow, which may carry oxygen, nutrients, food, and metabolic wastes around their body wall (Hickman et al. 2001). In high-current speed, the water flow containing oxygen and food is increased, speeding up the physiological growth process. When the current speed decreased, then the supply of oxygen and food through water flow also decreases, causing slower growth. The salinity and acidity of the environment were favorable for sponge growth.

Growth of *Clathria* Sponge on Biorock Structure

A study to observe the effect of Biorock on development and growth of *Clathria* transplanted inside and outside the Biorock structure was conducted in Gili Trawangan. Based on the results, the growth of *Clathria* sponges transplanted on Biorock was faster than those sponges transplanted outside the Biorock structures, which were on north and south blocks. But this was not the case for basal growth of sponges. The sponge tends to grow vertically, increasing length and branch

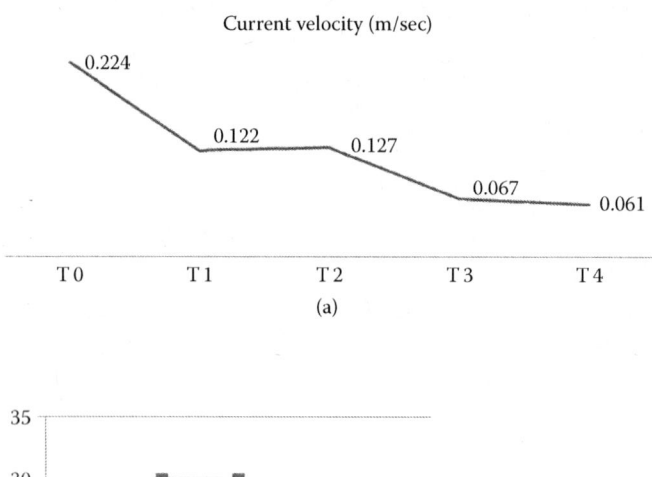

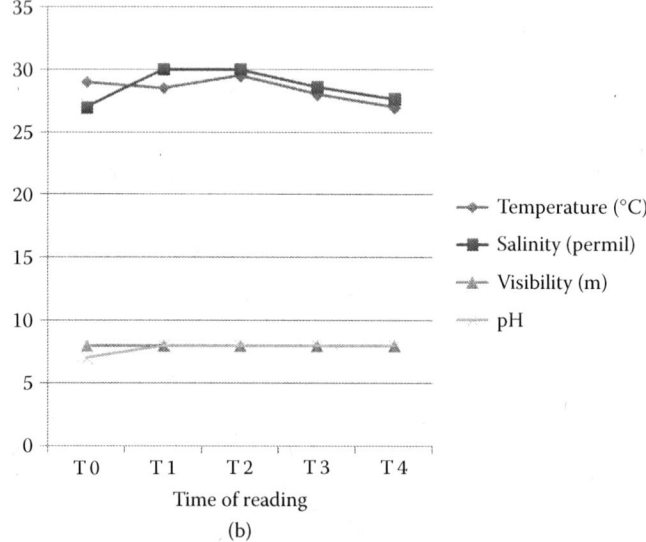

Figure 6.19 (a) Physical and (b) chemical conditions of aquatic environment for sponge study.

formation. Correlation analysis indicates that *Clathria* grew slower when they were transplanted farther from the Biorock structure. The survival rate of *Clathria* transplanted inside and outside Biorock was 100%.

Growth of Transplanted Sponge Fragments

The results showed that the growth rate of branching sponges two weeks after transplantation on Biorock was 2.55 cm and was the highest measured in the north block 1 followed by blocks 3, 4, 5, and block 2 (Figure 6.20). The same result was found in the south block. Growth rates were slower for sponges away from the Biorock. The slow growth of sponge at block 2 was probably caused by the sedimentation on the block due to diving activities around the Biorock. Sandy substrate on block 2 was monitored during the assessment. This sediment may inhibit the growth of sponge as it covers the surface, thus the ostium of sponge is closed, which may prevent the water flow that transports nutrients and oxygen required by the sponges.

Statistical analysis shows that the growth of sponge on the Biorock structure was significantly different than those sponges growing away from Biorock structure on both north and south blocks.

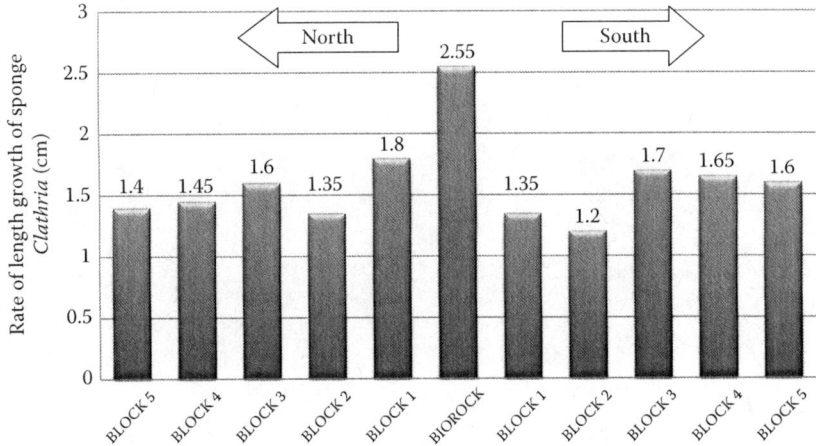

Figure 6.20 Rate of vertical growth of branching sponge inside and outside Biorock structure.

The correlation analysis gives a correlation coefficient value of −0.60. This means that the distance factor influenced the length growth. The closer the sponge from the Biorock, the faster the growth.

Rate of Basal Growth of Transplanted Sponge Fragments

In general, the basal growth rate of sponge over the substrate is low both inside and outside Biorock. This is normal, as sponge *Clathria* tends to branch vertically.

Figure 6.21 shows the rate of basal (encrusting) growth of sponges two weeks after transplantation inside and outside the Biorock structure. The distance factor influenced the circle growth rate of sponge. The growth of sponge on the Biorock is faster than any sponge grown in the south block, except for block 5. In the north block, the sponges transplanted outside the Biorock structure grew faster than the transplanted sponge inside the Biorock.

However, statistically there was no significant difference between the basal growth of sponge inside and outside Biorock. The correlation between distance factor and basal growth of sponge is positive, meaning that the farther the sponge from the Biorock, the higher the basal growth rate (0.16). This small coefficient value suggests that there is little influence of distance factor on the basal growth rate of sponge.

Rate of Survival of Transplanted Sponge

The survival rate of sponge fragments transplanted inside and outside Biorock structure was 100%. The sponges transplanted inside Biorock not only had high growth rate but also had high acclimatory ability. Furthermore, the sponges transplanted inside Biorock had no disease attack during the study, while the sponges transplanted outside Biorock were vulnerable to disease and predator attack (Figure 6.22).

It is clear that the Biorock influenced the development and growth of *Clathria*. The vertical growth of *Clathria* transplanted on the Biorock structure was significantly greater than the vertical growth outside the Biorock structure, but the basal growth was not.

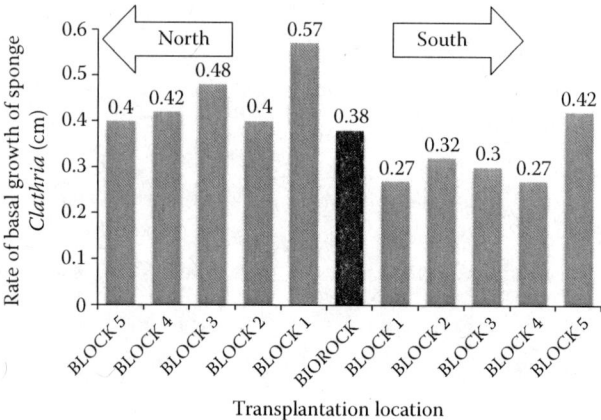

Figure 6.21 The rate of horizontal basal growth of sponge fragments inside and outside the Biorock structure.

Figure 6.22 Transplanted sponge away from Biorock suffering disease.

Density of *Pomacentridae* and *Labridae* Coral Fishes at Biorock Substrate

This study identified the composition, diversity, density, uniformity, and dominance of coral fishes of *Pomacentridae* and *Labridae* family around Biorock structures and determined the effect of Biorock on their population densities by direct monitoring in the field (Figures 6.23 and 6.24).

The *Pomacentridae* and *Labridae* around Biorock in Gili Trawangan consisted of 18 genera and 35 species. The highest density of *Pomacentridae* was represented by *Dasyllus aruanus*, while the *Labridae* was represented by *Thallashoma lunare*. Ornamental fishes of the genera *Amphiprion*, *Chrysiptera*, and *Pseudocheilinus* had low density.

The study observed that there was a significant difference in biological indices around the Biorock structure (Graph 4). Index of density, diversity, and dominance index around the Biorock structure were 1.54, 0.53, and 0.10, respectively. Whereas, in the control station the values were 0.97, 0.43, and 0.41, respectively. Hence, community structure at the Biorock is better than the community structure outside Biorock.

The result showed that Biorock was very effective in improving the population of coral fishes. The results indicate that the density of coral fishes found around the Biorock structure was six times greater than the density of coral fishes found outside the Biorock structure.

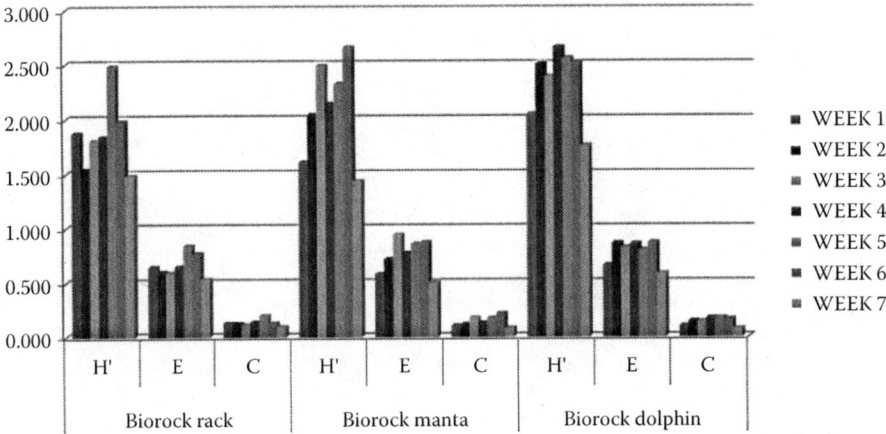

Figure 6.23 Indices of fish populations at Biorock measured at different weeks.

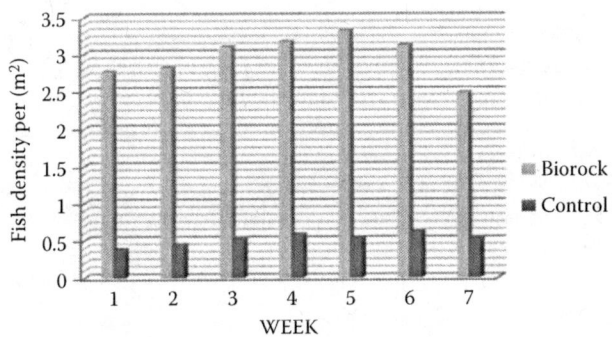

Figure 6.24 Density of coral fishes around the Biorock structure and control station.

CONCLUSIONS

The results of this study showed that the growth rate of *Acropora formosa* colonies transplanted on Biorock structure were four times higher than the growth away from the Biorock structure. *Acropora formosa* grew faster than *Montipora digitata* on Biorock, although control rates were similar. Growth of *Montipora digitata* colonies on Biorock was three times faster than *Montipora digitata* controls away from Biorock substrate.

The growth of *Acropora formosa* around Biorock was significantly greater than the growth at 5 and 8 m depths, which is possibly because of solar radiation intensity penetrating the water and influencing photosynthesis by symbiotic algae.

Based on the results, the vertical growth of *Clathria*, a sponge transplanted on Biorock, was faster than those sponges transplanted outside the Biorock structures. However, this was not the case for basal growth of sponges, which is normal, as the sponge tends to grow vertically, increasing length and branch formation.

This study showed that Biorock was very effective to improve the population of coral fishes. The result indicates that the density of coral fishes found around Biorock structure was six times greater than the density outside the structure. This study also indicates that the survival rate of all corals and sponges with mineral accretion and control corals was 100%.

This regular community-based Biorock reef restoration workshop/training should be continued to enable the Gili Island's stakeholders to share their opinions and vision toward sustainable ecotourism. It is clear that the Biorock structures installed in Gili Islands have contributed to the restoration of damaged coral reefs and repopulation of marine areas with many species of fishes and other sea organisms.

REFERENCES

Bengen, D. G. 2001. Ecosystem synopsis of marine and coastal natural Resources. Risert Center for Marine and Coastal Resources. IPB. Bogor.

Dahuri, R. 2003. *Marine Biodiversity-Sustainable Development Asset of Indonesia*. Gramedia Pustaka Utama. Jakarta.

Fauziyah. 2006. Growth Rate of *Acropora* sp. and *Hydnophora exesa* Transplanted in Pramuka, Seribu Islands. Universitas Sriwijaya.

Goreau, T. J. 2010. Coral reef and fisheries habitat restoration in the coral triangle: The key to sustainable reef management, invited keynote talk, world ocean congress, Manado, Sulawesi, Indonesia. In J. Jompa, R. Basuki, Suraji, M. Tesoro, and E. T. Lestari (Eds.) *Proceedings of the COREMAP Symposium on Coral Reef Management in the Coral Triangle*, pp. 244–253, Jakarta.

Goreau, T. J. and W. Hilbertz. 2005. Marine ecosystem restoration: Costs and benefits for coral reefs. *World Resource Review* 17: 375–409.

Goreau, T. J. and W. Hilbertz. 2007. Reef restoration as a fisheries management tool, in fisheries and aquaculture. In P. Safran (Ed.) *Encyclopedia of Life Support Systems (EOLSS), Developed under the Auspices of the UNESCO*. Eolss Publishers, Oxford. http://www.eolss.net. (accessed September 3, 2012).

Goreau, T. J. and W. Hilbertz. 2008. Bottom-up community-based coral reef and fisheries restoration in Indonesia, Panama, and Palau. In R. France (Ed.) *Handbook of Regenerative Landscape Design*, pp. 143–159. CRC Press, Boca Raton, FL.

Hickman, C. P., L. S. Roberts, and A. Larson. 2001. *Integrated Principles of Zoology* (11th ed.). McGraw-Hill, New York.

Krebs, J. 1989. *Ecological Methodology*. Harper and Collins, New York.

Nontji, A. 2005. *Indonesian Ocean*. Penerbit Djambatan. Jakarta.

Nybakken, J. W. 1992. Marine Biology as an Ecological Approach. In H. M. Eidman, Koesoebiono, D. G. Bengen, M. Hutomo, and S. Sukardjo (Eds.). PT. Gramedia. Jakarta.

Odum, E. P. 1996. *Fundamentals of Ecology*. Gadjah Mada University Press. Yogyakarta.

Robbe, D., R. Purnawadi, U. Ali, and A. Bakti. 2011. Gili matra, marine proteced area: Ecotourism and community management. In *Proceedings of the Second Coral Reef Management Symposium on Coral Triangle Areas*, 28–30 September 2011, Kendari.

Romimohtarto, K. and J. Sri. 2007. *Marine Biology: Knowledge for Marine Biota*. Penerbit Djambatan. Jakarta.

Sadarun. 1999. Stony Coral Transplantation in Seribu Islands of Jakarta Bay. Thesis Program Pasca Sarjana IPB. Bogor.

Soekarno, R., M. Hartono, M. K. Moosa, and P. Darsono. 1983. Coral Reefs in Indonesia, Resources, Problems and Management, Sumberdaya, Permasalahan dan Pengelolaannya. LON-LIPI. Jakarta.

Soesilo, I. and dan Budiman. 2006. *Science and Technology: Investigating Indonesian Ocean*. Sarana Komunikasi Utama. Bogor.

Suharsono. 2004. Types of Coral in Indonesia. Pusat Penelitian Oseanografi—LIPI COREMAP Program. Jakarta.

Suprjharyono. 2007. *Management of Coral Reef Ecosystems*. Penerbit Djambatan. Jakarta.

Thamrin. 2006. Corals, Reproduction Biology and Ecology. Minamandiri Pres. Pekanbaru.

Thornton, S. L., R. E. Dodge, D. S. Gilliam, R. DeVictor, and P. Cooke. 2000. Success and growth of corals transplanted to cement armor mat tiles in Southeast Florida: Implications for reef restoration. In *Proceedings of the Ninth International Coral Reef Symposium*. Bali.

CHAPTER 7

Electrical Current Stimulates Coral Branching and Growth in Jakarta Bay

Neviaty P. Zamani, Khalid I. Abdallah, and Beginer Subhan

CONTENTS

Introduction 81
Materials and Methods 82
    Study Area 82
    Environmental Parameters 83
    Experimental Design 83
    In Situ Biological Observations 83
    Data Analysis 83
        Environmental Parameters 83
        Survival Rate 83
        Test of Significance 84
Results and Discussion 84
    Environmental Parameters 84
    Growth Rate and Branch Number of Coral Fragments 85
    Survival Rate 86
Conclusion 88
References 88

INTRODUCTION

Coral reefs are biologically diverse ecosystems built by the accumulation of calcium carbonate ($CaCO_3$) produced by reef-building Scleractinian corals and calcareous algae (Nybakken 1988). Corals consist of two different groups, hermatypic (or reef building) with special symbiotic relationship between polyps and zooxanthellae dinoflagellates (*Symbiodinium*) (Veron 2000; Castro and Michael 2003). Coral reefs serve as the venue for the cycle of biological, chemical, and physical as well as for wave-surge protection (Winarso et al. 1999).

According to Suharsono (1998), the percentages that cover live coral of the western part of Indonesia is 3.93% excellent, 19.10% good, 28.09% fair, and 48.88% poor. The central part is 7.09% excellent, 22.70% good, 33.33% fair, and 36.88% poor. The eastern part is 9.80% excellent, 35.29% good, 25.49% fair, and 29.42% poor. There are five main threats leading to coral reef deterioration in Indonesia: (1) poison fishing: cyanide is squirted on coral heads not only to stun and capture live aquarium and food fish but also for killing the polyps; (2) blast fishing: small bombs are detonated in shallow reef areas, not only killing targeted schools of fish but also killing larvae, juveniles, and corals; (3) coral mining: corals are collected and smashed for house construction and live sale for the aquarium trade; (4) sedimentation and pollution from logging, erosion, untreated sewage, and industrial discharges, which kill the corals; and (5) overfishing, which does not destroy corals directly but reduces an abundance and diversity of fish and invertebrates (Tomascik et al. 1993).

Because of its close proximity to Jakarta, a metropolitan city of about 12 million people, Seribu Islands Marine National Park serves as a holiday destination for people in Jakarta and vicinity (Fauzi and Buchary 2002). The islands are subjected to considerable pressure from human use and pollution, which make the coral reefs in Seribu Islands increasingly vulnerable to domestic sewage, industrial wastes, and destructive fishing, including cyanide bombing and toxic materials (Bryant et al. 1998; Erdmann 1996).

An artificial reef is one or more objects, natural or man-made, deployed purposefully on the seafloor to influence, physical, biological, or socioeconomic processes related to living marine resources. Artificial reefs are defined physically by the design and arrangement of materials according to their purpose (Bohnsack and Sutherland 1985; Carr and Hixon 1997; Seaman and Jensen 2000).

The Biorock method, invented, developed, and patented by the late Prof. Wolf Hilbertz and Dr. Thomas J. Goreau, uses low-voltage direct current (between 1.2 and 12 V) to grow solid limestone minerals on conductive substrates. The minerals grown are naturally present in large amounts in seawater but do not crystallize by themselves. These current are safe to humans and all marine organisms. There is no limit in principle to the size and shape of Biorock structures.

Hard coral typically grow 2–6 times faster on Biorock structures than on controls (depending on species and conditions), show dense branching, have 16–50 times higher survival rates after severe high temperature stress, and show rates of new coral recruitment hundreds to thousands of times higher per unit area per unit time than recorded in the literature (Goreau 2008).

The hypothesis of this study is that the Biorock process can increase the health, growth, and survival rates of hard corals and also can influence the growth forms of corals. The main purpose of this study was to (1) estimate the growth and survival rates of corals (*Acropora* sp.) on Biorock artificial structures, (2) compare them with survival and growth rates on the non-Biorock artificial reef, and (3) ascertain the effect of limestone deposited on corals' growth morphology. The main aim of this study is to assess Biorock artificial reef as an alternative way for reef rehabilitation in Jakarta Bay by providing a natural limestone substrate produced *in situ*. This kind of artificial reef will be used for a variety of purposes, such as for aquaculture and breakwaters, to support conservation of biodiversity, to test ecological theories, to make available to the public as alternative diving sites, and may reduce human pressure on nearby natural reefs and therefore facilitate their rehabilitation.

MATERIALS AND METHODS

Study Area

This study was conducted for seven months (April through November 2009) as a field experiment at depth of −6 m at Pramuka Island in the Seribu Islands (Jakarta Bay, Indonesia). The Seribu Islands (5°24′–5°45′ South, 106°25′–106°40′ East) are scattered around the northern part of the Jakarta Bay in the southwest Java Sea and consist of 78 small islands. Because of its close proximity to Jakarta, Indonesia's major population center, the islands are subjected to considerable pressure from human use and pollution.

Environmental Parameters

The *in situ* environmental parameters measured monthly were temperature, pH, salinity, water current speed, and dissolved oxygen. Water samples were collected (underwater sampling at a depth of −6 m) from each study site. For long storage, water samples were preserved with 2 mL H_2SO_4/L and stored at 4°C (APHA 2005). Other parameters measured in Proling Laboratory (Bogor Agricultural University, Faculty of Marine Science and Fisheries) were turbidity, total suspended solids (TSS), nitrate (NO_3-N), and phosphate (PO_4-P) using the nephelometric method, gravimetric method, colorimetric method, and ascorbic acid method, respectively (APHA 2005). The purpose was to check whether the changes in growth and survival rates were due to the experimental treatment or changes in the environmental conditions.

Experimental Design

Biorock submerged structures were constructed from welded steel cathodes with a positively charged anode (titanium mesh). A low electric direct current (3 V and 6 A) flows between them provided by a DC power supply. Calcium ions combine with carbonate ions and adhere to the structure (cathode), resulting in the growth of calcium carbonate over the steel. Coral fragments adhered to $CaCO_3$ and grew quickly. Two species, namely, *Acropora tenuis* (branching) and *A. cytherea* (tabulate) were used for the experiment. Corals fragments collected from study area reef were cut as a thumb-sized tips and then attached to iron structures, both electrically charged Biorock structures, and identical uncharged control structures. Sixty coral fragments of each species were attached to the Biorock and control structures. The study was divided in two phases: the first phase of three months and then continued for the next 4 months (this is still in progress, to be published in details by Abdallah and coworkers) as the second phase.

In Situ Biological Observations

The submerged Biorock structures were observed every month; coral colony branch lengths were measured using a caliper with 0.05 mm scales, and other biological parameters were also noted. SCUBA diving gear and documentation equipment (writing board, underwater digital camera) were used (Figures 7.1 and 7.2).

Data Analysis

Environmental Parameters

To test the significant effect on coral growth and branch numbers of the environmental parameters, analysis of covariance was computed using alpha 0.05 (Zar 1984), where the electrical current was the main variable, the environmental parameters were the covariates, and mean growth and increase of branch numbers were dependent variables.

Survival Rate

In biostatistics, survival rate is a part of survival analysis, indicating the percentage of samples in a study or treatment group that are alive at a given period of time. Survival rate in this study was determined by dividing the number of live coral colonies at the end of the study with the number of coral colonies at beginning of the study, multiplying by 100.

Figure 7.1 Control corals (a) before and (b) after the experiment.

Figure 7.2 Biorock corals (a) before and (b) after the experiment.

Test of Significance

To investigate the statistical significance of the difference between Biorock and control (μ_1 and μ_2), means of coral growth rates, increase in branch number rates, and survival rates (t-test) have been used (Kanji 2006).

SPSS release 11.5.0 (September 6, 2002) and Microsoft Office Excel 2007 statistical software were used for collection, management, and analysis of the data. Textbooks were used for the identification of corals (Suharsono 1996; Veron 2000).

RESULTS AND DISCUSSION

Environmental Parameters

In general, water quality at the study site was suitable for coral growth. Water temperatures ranged from 23–30°C, salinity 30–35‰, pH 7.6–8.3, turbidity 0.50–1.18 NTU, TSS 4–7 mg/L, dissolved oxygen 6.4–7.4 mg/L, current speed 0.05–0.09 m/S, PO_4-P 0.02–0.09 mg/L, and NO_3–N 0.002–0.022 mg/L.

Analysis of covariance on the effect of the environmental parameters showed that none of the environmental parameters had a significant effect on coral (*A. tenuis* and *A. cytherea*) growth rates or branch numbers. Differences in growth rates and branch numbers were due to the electricity.

Covariance analysis showed that there is no statistically significant effect of environmental parameters, and electricity ($p = .548$) helped in increasing the branch numbers of *A. cytherea*. This makes sense, because *A. tenuis* has three-dimensional open branching, while *A. cytherea* is tabulate and adds new polyps only at the periphery.

The deposition of minerals onto a cathode substrate through seawater electrolysis presumably enabled the increase in the concentration of mineral ions in a small boundary layer above the cathode that could be utilized by the coral for skeleton formation (Hilbertz and Goreau 1996).

Growth Rate and Branch Number of Coral Fragments

By the end of the first three-month duration, our results estimated the mean growth rates for *A. tenuis* as 1.8 cm on Biorock and 2.4 cm on control, and for *A. cytherea* as 0.57 cm on Biorock and 0.74 cm on control, while for the second phase (next four months), the mean growth rates for *A. tenuis* were estimated as 5 cm on Biorock and 3 cm on control, and for *A. cytherea* as 1 cm on Biorock and 0.3 cm on control (Figure 7.3).

The lower growth rates of fragments transplanted in Biorock structure compared with those transplanted in uncharged control structure (Figure 7.3a) was apparently due to the electrical power stability problem and irregular charging during the first three months, but when the power supply was stabilized during the second phase, this resulted in higher growth rates of corals transplanted on Biorock compared to controls (Figure 7.3b). This could indicate that the main key for success of mineral-accretion artificial reefs is the electrical current stability. Electric fields allows accretion of carbonate because a low current causes elevation of pH in the immediate vicinity of the coral, increasing natural calcification, and from excess production and release of electrons, because the electrochemical processes at the cathode may affect the electron transport chain for ATP production, where excess energy can be used for growth promotion (Hilbertz and Goreau 1996; Sabater and Yap 2004; Schuhmacher et al. 2002; Tambutte et al. 1995, 1996).

The growth rate reported in our study reflects a particular level of voltage and current (3 V and 6 A). Growth rates may differ under different electrical regimes. Previous studies that used this method had voltages varying from 18 V and 4.16 A (Sabater and Yap 2002, 2004), 1 to 24 V (van Treeck and Schuhmacher 1997), 8 to 12 V (Schuhmacher and Schillak 1994), and 6 to 12 V (Hilbertz and Goreau 1996). Our observations showed a large variation of increase of coral branches during two different phases. In the first phase, the mean increase in branches for *A. tenuis* was four branches on Biorock and three branches on control, and for *A. cytherea* it was six branches on Biorock and nine branches on control. During the second phase, the mean of increasing branch number for *A. tenuis* was eight branches on Biorock and five branches on control, and for *A. cytherea* it was 14 branches on Biorock and 11 branches on control (Figure 7.4).

Fragments of corals transplanted on Biorock had unstable electrical power during the experimental first phase and this resulted in no significant difference in increasing branch numbers (Figure 7.4a), while during the second experimental phase with stable power, the coral fragments on the charged structure showed a significant increase in branch numbers compared with those transplanted on uncharged control structures (Figure 7.4b).

Increasing pH in the vicinity of Biorock transplants may create a higher concentration gradient between the water outside the coral polyp and coelenteric cavity. This can lead to diffusional influx of mineral ions into the coelenteron via the "leaky" nature/permeability of *cnidarian* epithelium (Benazet-Tambutte et al. 1996; Tambutte et al. 1995, 1996). This could increase the availability of ions for active uptake via transcellular route of calcification and would explain the significant increase in the number of branches and growth rates (Hilbertz and Goreau 1996). The higher rates of increase in branch numbers than in the linear coral fragment growth implies that the coral's tissue growth is benefitting more than the skeleton growth.

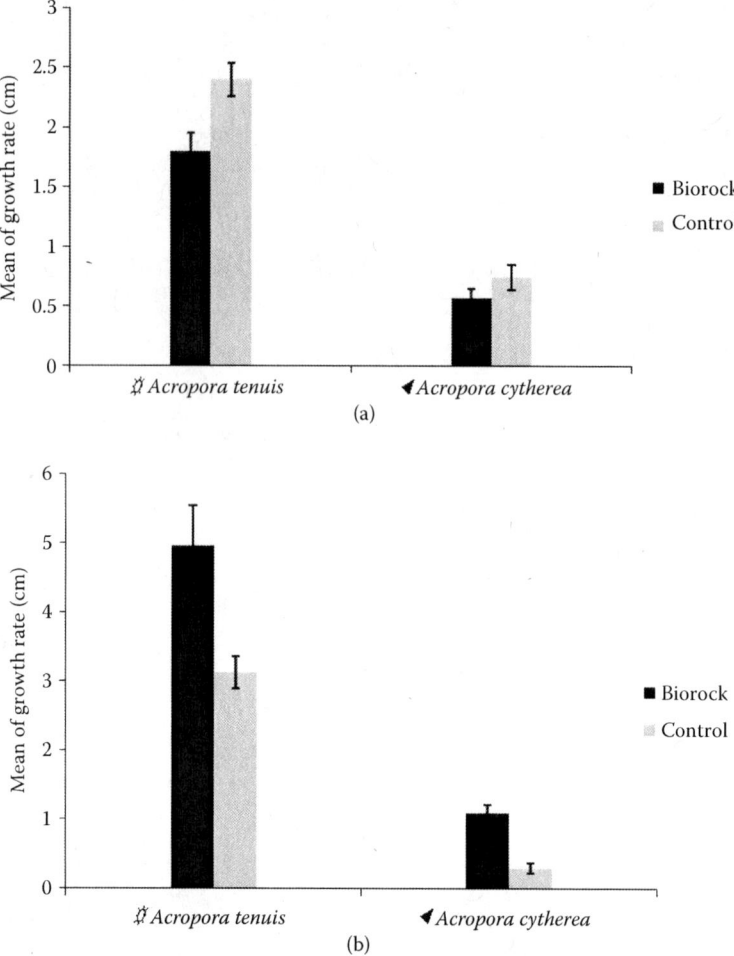

Figure 7.3 Mean of growth rates (fragment length) for *A. tenuis* and for *A. cytherea* on all the study sites for the whole seven months: (a) first three months and (b) second four months.

Survival Rate

Survival rate of corals reflects their ability to adapt/adjust to the new environment. All transplanted fragments were collected from around the experiment site and are influenced by transplantation techniques. Fragment size plays a role in transplantation success, where large fragments have a higher chance to survive more than small ones.

The data presented refer to survival rate as numbers of fragments found in original position and alive on the respective structures. Figure 7.5 shows the higher survival rates on Biorock (100%) than on control (73.3% and 83.3%) of the two transplanted corals species (*A. tenuis* and *A. cytherea*). The difference between Biorock and control may be due to the influence of electrical field that inhibited the settlement of red filamentous and fleshy (*Caulerpa racemosa* and *Padina* sp.) algae. Losses were probably due to wave action, grazing activities, predation, and death caused by intrinsic factors.

We found some of the missing nubbins on the seabed close to the structures. Transplanted fragments were not all found dead on the structures but some of them disappeared due to wave action

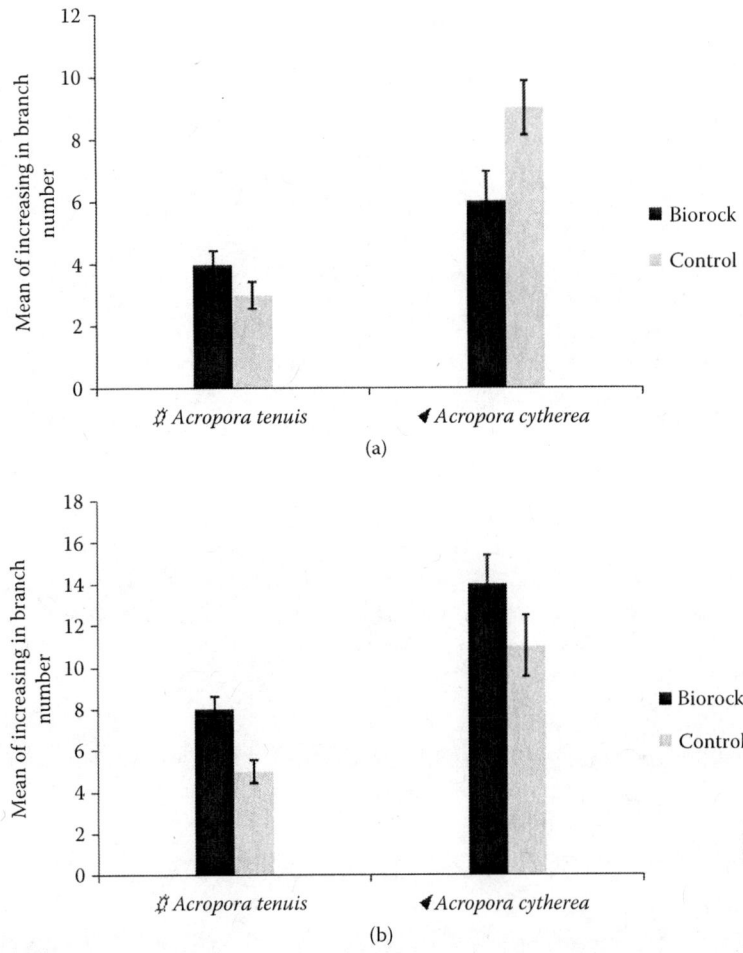

Figure 7.4 Mean of increasing in branches number for *A. tenuis* and for *A. cytherea* on all the study sites for whole seven months. (a) First three months, with no reliable power and (b) second four months, with reliable power.

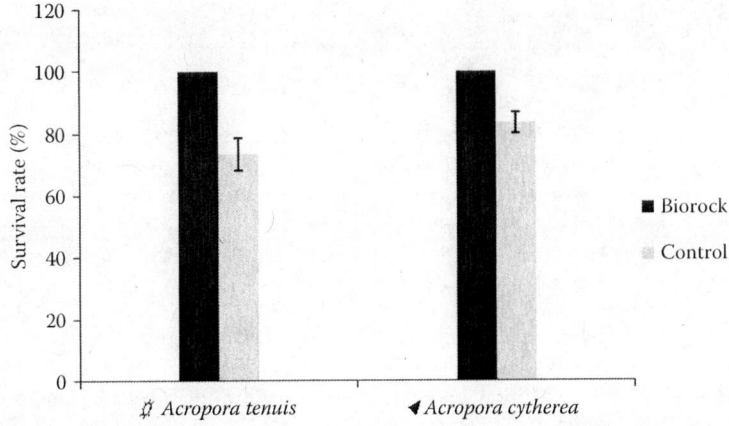

Figure 7.5 Mean of survival rates for *A. tenuis* and for *A. cytherea* on all the study sites for whole seven-month duration of the study ($p = .04$, ±SE).

or biogenic disturbances such as fish grazing. Van Treeck and Schuhmacher (1997) observed fishes (Labrids, e.g., *Thalassoma* spp., *Cheilinus abudjubbe*) moving some coral fragments while searching for invertebrates living in the skeletons of fresh fragments.

CONCLUSION

Electrical stimulation appeared to increase budding and branching of corals even faster than growth, suggesting that the positive effects of the electrical field on tissue-cell growth and division are greater than on the skeleton growth. The effects were greater on corals with three-dimensional branching than on table corals, where new polyps are limited to the periphery. By increasing the growth and survival rates of hard corals by stimulating coral branching and fragment length, Biorock reef restoration could be introduced as an alternative way for reef rehabilitation by providing a natural-like substrate with limestone character generated *in situ*, which will be useful for a wide variety of coastal management purposes.

REFERENCES

American Public Health Association (APHA). 2005. *Standard Methods for the Examination of Water and Waste Water*, 21st Ed. APHA, AWWA (American Water Works Association), and WEF (Water Environment Federation). Washington, DC.

Bohnsack, J. A., and D. L. Sutherland. 1985. Artificial reef research: A review with recommendations for future priorities. *Bull. Mar. Sci.* 37: 11–39.

Bryant, D., L. Burke, J. McManus, and M. Spalding. 1998. *Reefs at Risk: A Map-Based Indicator of Threats to the World's Coral Reef*. World Resources Institute. Washington, DC.

Carr, M. H., and M. S. Hixon. 1997. Artificial reefs: The importance of comparisons with natural reefs. *Fisheries* 22: 28–33.

Castro, P., and E. Michael. 2003. *Marine Biology*, 4th Ed. Mc Graw Hill. Boston, MA.

Erdmann, M. V. 1996. Destructive fishing practices in the Pulau Seribu Archipelago. Report on the Coral Reef Management Workshop for Pulau Seribu. Jakarta. No.10.

Fauzi, A., and Buchary, E. A. 2002. A Socioeconomic Perspective of Environmental Degradation at Kepulauan Seribu Marine National Park, Indonesia Coastal Management, 30: 167–181.

Goreau, T. J. 2008. Brief summary of biorock habitat restoration methods. The 6th Indonesian Biorock Training Workshop. Gili Trawangan. Indonesia.

Hilbertz, W. H., and T. J. Goreau. 1996. Method of enhancing the growth of aquatic organisms, and structures created thereby. http://www.patentsbase.com. (accessed August 13, 2012).

Kanji, K. 2006. *100 Statistical Tests*. 3rd Ed. SAGE Publications. London.

Nybakken, J. W. 1988. *Biologi Laut: Suatu Pendekatan Ekologis*. Terjemahan. PT. Gramedia Pustaka Umum, Jakarta. Harper and Row. New York.

Sabater, M. G., and H. T. Yap. 2002. Growth and survival of coral transplants with and without electrochemical deposition of $CaCO_3$. *J. Exp. Mar. Biol. Ecol.* 272: 131–146.

Sabater, M. G., and H. T. Yap. 2004. Long-term effects of induced mineral accretion on growth, survival and coralite properties of *Porites cylindrica* Dana. *J. Exp. Mar. Biol. Ecol.* 311: 355–374.

Schuhmacher, H., and L. Schillak. 1994. Integrated electrochemical and biogenic deposition of hard material—a nature-like colonization substrate. *Bull. Mar. Sci.* 55: 672–679.

Schuhmacher, H., P. van Treeck, M. Eisenger, and M. Paster. 2002. Transplantation of coral fragments from ship groundings on electrochemically formed reef structures. Proceedings of the 9th International Coral Reef Symposium. October 23–27, 2000. Bali, Indonesia.

Seaman, W., and A. C. Jensen. 2000. Purposes and practices of artificial reef evaluation. In: Seaman W. (ed.) *Artificial Reef Evaluation with Application to Natural Marine Habitats*, 2–19. CRC Press. Boca Raton, FL.

Suharsono. 1996. *Common Coral Species in Indonesian Water*. Oceanology Publications. LIPI, Jakarta, Indonesia.

Suharsono. 1998. Condition of coral reefs resources in Indonesia. *Journal of Coastal and Marine* 1(2): 44–52.

Tambutte, E . Allemand, D., Bourge, I., and Gattuso, J. P., 1995. An improved 45Ca protocol for investigating physiological mechanisms in coral calcification. *Mar. Biol.* 122: 453B–459.

Tambutte, E., Allemand, D., Mueller, E., and Jaubert, J., 1996. A compartmental approach to the mechanism of calcification in hermatypic corals. *J. Exp. Biol.* 199: 1029–1041.

Tomascik, T., Suharsono, and A. J. Mah. 1993. Case histories: A historical perspective of the natural and anthropogenic impacts in the Indonesian Archipelago with a focus on the Kepulauan Seribu, Java Sea. Proceedings of the Colloquium on Global Aspects of Coral Reefs: Health Hazards and History. University of Miami, FL.

Van Treeck, P., and H. Schuhmacher. 1997. Initial survival of coral nubbins transplanted by a new coral transplantation technology: Options for reef rehabilitation. *Mar. Ecol. Prog. Ser.* 150: 287–292.

Veron, J. E. N. 2000. *Corals of the World*. Vols. 1–3, Australian Institute of Marine Science. Townsville, Australia.

Winarso, G., B. S. Tejasukmana, and B. Irianto. 1999. Analysis of Landsat-TM for distribution and area of coral reef at Spermonde Island South Sulawesi. *Majalah LAPAN, Remote Sensing Edition* 1(1): 56–62.

Zar, J. H. 1984. *Bio-Statistical Analysis*, 2nd Ed. Prentice-Hall. Englewood Cliffs, NJ.

CHAPTER **8**

Electricity Protects Coral from Overgrowth by an Encrusting Sponge in Indonesia

Jens Nitzsche

CONTENTS

Introduction	92
Materials and Methods	93
Experimental Design	93
Study Site	93
Experimental Organisms	93
Setup of the Experiment	94
Method of Attachment	95
Electricity Supply	95
Data Collection Time Frame	96
Measurements and Documentation	96
Data Selection	96
Sponge-Infested Sample Corals	96
Uninfested Sample Corals	97
Results	97
Sponge Extension Increase Experiment	97
Acropora microphthalma Growth Rate	97
Relation of Sponge Extension Increase and Distance to Electrical Source	97
Comparison of First and Second Measurement Period Regarding Their Respective Sponge-Extension Increase	99
Discussion	99
First and Second Growth Period	99
Coral Growth	99
Negative Impact of Electricity on Sponges	100
Reasons for Sponge Extension Increase on Electrified Structures	100
Conclusions	102
Acknowledgments	102
References	103

INTRODUCTION

Corals growing on Biorock reefs typically grow 2–6 times faster than controls and have 16–50 times higher survival after severe bleaching (Goreau and Hilbertz 2005, 2008). Nevertheless, sponge overgrowth of some species of corals by certain sponges has occurred on some electrical coral-reef restoration projects in the Maldives and in Bali and Lombok Indonesia. This project determined the growth rates of corals and sponge overgrowth as a function of distance from the electrical field to determine whether electricity affected the sponge–coral interaction. Due to the higher levels of sponge overgrowth on some structures than on nearby natural reefs, it was initially hypothesized that sponge growth might be stimulated more than coral growth, favoring sponge overgrowth (Figures 8.1 through 8.3).

Figure 8.1 Biorock corals resistant to red encrusting sponge overgrowth.

Figure 8.2 Pink (left) and red (right) encrusting sponges overgrowing shaded substrate on weakly charged Biorock structure.

ELECTRICITY PROTECTS CORAL FROM OVERGROWTH

Figure 8.3 Coral being overgrown by encrusting sponge.

MATERIALS AND METHODS

Experimental Design

Seven measurement sites, two on electrically connected Biorock structures and five at different distances from the electrical structures, give insight into the question whether sponges or corals gain advantages through electrical sources and how this decreases with distance from the electrical field.

Study Site

The experiment was carried out on the Karang Lestari Coral Reef Restoration project in Pemuteran, Bali, Indonesia. The village of Pemuteran (8°02' South; 114°39' East) faces the Bali Sea near the northwest corner of the island. Here, around 50 Biorock structures (2008 status) are located inside a village No Fishing protected area. This zone is parallel to the coastline with a total length of around 300 m. The structures are mostly 50–80 m from the shore at depths between 3 and 7 m.

Experimental Organisms

While they are hard to find in the natural reef surrounding the structures, several species of sponges, mainly red-, orange-, and pale pink-colored species seem to be a problem due to their high abundances on certain structures. They chiefly grow along the underside of the horizontal rails of a structure or fully along the vertical rails of structures, and when they reach a coral of certain species they start to overgrow the coral, whereby the coral often dies, with some species being far more susceptible to overgrowth than others. Above all, the extension of a pale pink sponge involves encroachment over living coral tissues. The most abundant species of sponge on the projects, an unidentified pink-colored encrusting sponge, often with a whitish surface coating and fleshy red tissue, specifically attacks *Acropora microphthalma*, a fragile but common species on the projects. After the coral has been overgrown and killed, it eventually breaks off, and the sponge appears to weaken the dead coral skeleton.

The sponge usually occurs as a few millimeters-thick encrusting layer, but on some spots it occurs as thin, fingerlike growths up to 20 cm long along the substrata, with connected branches growing out from an encrusting base. During a six-month period, these sponge branches were noted reaching from one overgrown coral branch to another branch to another overgrown coral branch. The sponge grew together, filling the gap between the coral branches.

Setup of the Experiment

The experiment, which lasted for 88 days, was set up at the western end of the project and up to 100 m westward of it. In order to determine if the strength of the electrical field affected coral and sponge growth rates, specimens whose growth was measured were placed on top of a structure—directly under a structure but not in physical contact with it—and at distances of 3, 10, 30, and 100 m away from a structure. Structure 22 and Fish Structure 3 were used for the experiment.

The measurement sites are named in compliance with their distance to the closest structure. Therefore, the seven measurement sites are called 3 m, 10 m, 30 m, and 100 m; and accordingly, samples under Structure 22 are called uST22 and measurement sites with samples directly on Structure 22 or Fish Structure 3 are called ST22 or Fish3.

Both structures were used for the experiment, as they showed interesting differences although located near to each other. With its simple design and its few struts, Structure 22 is one of the most successful structures of the project in building up limestone. Around the construction steel bars of diameter 1–1.2 cm, up to 4–8 cm diameter of limestone has accumulated. That structure has very little sponge growth. Fish Structure 3 is made from a construction-wire mesh with a diameter of 0.6–0.8 cm. The material itself and its many welds diminish conductivity. Additionally, the cables of this structure have to be some 15 m longer, so that there is more voltage drop. These factors give reasons for less power flow through Fish Structure 3, and so for a diameter increase has been much slower, only 2 cm. The essential point is the divergence in sponge infestation, which is highest on the structure that is receiving less power and growing more slowly. While rapidly growing Structure 22 (Figure 8.4) and its corals show only little sponge presence, slowly growing Fish Structure 3 (Figure 8.5) was almost completely covered with sponge, and sponge infestation on the corals growing on the structure is intense.

Figure 8.4 Structure 22.

Figure 8.5 Fish Structure 3.

Method of Attachment

Two options of material for the coral-attaching devices were chosen: thirty 2 cm diameter and 50 cm long plastic tubes stabilized by cement pedestals, and thirty 1 cm diameter and 1 m long iron rods that simply could be hammered into the ground.

Around each measurement site mark, six iron rods were driven into the sand, and six plastic cement devices were placed. These two sorts of attaching devices were mixed-positioned, so there was no group forming, and effects of location were minimized. Although all the attaching devices were positioned as close as possible to minimize divergences in distance to the electrical source and depth, an inevitable distance was kept to prevent any contact of the samples with each other during the experimental period. In the case of uST22, an adequate distance to prevent contact to the structure was included.

All corals on the attaching devices, including those on Structure 22, are around 50 cm above the bottom. The corals on Fish Structure 3 are around 2 m above the bottom. All corals were located at depths of 3.9 to 4.4 m below the surface. The specimens were tied on with plastic cable ties.

On the five measurement sites that are not directly on the two structures, namely sites uST22, 3 m, 10 m, 30 m, and 100 m, three corals without any sponge and three sponge-infested corals were fixed onto iron rods and the same numbers onto the plastic tubes.

On Fish Structure 3 and on Structure 22, four corals without any sponge and four sponge-infested corals were fixed on to each of them.

Coral samples, after a few days of attachment at all measurement sites, appeared to be colorful and healthy-looking.

Electricity Supply

The power supply is switched on for around eight hours per day.

To check that the electrolysis process is working, structures were observed to see if hydrogen gas bubbles were rising and the anodes were clean and working normally.

During the running time of the experiment, voltage values fluctuated between 11 and 13 V. Amperes on the measured cables connected to the anodes placed around Structure 22 and Fish Structure 3 varied from 7.2 to 14.5 A, with an average of 9.6 A.

Data Collection Time Frame

Growth measurements were carried out three times. The process of data collection was repeated every time in the same chronological order. The first images of the samples were taken just after the setup of the experiment on April 17, 2008; the first repetition was right in the middle of the running period on May 30, 2008; and the last one was 88 days later on July 12, 2008. Each particular data collection was carried out after the fifth day with the measurement of the second half of the sample corals. The experiment ended on July 16, 2008.

Measurements and Documentation

To record data about sponge extension increase, photographs were taken for each coral branch infested with sponge.

A ruler was held on the same spatial level as the branch. This provides an equivalent scale for the measurement software ImageJ 1.40g with which the sponge expansion was ascertained.

Coral growth was documented by measuring height, width, and depth of each sample coral with a vernier caliper.

Data Selection

Sponge-Infested Sample Corals

Excluded Sponge Growth Values

At the collection of data, the increasing extension of sponge on all infested branches of the sample corals was measured. Sixty-six separate values were recorded.

Thirty-five of them were used for further analysis, while 31 values were excluded. The reasons for their exclusion are listed in Table 8.1

After Fish Structure 3 was knocked over by an exceptionally rough sea, samples on it were excluded from statistical interpretations. Other reasons for sample loss, like break-offs, overgrowth or bleaching are documented in Table 8.1.

Remaining Sample Distribution

How many samples on each measurement site were taken into consideration and how many samples got lost on each site is shown in Table 8.2.

Table 8.1 Reasons for Sample Loss

Reason	Sponge extension on bleached branches	Fish structure collapsed, samples were not taken into account	Sample coral broke off the attaching device	At second measurement already totally overgrown branch	At third measurement already totally overgrown branch	Angle of taken pictures at different measurement periods did not conform
Number	4	5	4	6	9	3

Table 8.2 Samples Lost on Each Site

Measurement Site	ST22	uST22	3 m	10 m	30 m	100 m
Lost Samples	0	2	1	2	1	3
Remaining Samples	4	4	5	4	5	3

Uninfested Sample Corals

Out of the 38 attached samples, the data of 27 samples could be used for further interpretations. Eleven samples had to be excluded. The exclusion of samples is attributed to three reasons. Fish Structure 3 was knocked over during exceptionally rough seas, and therefore the data of Fish Structure 3 were not used anymore for a comparison of the various measurement sites. Furthermore, two samples together with their spear qualified iron rod attaching device disappeared at measurement site 100 m. The remaining five losses are attributed to reductions of the volume of the sample coral. This reason for exclusion is caused by fish bites or somehow incurred break-offs. These are equally distributed among the measurement sites; they all have one loss because of volume reduction.

RESULTS

Sponge Extension Increase Experiment

No differences in growth, health, or sponge extension increase are found due to the use of iron rods compared to the use of plastic tubes.

Acropora microphthalma *Growth Rate*

Considering the total experimental time frame, a significant difference between two groups could be detected (ANOVA; $P = 0.004$). There is accelerated coral growth on Structure 22, under it, and near to it, compared to the other measurement sites, which are further away. This beneficial effect seems to be the result of the electrical field around the structures and extends at least 3 m, but less than 10 m, from it.

However, it is important to note that coral growth was fairly variable between the two time intervals (Figure 8.6). In general, the values of the second period are closer together than those of the first period. The near the structure, sites uST22 and 3 m grew faster the first 44 days than the second 44 days. The other four sites had faster growth rate in the second period than the first. The majority of individual growth values of ST22 at the first growth period rank behind those of 3 m and uST22 and ahead of those at the further measurement sites, but at the second growth period, the single values of ST22 are like those of 3 m and uST22: not very outstanding compared to the other measurement sites. Group forming could be determined in the first time interval (ANOVA; $P = 0.000$) but not for the second (ANOVA; $P = 0.473$).

Relation of Sponge Extension Increase and Distance to Electrical Source

Sponge extension increase is higher on weak corals than on vital ones. For example, if a branch is already bleached, there is hardly any defense against the sponge anymore. So the sponge growth

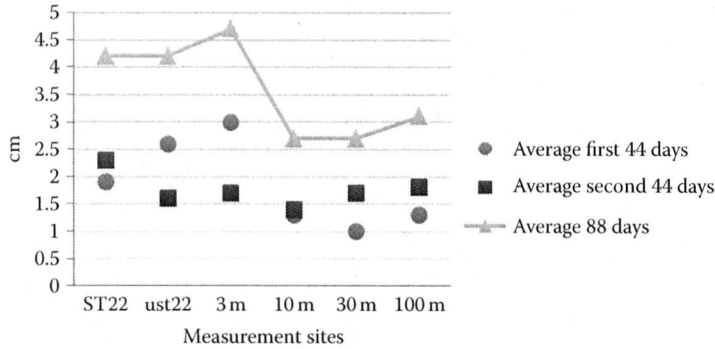

Figure 8.6 *Acropora microphtalma* growth rate.

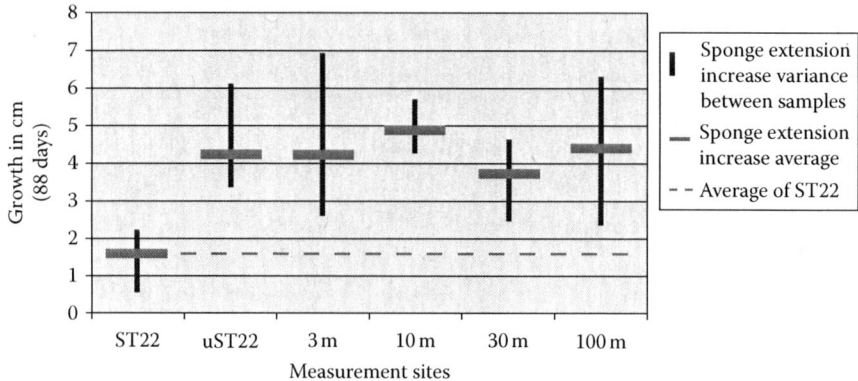

Figure 8.7 Sponge extension increase on each particular measurement site.

rate over bleached branches is clearly higher. Average sponge growth over bleached branches was 8.4 cm, about twice as high as on healthy corals (i.e., not bleached). Many corals being overgrown by sponges were observed bleached, although a thermal bleaching event was not under way. It is possible that this bleaching is a stress response to allelochemicals released by the sponge. This needs to be tested.

The samples on ST22 are the strongest in contrast to the bleached samples. Three of the four lowest sponge growth rates, below 2 cm, belong to the three samples on ST22. The corals on the structure appear to be more resistant to sponge overgrowth (Figure 8.7).

Sponges clearly grew more slowly, by about 62%, on corals on Structure 22 than they did anywhere near or far from them. Considering the total experimental time frame, a significant difference between two groups, ST22 versus all other measurement sites, could be detected (ANOVA; $P = 0.004$). The role of electricity, therefore, appears to inhibit overgrowth of living corals by encrusting sponges on the electrified reef in Pemuteran, providing an advantage to the corals.

Further, it is noteworthy that as mentioned, only the samples at ST22 seem to be influenced. The two nearby sites uST22 and 3m, which showed elevated coral growth during the first measurement period, do not stand out from those farther away with regard to sponge growth. Both uST22 and 3 m have the same average of 4.2 cm, which hardly differs from the total average of 4.3 cm of the three measurement sites 10 m, 30 m, and 100 m.

ELECTRICITY PROTECTS CORAL FROM OVERGROWTH

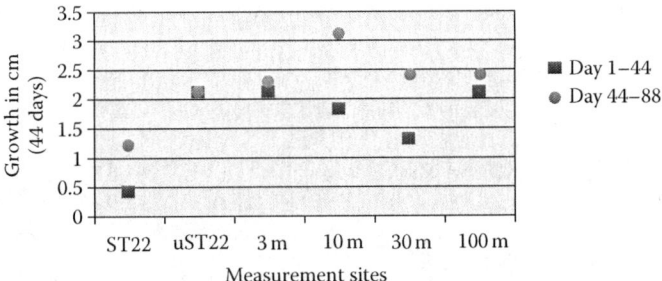

Figure 8.8 Average growth during first and second measurement intervals.

Comparison of First and Second Measurement Period Regarding Their Respective Sponge-Extension Increase

The outstanding character of ST22 exists in both periods. After the first period, sponge extension increase is on average 1.5 cm slower on ST22 than on the other measurement sites, and with 1.3 cm slower increase after the second period.

However, during the second 44 days, sponge-extension increase was greater than during the first 44 days. At the second measurement on some infested branches of the samples at ST22, the sponge extension even declined. Such developments could not be registered after the third measurement.

On the other measurement sites, there is no such sponge decrease. On comparing the two measurement periods, the averages of 10 m and 30 m had gone up, respectively, 1.1 and 1.3 cm. The averages of the other three measurement sites went up only a little or not at all. They already had shown high increase values at the first period. Like the *A. microphthalma*, growth measurements during the first period show higher results than the during second period. Over the first period, group forming (ST22 vs. all other measurement sites) can be confirmed (ANOVA; $P = 0.039$). Over the second period, it was not significant (ANOVA; $P = 0.878$). The different character of ST22 can only be guessed. However, as shown in Figure 8.8, ST22 has the slowest growth values in the first and second periods.

DISCUSSION

First and Second Growth Period

The difference between growth values of samples on the structure and other measurement sites for coral growth and for sponge growth is more significant at the first period. The cause of this difference cannot be detected with certainty. Only assumptions can be made. One reason which would confirm the results of this work would be if, at the second measurement period, the power supply would have been interrupted or off for various days unobserved. Another possibility is seasonal variations in zooplankton food supply for corals and bacterial food for sponges. Higher sample numbers and more measurements to detect seasonal changes would have brought more certainty.

Coral Growth

The stimulating influence of the electricity on the growth of *A. microphthalma* on the structures and in the area around the structures seems to be obvious. During the first growth period, growth-rate ranking of samples is divided, with hardly any exceptions, into two groups. Samples on Structure 22, under the structure and 3 m away, grew faster than samples on the other measurement sites farther away. Curiously, during the second growth period, there were smaller differences. For future research on coral growth acceleration, the distance and position to the anode should be taken into consideration, since at the first growth period, corals had high growth values, especially corals on measurement site 3 m, which is located in between Anode 7 and Structure 22.

Negative Impact of Electricity on Sponges

As the results of the sponge extension increase experiment show, electrical current does not offer any growth advantage for the pale pink sponge already infested on the coral *A. microphthalma*; in fact, it reduces sponge growth. Although the coral could not effectively reject the sponge, sponge growth was even slower on sample corals attached to the structure than on those attached to devices of the other measurement sites.

Since electricity decreased sponge growth rates, and since the structures having fast growth due to higher electrical current have few sponge problems compared to those that are growing slowly, we decided to test if increasing the current on a slow-growing, sponge-infested structure would get rid of the sponges. Fish Structure 3, which was heavily sponge-infested, was used for this experiment. The power was about doubled, and the mineral growth on the structure immediately increased. Sponge cover, which covered almost all the surface of the structure at the start, steadily decreased until only small isolated patches of sponges remained several months later. Therefore, structures that have sponge problems can get rid of them by increasing the power. The result of this attempt confirms the results of the sponge extension increase experiment.

Why sponges should grow more slowly on the structures while hard corals grow faster—and soft corals, tunicates, and bivalves also appear to—is curious. Since this is a function of being on the structure itself, and not the electrical field that stimulates corals in the vicinity of the structures, a likely explanation is the high pH generated by electrochemical reactions at the growing surface. While this makes limestone less insoluble, causing it to precipitate and grow from seawater, the silica material that makes up sponge skeletal spicules has the opposite behavior: it dissolves under high pH. Loss of the skeletal material may be the cause of reduced growth, and histological studies would be needed to clarify this.

Reasons for Sponge Extension Increase on Electrified Structures

Sponge extension on the slow-growing electric reefs is shown by close encrusting growth over the frame of the slow-growing structures, but fast-growing structures and the natural reef surrounding the structures are not affected by excessive sponge abundance. If it was only a problem of nutrition or the absence of sponge growth-controlling fish, the reef straight under and beside the structure would be impacted too.

Sponge-feeding fish seemed pretty conspicuous around the artificial reef, but a further examination of the pale pink sponge on the structures as well as on those measurement sites away from any electrical influence indicated very little physical damage or bite marks on sponges, as injuries of the pale pink tissue would have been obviously noticeable through the underlying bright red tissue beneath a pale surface layer.

The high abundance of sponges on the structures is not due to increased growth rates, since the sponge actually grows more slowly under electricity. Presumably higher recruitment and settling of sponge larvae on slow-growing structures and unhindered sponge extension on the preferred shadowed underside areas or the superior location of the three-dimensional structure—which allows filter feeders to be right in the middle of water movement, offering increased food supply—may be reasons for the very successful growth of this species.

In the following, various underlying possibilities for excessive sponge settlement will be further discussed:

According to a progress report from 2001 about the attachment of the first corals on, at this time, still-bare frames, "In order to avoid damaging healthy corals but to rescue broken-off specimens, fragments lying on reef slopes or buried in sand were selected for transplantation." Since "large numbers of corals were found loose, often having had their bases undermined by boring worms, clams and sponges" and even "a few totally dead corals were attached because some divers involved in collection were insufficiently experienced at identifying live corals" (Hilbertz and Goreau 2001), the possibility that sponges were widely transplanted together as overgrowths with the first corals is quite likely.

Additionally, another human activity, the attempt to get rid of the recognized threat to the coral garden by scrubbing sponges off the bars of the structures, may again have caused further spreading. Even though regeneration from removed tissue is slow—and sometimes, when large parts are removed, is irreparable for some encrusting species—other sponges are rapidly regenerating. Some encrusting species even stop undisturbed tissue growth when regenerating, and patches where almost all living tissue was scraped off rocks can recover rapidly (Ayling 1983). The futility of the attempt to free the structures from sponge cover by scrubbing them and assuming that encrusting sponge regeneration is inversely related to resistance to damage (Wulff 2006) suggests that the soft, easy-to-injure pale pink sponge can rapidly regenerate in supposedly cleaned areas. Torn fragments may rapidly reattach to the available substrata (Rützler 2), and probably all sponges are capable of regenerating viable adults from fragments (Brusca and Brusca 2003), so there is a high risk to support asexual spreading. Experiments with a sample sponge of the same genus underline the impressive survival strategies of some sponges. The cells of sponges pressed through fine cloth immediately began to reorganize themselves and reformed a functional sponge within two to three weeks (Brusca and Brusca 2003). Additional asexual processes, such as the formation of reduction bodies with omnipotent cells that were spread during the attempt to clean the structures from sponge cover and could hatch out on numerous new spots of available substrata, may aggravate the sponge problem.

Although transplanting infested samples or scrubbing off sponges may aggravate the situation, it is probably not responsible for it, as the moment of the first attempt to clean structures was after obvious infestation and sponge spreading and did not result in large new colonization adjacent to the project.

In the first place, the spreading of the pale pink sponge at the artificial reef has very likely something to do with natural recruitment and with characteristics of the structure, including the availability of substrata, the location in the middle of water movement, or even the electrical attraction of the larvae. This needs to be directly tested.

The artificial reefs provide new suitable substrata, which allow opportunities for the dominant fastest-spreading species. Some corals may even be successful against sponge encroachment; sponge recruits will develop faster, and sponge is occupying space, preventing coral recruitment. The shaded underside of the bars is preferred by the sponge.

Furthermore, the structures may have a filter function. Standing in the movement of water, structures are offering substrata for sponge larvae or reduction bodies brought along. Particularly, strong sponge extension on more vertical structures with relatively close spacing of structural elements and slow mineral aggregation rates support the argument.

The most sponge-covered structures are constructed out of construction-wire mesh, with smaller diameter, inferior material composition, many welds, and lower electrical conductivity. The most infested of all structures inside the Karang Lestari project area is Fish 3 which is furthermore one of the structures with the longest distance to the land-based chargers and therefore has higher voltage drop because of longer cables.

Since the sponge is growing more slowly, higher sponge abundance must be due to greatly increased sponge-settlement rates, perhaps because the sponge larvae are attracted to low electrical fields. Whether they settle more or less on rapidly growing structures is unclear, as the smaller number of sponges apparent could also be due to the electrical inhibition of growth. Direct measurements of the effect of electrical fields on sponge settlement are needed.

CONCLUSIONS

Corals growing on electrified Biorock reef-restoration projects have higher growth and survival than control corals, yet on some projects in the Maldives and the Indonesian Islands of Bali and Lombok, sponge overgrowth of corals by several sponge species has been a problem, raising the question whether the sponge growth rate was accelerated more than the coral growth by the electrical current, thereby shifting the competitive advantage to the sponge. Direct measurements were made of the growth rate of a highly susceptible corals species, *A. microphthalma*, and of sponge overgrowth as a function of distance from the electrical field. The results showed that coral growth was stimulated in the vicinity of the Biorock structure, not only on it, but around it up to at least 3 m away from the structure, close to the anode. In contrast, sponge growth was sharply reduced, but only directly on the structure itself, and not in the vicinity, where it was much higher. Therefore, the electrical current represses sponge growth and favors coral survival over sponges on the electrified structures. The high abundance of overgrowth on *A. microphthalma* must, therefore, be due to much higher settlement of sponge larvae on the structure due to the attraction to the current and/or unhindered growth on seemingly preferred shadowed areas on the underside of a structure bar, wherefrom the sponge starts to attack corals attached on top of a bar.

ACKNOWLEDGMENTS

First of all, I would like to thank the two supervisors of my bachelor thesis (Coral Reef Management—The Role of Electricity on Coral overgrowth by Encrusting Sponges on a Biorock Artificial Reef Project) who supply the basis for this chapter, Dr. Thomas J. Goreau (Global Coral Reef Alliance, USA) for making the first contact to Pemuteran and for his constant support and engagement during all steps of this work, and Prof. Dr. Martin Welp (FH Eberswalde, Germany) for his support and guidance along the process of my bachelor thesis. I deeply appreciate the support of the following individuals who contributed to the achievements of the field trip and/or made the six-month stay in Pemuteran an unforgettable experience: Nara Narayana and, Rani E. Morrow-Wuigk (both in multiple ways); Pak Agung Prana; Michael Cortenbach from Bali Hai Diving Academy for placing free diving equipment at my disposal; the friendly staff of Bali Hai; Putu Yasa, Komang, and Putu from the Biorock Centre; the whole "Badini family" for being my home away from home; as well as Komang, Sandi, and Bagong. I also thank Stephanie Abel for editing grammar and style; Eric Fee for making scientific papers available to me; Dr. Carden Wallace for sample coral identification; Dr. Christine Schönberg; and Dr Nicole J. de Voogd for communication about the project. Last but not least, I thank my family and friends for their support and patience, and Melanie Adam for not only being by my side in the Bali Sea, taking the photographs, assisting in setting up and documenting the experiment, and sharing great impressions and experiences, but also for being in my life. My apologies to everyone I should have mentioned here but omitted to do so.

REFERENCES

Ayling, A. L. 1983. Growth and regeneration rates in thinly encrusting demospongiae from temperate waters. *Biological Bulletin* 165: 343–352.

Brusca, R. C. and G. J. Brusca. 2003. *Invertebrates*. Second Edition. Sunderland. Sinauer Associates.

Goreau, T. and W. Hilbertz. 2005. Marine ecosystem restoration: Costs and benefits for coral reefs. *World Resource Review* 17: 375–409.

Goreau, T. and W. Hilbertz. 2008. Bottom-up community-based coral reef and fisheries restoration in Indonesia, Panama, and Palau. In *Handbook of Regenerative Landscape Design*. R. L. France (Ed.), pp. 143–159. Boca Raton, FL. CRC Press.

Hilbertz, W. and T. J. Goreau. 2001. Pemuteran Coral Reef Restoration Project Progress Report. http://www.globalcoral.org/pemuteran_coral_reef_restoration.html (accessed August 2008).

Rützler, K. 2004. Sponges on coral reefs: A community shaped by competitive cooperation. In Pansini M, Pronzato R, Bavestrello G, Manconi R (eds). Sponge science in the new millennium. *Bulletin Museum Institute of Biology,* Genoa 68: 85–148.

Wulff, J. L. 2006. Resistance vs recovery: Morphological strategies of coral reef sponges. *Functional Ecology* 20: 699–708.

CHAPTER 9

Gorgonian Soft Corals Have Higher Growth and Survival in Electrical Fields

Diannisa Fitri and M. Aspari Rachman

CONTENTS

Introduction 105
Materials and Methods 106
    Biorock Design and Setting 106
    Maintenance of Animals 106
    Growth and Survival 107
    Physical Parameters 107
Results 108
Discussion 109
Conclusion 110
Acknowledgments 110
References 110

INTRODUCTION

    The majority of reef corals, as well as other reef-dwelling cnidarians, are inhabited by symbiotic dinoflagellates of the genus *Symbiodinium* (Muscatine 1990). The gorgonian soft corals (*Isis*) also have *Symbiodinium* in their tissue (Fabricius and Alderslade 2001). The dependence of the coral animals upon the photosynthetic activities of endosymbiotic *Symbiodinium* living within their endodermis tissue (Barnes and Chalker 1990) restricts the distribution of these coral animals to sunlit waters (Veron 1993). The photobionts supply nutrients to their hosts, which allow them to colonize habitats they normally could not because of limited supplies of food (Paracer and Ahmadjian 2000).
    Gorgonian soft corals, *Isis* (Isididae), are relatively new to the list of reared marine species under controlled environment. The present work focuses on the enhancement of growth and survival of gorgonians in raceway tanks by comparing the growth in electrical fields and uncharged controls.
    Isis hippuris has been exploited to the extent of a thousand tons from the reefs of the Eastern Indonesia Archipelagos for their skeletal materials, which have been exported to Singapore, Malaysia, and Hong Kong.
    Highly bioactive compounds have been extracted from *I. hippuris* known as "hippuristanol." The kind of *Isis* grown contains Hippuristerone A, polyoxygenated steroids, hippuristerones J-L,

and hippuristerols E-F (Chao et al. 2005; Sheu et al. 2000). Their medical applications have been found to exhibit antiviral activity (Chen et al. 2011).

They would appear to have great potential for mariculture and to be well suited for restoration of depleted reefs. The purpose of this study is to document the changes in growth and survival of the gorgonian, *I. hippuris*, in electrical fields.

MATERIALS AND METHODS

Biorock Design and Setting

The Biorock design and setting are illustrated in Figure 9.1a. The cathode compartments consisted of three trapezoidal structures constructed from steel bars (8 mm diameter) and a plate iron (3 mm thick) as frame. Chicken-wire mesh of 4 cm mesh size and rectangular cement plates were attached (alternately) to both sides of structures. Each trapezoid length was 2 m, base width 0.85 m, and height 0.50 m with 45° side slope. The cathodes were placed in flow-through concrete raceway tanks (8 × 1.5 × 1.2 m). Anodes were made of coated titanium wire meshes and were installed inside the outlet polyvinyl pipe (size 2 in. diameter and 0. 95 m long). The anode pipes were inserted into the outflow polyvinyl pipe (size 3 in. diameter and 1 m long). A DC power supply (RTVC, Volta Electronic) was placed on the upper supporting woods of building above the Biorock tank. An electric current of 5 V and 5 A was applied for about 12 hours during the day. At night, the electric current was turned off. Biorock design and construction was advised by T. J. Goreau.

Maintenance of Animals

These experiments were conducted at Barrang Lompo Island Marine Field Station of Hasanuddin University, Makassar, South Sulawesi, Indonesia. Three colonies of the gorgonian (*I. hippuris*) were collected from their natural habitats, at a depth of 6 m on the reef of Bonetambung, western Barrang Lompo Island, and transported in containers of seawater to the islands of Barrang Lompo. These colonies were growing in close proximity and all appeared to be of the same phenotype (Figure 9.2). All were kept in running seawater raceways (~1 m deep) for 1 week before being used in experiments. Only healthy-looking colonies with normal polyps expanded were used by carefully cutting the branch (approximately 6–7 cm in length) with a sterile stainless scissor and then attaching it directly to the artificial substrate. All of the transplantation processes were done underwater to

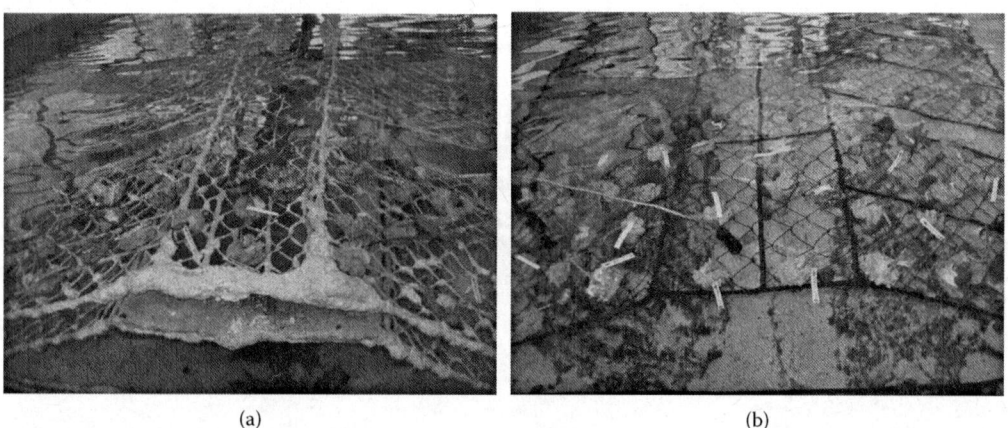

Figure 9.1 The transplanted fragments of *Isis*: (a) Biorock, (b) controls.

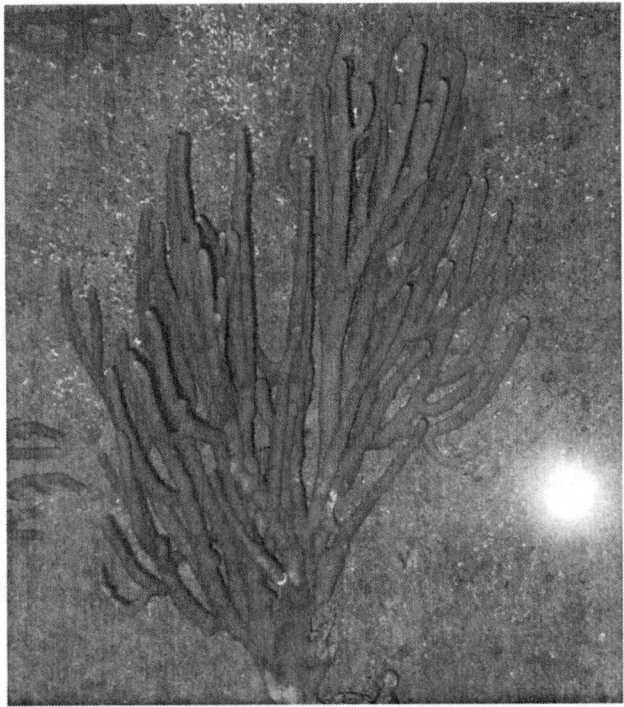

Figure 9.2 One of the healthy *Isis* mother colonies.

prevent stress to the animals, because the gorgonians must be carefully handled. As it is very sensitive to handling. Height measurements were made only at the start of 8 weeks, to minimize handling stress. A total of 40 nubbins were placed on the cathode wire mesh and on the control uncharged wire-mesh structure. Two concrete raceway tanks (8 × 1.5 × 1.2 m) were used for this experiment. Sunlight transmission was reduced to about 85% under a translucent fiberglass roof canopy. The flow-through system was used with the flow rate of about 60 L/min. Water overflowed from inlet PVC pipe by gravity into the experimental tanks, with water height maintained with a moveable outlet PVC pipe for the outflow. Aeration was provided inside the experiment tanks. A vernier caliper (0.05 mm) was used to measure gorgonian body lengths, and the measurement was held underwater. A light meter (Lutron LX101) and thermometer were used to measure light intensity and temperatures during the experiment.

Growth and Survival

All of the gorgonian lengths were measured at the start and the end of the experiments. Survival was recorded at weeks two, four, six, and eight by determining the proportion of live and dead animals from a total of 40 animals. Growth increments were analyzed using *t*-test, with the probability of type 1 error set at $p = 0.05$.

Physical Parameters

Temperatures were recorded every two weeks with thermometers in two concrete raceways tanks. Light readings (lux) were taken every two weeks, around midday, in two concrete raceways tanks at the surface of seawater.

RESULTS

Growth of gorgonians in Biorock and control tanks are shown in Table 9.1 and Figure 9.3. The growth increment of gorgonians in Biorock and control tanks were 0.59 cm and 0.25 cm, respectively. Growth of gorgonians on Biorock was significantly (t-test, $p < .01$) higher than controls, with the growth on Biorock 2.68 times higher than control.

Survival of the gorgonians in Biorock and control tanks are given in Table 9.2 and mortality in Figure 9.4. The mortality in the control tank was 1.88 times higher than in the Biorock tank after the end of the experiment. Mortality in the Biorock tank was slightly higher than controls at week two, but at weeks four, six, and eight consistently higher in the control tank compared to the Biorock gorgonians. The Biorock gorgonian mortality stopped after six weeks, but controls continued to die.

Light intensity values are given in Figure 9.5. Light intensity ranged from 8,000 to 20,000 lux both in Biorock and in control tank from October to December (rainy season). The temperatures in Biorock and control tank ranged from 26 to 30.5°C and 26–29.5°C, respectively, over the experiment, with temperatures in Biorock slightly higher than control tank (Figure 9.6).

Table 9.1 Statistical Analysis of the Gorgonian Lengths, *I. hippuris*, in Biorock and Control Tanks

Treatment	Start of Experiment	End of Experiment	t-test of the Length at the End of Experiment	
	Mean Length (cm)		p Value	Significance
Biorock	7.01 ± 0.38	7.60 ± 0.21	0.00	*
Control	6.87 ± 0.30	7.09 ± 0.31		

* Significant at 0.01 level.

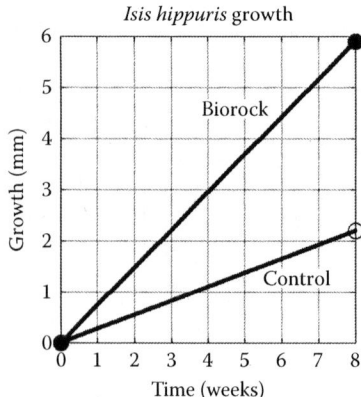

Figure 9.3 Growth (mm) of the gorgonians, *I. hippuris*, at the end of the experiment in Biorock and control tanks.

Table 9.2 Survival (%) of Gorgonians, *I. hippuris*, Over Eight Weeks of Experiment in Biorock and Control Tanks

	Week Interval				
Treatment	0	2	4	6	8
Biorock	100	87.5	82.5	80	80
Control	100	92.5	77.5	67.5	62.5

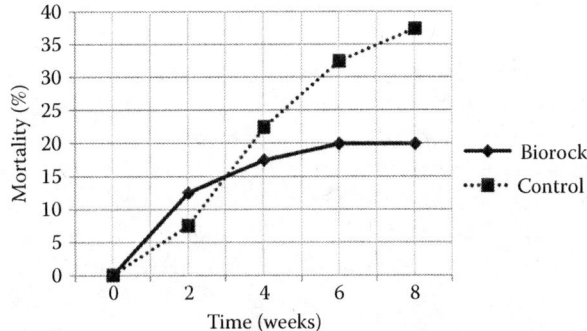

Figure 9.4 Percent mortality of the gorgonians, *I. hippuris*, over eight weeks in Biorock and control tanks.

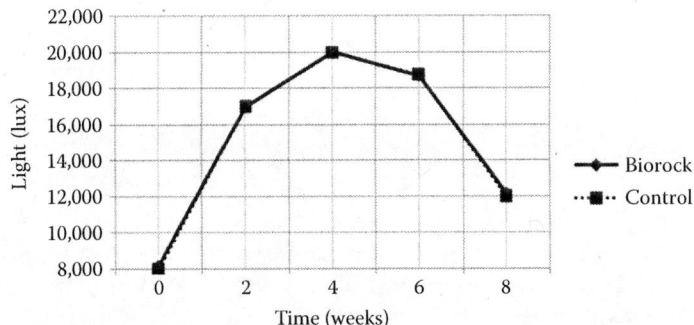

Figure 9.5 Light intensity (lux) in Biorock and control tanks over the period of the experiment.

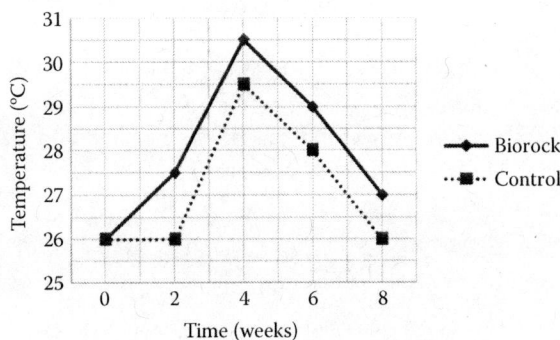

Figure 9.6 Temperature (°C) in Biorock and control tanks over the period of the experiment.

DISCUSSION

These differences in growth and survival of the gorgonians correlated with and may be attributed to the electric-field effects from the Biorock process. Goreau and Hilbertz (Chapter 4) stated that average hard-coral growth on Biorock structures is faster than controls.

The Biorock process increases calcification by providing alkalinity for skeletal growth and metabolic energy for proton and calcium pumping, growth, reproduction, and resisting environmental stresses (Goreau et al. 2004; Goreau and Hilbertz, unpublished data). Soft corals, including gorgonians (*Isis*, *Rumphella*, and *Anthogorgia*) grew on Biorock with normal appearance, but with faster basal growth and more new budding (branches).

It is difficult to isolate the importance of any single factor in the ecophysiology of corals (Falkowski et al. 1990). The variation in size and shape between colonies of *I. hippuris* from different environments and geographic regions is very large (Fabricius and Alderslade 2001).

Although temperatures were slightly higher in the Biorock tank than the controls, both showed a normal polyp extension, but polyp color was darker on Biorock compared to controls. High temperatures could cause *Symbiodinium* expulsion from the coral hosts, and then corals become bleached (Tomascik et al. 1997). Corals on Biorock nurseries in the sea bleached, but had 16–50 times greater survival after severe bleaching (Goreau et al. 2000).

Tissues from two dead corals prior to falling off the skeletons from Biorock and control structures were examined under a microscope. Pale-colored *Symbiodinium* were still intact in the host, but *Symbiodinium* became smaller as if cysts developed and were not normal yellow-brown in appearance. Rachman and Fitri (2011) found that *Symbiodinium* isolated from *I. hippuris* showed a normal germination and grew when recultured under laboratory condition after being stressed with extreme temperatures (0 and 80°C).

CONCLUSION

The electrical fields from Biorock lead to 2.68 times increased gorgonian soft-coral growth compared to controls in the raceway tank conditions. The controls had 1.88 times higher mortality than electrified corals. These results suggest that Biorock will result in superior growth and survival for mariculture of these species. Based on the pilot project results reported here, more detailed work is currently under way to find the optimum conditions for *I. hippuris* and other species.

ACKNOWLEDGMENTS

We especially thank Dr. Tom Goreau for kind support with experimental setup and editing this chapter; Dr. Noel Janetski, Dr. Sven Blankenhorn (Mars Symbioscience), and Dr. Idrus Patturusi (Rector of Hasanuddin University) for facilitating this experiment; and all the staff of Barrang Lompo Marine Station for their help during experiments.

REFERENCES

Barnes, D. J. and B. E. Chalker. 1990. Calcification and photosynthesis in reef-building corals and algae. 109–127 in: Z. Dubinsky (Ed.), *Ecosystems of the world 25*. Coral Reefs. Elsevier, Amsterdam.

Chao, C. H., L. F. Huang, S. L. Wu, J. H. Su, H. C. Huang, and J. H. Sheu. 2005. Steroids from the Gorgonian Isis hippuris. *Journal of Natural Products* 68:1366–1370.

Chen, W. H., K. W. Shang, and Y. D. Chang. 2011. Polyhydroxylated steroid from the Bamboo Coral *Isis hippuris*. *Marine Drugs* 9:1829–1839.

Fabricius, K. and P. Alderslade. 2001. *Soft corals and sea fans*. Australian Institute of Marine Science, Melbourne, Australia.

Falkowski, P. G., P. L. Jokiel, and R. A. Kinzie III. 1990. Irradiance and corals. 89–105 in: Z. Dubinsky (Ed.), *Ecosystems of the world 25*. Coral reefs. Elsevier, Amsterdam.

Goreau, T. J., J. M. Cervino, and R. Pollina. 2004. Increased zooxanthella numbers and nitotic index in electrically stimulated corals. *Symbiosis* 37:107–120.

Goreau, T. J., W. Hilbertz, and A. A. Hakeem. 2000. Increased coral and fish survival on mineral accretion reef structures in the Maldives after the 1998 Bleaching Event. http://www.globalcoral.org/Increased%20Coral%20and%20Fish%20survival%20. (accessed August 25, 2012).

Muscatine, L. 1990. The role of symbiotic algae in carbon and energy flux in reef corals. 75–85 in: Z. Dubinsky (Ed.), *Ecosystems of the world 25*. Coral reefs. Elsevier, Amsterdam.

Paracer, S. and V. Ahmadjian. 2000. *Symbiosis: An introduction to Biological Associations*, 2nd edition, Oxford University Press, New York.

Rachman, M. A. and D. Fitri. 2011. Testing resistances of zooxanthellae under laboratory conditions. Presented at the International seminar on the marine and global climatic change. Udayana University, Mar. 10–11, 2011. Denpasar, Bali, Indonesia.

Sheu, J. H., S. P. Chen, P. J. Sung, M. Y. Chiang, and C. F. Dai. 2000. Hippuristerone A, a novel polyoxygenated steroid from the gorgonian *Isis hippuris*. *Tetrahedron Letters* 41:7885–7888.

Tomascik, T., A. J. Mah, A. Nontji, and M. K. Moosa. 1997. *The Ecology of Indonesian Seas, Part I. The Ecology of Indonesia Series*, Volume VII. Periplus Editions, Singapore.

Veron, J. E. N. 1993. *Corals of Australia and the Indo-pacific*. University of Hawaii Press, Honolulu.

CHAPTER 10

Suitability of Mineral Accretion as a Rehabilitation Method for Cold-Water Coral Reefs*

Susanna M. Strömberg, Tomas Lundälv, and Thomas J. Goreau

CONTENTS

Introduction 113
Materials and Methods 115
    Experimental Design 115
    Tanks and Substrates 116
    Feeding and Monitoring 116
    Growth Measurements 118
    Statistics 118
Results 119
    Mineral Accretions 119
    Coral Response to the Treatments 120
Discussion 124
    Comparisons with Previous Studies 125
Summary 127
Acknowledgments 128
References 128

INTRODUCTION

The cold-water scleractinian coral *Lophelia pertusa* (Linneaus 1758) is the main ecosystem engineer in the northeast Atlantic (Freiwald et al. 2004) building important habitats for fish and invertebrates (Costello et al. 2005, Mortensen et al. 1995, Reed 2002, Ross and Nizinski 2007). Coral presence increases diversity threefold compared to surrounding soft-sediment habitat (Gage and Roberts 2003, Henry and Roberts 2007, Husebø et al. 2002, Jensen and Frederiksen 1992, Jonsson et al. 2004, Jonsson and Lundälv 2006), and loss of coral habitat adversely affects local fisheries (Fosså et al. 2002, Koenig et al. 2000). It is estimated that 30% to 50% of the Norwegian reefs have been damaged by bottom-trawling (Fosså et al. 2002). Likewise, the coral coverage

---

* This chapter was reprinted from *Journal of Experimental Marine Biology and Ecology*, Vol. 395, S. M. Strömberg, T. J. Goreau, T. Lundälv, Suitability of mineral accretion as a rehabilitation method for cold-water coral reefs, pp 153–161, copyright 2010, with permission from Elsevier.

and habitat complexity of the *L. pertusa* reefs in the Skagerrak have been severely reduced. In the Swedish part of Skagerrak, only one small live reef, consisting of small detached and scattered colonies, remains, whereas six hitherto known reefs are extinct and consist only of dead coral rubble. Rubble fields are known to give a poor substrate for recolonization, with little or no recovery in both high and low wave-energy environments, and, rather than recovery, a further deterioration has been the case (Brooke et al. 2006, Clark and Edwards 1994). Furthermore, due to geographic isolation of reef sites in the area, the natural recolonization by coral larva from neighboring reefs is unlikely. For example, dispersion probabilities using a Lagrangian model shows a mere 4–9% probability of larval recruits from the nearest Norwegian reef (Tisler) reaching Saekken (Ericson and Ljunghager 2006). These results are corroborated by genetic data showing high genetic differentiation between reefs in Skagerrak (Broberg 2006) and high levels of clonality within reefs (Dahl 2006). Hence, it seems that rehabilitation efforts by means of deployment of artificial reefs with coral transplants are necessary to restore cold-water coral cover in the area.

The method of mineral accretion through electrolysis in seawater for the purpose of coral reef rehabilitation was developed by Wolf Hilbertz and Thomas Goreau (1996) and has been experimentally tested by several workers (Borell et al. 2009, Eisinger 2005, Goreau et al. 2004, Romatzki 2009, Sabater and Yap 2004, Schuhmacher et al. 2000, van Treeck and Schuhmacher 1997). The results have widely varied between experiments and between species of corals, and although measurements from rehabilitation projects around the tropics find growth rates two to six times higher than natural coral colonies, the scientific reports published so far have more ambiguous results to present. Few of the studies from rehabilitation projects have yet been published except as theses in Indonesian. It has nevertheless been substantiated that there is an increased survival rate of coral transplants growing on cathodes (Sabater and Yap 2002, van Treeck and Schuhmacher 1997). The method has thus far been used in the rehabilitation of shallow tropical coral habitats, and has never been tested in deeper temperate habitats.

The occurrence of *L. pertusa* on oil rigs (Gass and Roberts 2006) and the use of electrical stimulation in the form of microcurrent electrical therapy (MET) in human and animal medicine with several positive effects observed in different studies (Cheng et al. 1982, Kloth 2005) suggest the possibility of using metal structures and trickle currents to enhance survival and growth in transplanted corals. Gass and Roberts (2006) reported high growth rates of corals growing on oil rigs; trickle currents are used to protect the metal from corrosion and could explain the positive effects on growth. Cheng et al. (1982) observed an increase in adenosine triphosphate (ATP) production in skin-tissue samples exposed to microcurrents ranging from 50 to 1000 µA; thus there is a substantiated positive effect of microcurrents on the ATP production. Furthermore, the application mimics the natural mode of mineralization. Corals produce an alkaline environment in an enclosed space between the calicoblastic cells and the substrate that promote the spontaneous precipitation of aragonite crystals, with the deposited organic matrix working as a primer. Electrodeposition produces the same type of orthorhombic aragonite crystals, thus growing a seminatural substrate on the metal conductor. The proposed benefits for corals (and other calcifying organisms) growing on the cathodes are as follows: (1) supersaturation of calcium and carbonate ions in the vicinity of the cathode, (2) increased efficiency of cation uptake and transport due to the availability of electrons, and (3) increased metabolic efficiency because free electrons are available for ATP production (Hilbertz and Goreau 1996).

The aim of this study was to evaluate the potential of mineral accretion for the rehabilitation of cold-water coral habitat. More specifically, a predeposited mineral layer was tested with galvanic elements (GEs) (~0.4 V) and three levels of applied electrode potential (2.0 V, 2.5 V, and 3.0 V) without applying electric current to find the optimal level of current density, considering survival and growth of the coral transplants, or if the substrate per se without applied electrical current would have a positive effect. In addition, the response in polyp behavior (degree of extension and activity) and budding frequencies were evaluated.

MATERIALS AND METHODS

The experiment was conducted at the marine research station in Tjärnö (Department of Biological and Environmental Sciences, University of Gothenburg), at the Swedish west coast facing the Skagerrak. The research facility has a complete flow-through saltwater system with deepwater of similar composition (chemistry, salinity, etc.) as the ambient water of the nearby reefs of *L. pertusa*. The experiment was launched on September 3, 2008, and ran for six months.

Experimental Design

Twenty-four aquarium tanks were set up in a constant temperature room with the inflowing water temperature set at 8°C to imitate the *in situ* conditions for local coral populations (natural range: 4–10°C). The water intake for the deepwater flow-through system is at 40–45 m depth in the adjacent Koster Fjord. The tanks were assigned either one of the six different treatments to be tested, four replicate tanks per treatment with four coral pieces in each, making a total of 96 coral fragments (mean length ± SD: 47.2 ± 12.6 mm and size range: 22.2–85.4 mm). The coral pieces were randomly distributed between tanks. The number of calices on each fragment ranged from 3 to 20. Three different morphological types were recognized: white slender (45 pieces), compact white corals (31 pieces), and a red morph (20 pieces) with smaller calices and a thin, richly branched skeleton. The red morph was spread out to have not more than one representative in each tank. The corals were collected with a remotely operated vehicle (ROV) (Sperre subfighter 7500 DC) equipped with a manipulator arm at ~100 m depth at the Tisler reef (58°59.70′N and 10°58.00′). Six different treatments were applied as follows:

C:	Controls; corals on plastic mesh
AS:	preprecipitated Aragonite[a] Substrate on cathodes, no voltage applied
GEs:	Galvanic Elements (FeIZn), galvanic potential in seawater ~0.4 V
LI:	Level 1; 2.0 V and 0.01 A, end value: 0.00 A
LII:	Level 2; 2.5 V and 0.12 A, end value: 0.05 A
LIII:	Level 3; 3.0 V and 0.35 A, end value: 0.21 A

[a] Comment: The precipitated minerals were initially very porous and thus consisted mainly of brucite. When the experiment was terminated, however, the minerals had hardened by a gradual deposition of aragonite.

The AS treatment could be viewed both as a procedural control and a zero-voltage treatment, testing the substrate per se. The GEs with steel cathodes and zinc anodes have a galvanic potential of ~0.4 V in seawater, and thereby offers a lower current alternative to the three treatments with applied DC current. The different voltage levels were applied by three EP-613 DC sources with three-digit LCD displays (0–30 V and 0–3 A, Manson Engineering Industrial Ltd). Connections were made with the four replicates of each treatment in parallel from one DC source. The voltage was monitored and kept constant throughout the experiment, while the amperage decreased over time. The different treatments were randomly distributed between the tanks as far as possible to avoid artifact effects of position in the room. The cables and connections, however, made it impossible to fully randomize, and therefore all the AS treatments were assembled on the same bench together with one control and one GE tank. The current density (A m^{-2}) was calculated for comparisons with other studies where A m^{-2} has been used.

Tanks and Substrates

The tanks were constructed with two separate main chambers: one that would contain the corals and one that would hold the anodes or left empty. Perforated partition walls guaranteed that there was no back-mixing between the chambers to avoid exposure to the lower pH or chlorine gas produced at the anodes. All the tanks were given the same design to have equal flow regimes, and all anode chambers were covered with plastic and sealed with duct tape. The sealing of the anode chambers were done to retain chlorine gas within the water to purify and neutralize the outflowing water in filter trays containing activated carbon and oyster and mussel shells prior to release.

The cathodes were cut to 110×160 mm pieces from a steel mesh with a metal surface area calculated to 0.04 m^2. Average weight of the cathodes was 253 g. The anodes were made of a 1×1 mm angular wire mesh of coated titanium (150×200 mm and average weight 12 g). The GEs were produced by attaching zinc anodes to one end of the cathodes, ~43.8 g of zinc on each GE. In the controls, the corals were placed on a plastic mesh of the same size as the cathodes; mounted on a small, flat stone; and affixed with aquarium silicon.

All materials were conditioned in seawater for five weeks before mounting the corals. The AS treatment was run on electrolysis (4.0 V) during these weeks to get a mineral cover; however, the deposited minerals consisted mainly of brucite and were very porous. Additional short periods of DC currents (3.0 V) were given on six occasions during the first three weeks of the experiment, to counteract corrosion on cathode surfaces that had its mineral layer scraped off while mounting the corals. When the experiment was terminated, the mineral layers were measured with a pair of vernier calipers, with five replicate measurements at each end of the cathode (close or far end relative to the anode).

Feeding and Monitoring

During the experiment, the corals were fed with *Artemia salina* nauplii (brine shrimp) on three occasions per week. The *Artemia* were hatched over 48 hours and reared for another 48 hours while fed with microalgae (*Isochrysis* sp.) to the increase nutrient value. The food was supplied via the water flow-through system, to be evenly distributed. Four plastic containers were used to buffer water, so that the water levels in the containers and thereby pressure and flow rate to the tanks could be kept equal. Each container supplied six tanks each with water. Flow rates were measured on three occasions during the experiment with three replicate measurements from each tank.

Water temperature and pH values were monitored weekly during the experiment with a handheld digital pH meter (Waterproof pH Testr 30, accuracy; ± 0.001 pH units and $\pm 0.5°C$). Measurements were taken 1 cm above the substrates and randomized between the tanks. The pH were between 8.06 and 8.10 (see Table 10.1 and Figure 10.1). Water temperature was $7.07°C \pm 1.45°C$ (mean $°C \pm SD$), starting around $10.0°C$ and leveling off at around $5.5°C–6.0°C$ over several weeks in the second half of the experimental period. During the four last months of 2008, the measured salinity at 30 m depth (close to the depth of water intake) in the Kosterfjord was 32.5–33.8 psu (SMHI Report No. 2009–6).

The health status of the corals was likewise monitored on a weekly basis; the corals were given points based on the degree of extension of tentacles. Fully withdrawn = 0, partially withdrawn = 1, extended but slack = 2, intermediate = 3, extended and vivid (stiff) = 4, and actively moving tentacles = 5 points. The health-status observations were not randomized due to the corals' sensitivity to vibrations that caused them to withdraw at the slightest disturbance. Instead, the observations were made in the same order, from tank 1 to 24, as swiftly and quietly as possible, and repeated in the opposite direction for another day of the week to elucidate whether there would be a difference in extension rates due to procedures.

MINERAL ACCRETION AS A REHABILITATION METHOD FOR COLD-WATER CORAL REEFS

Table 10.1 Average pH Values (mean ± SD) over the Different Treatments. The Measures from the First Five Weeks (during the Initial Larger Fluctuations) Were Excluded

Treatment	pH	SD
C	8.06	±0.024
AS	8.08	±0.030
GE	8.07	±0.029
LI	8.07	±0.021
LII	8.10	±0.032
LIII	8.07	±0.055

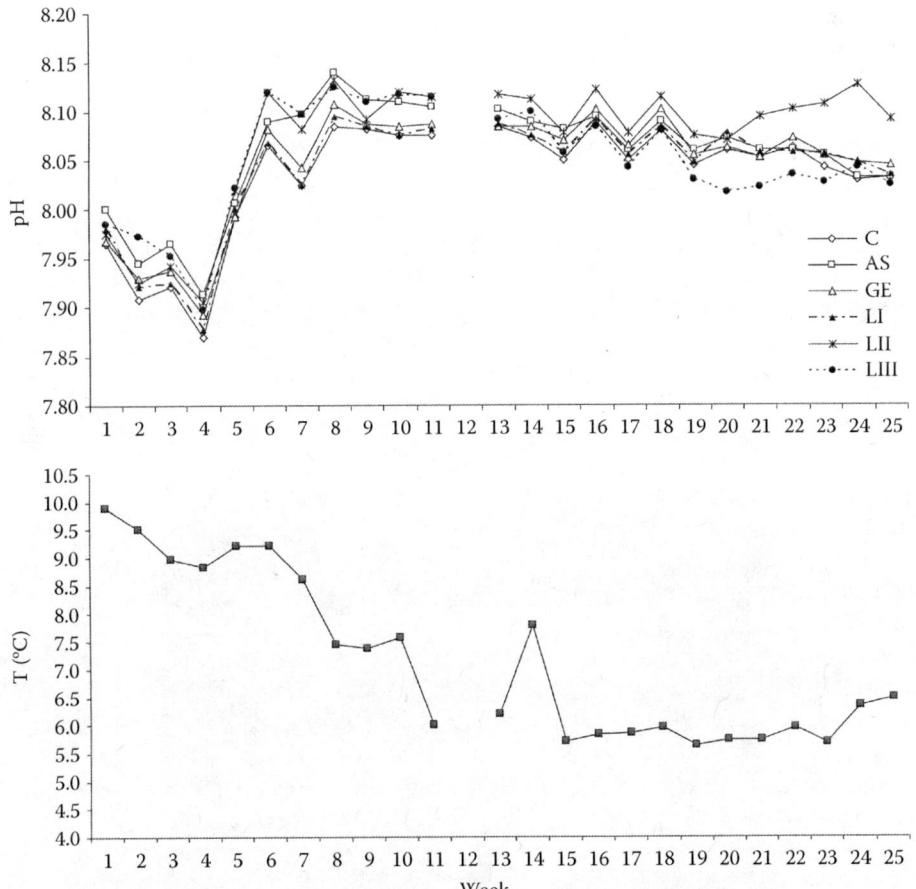

Figure 10.1 General conditions; pH differed slightly between treatments, that is, the AS, LII, and LIII treatments had higher pH levels, and the lowest current density treatment (LI) and the noncharged treatments (C and GE) had lower pH. The pH levels decreased over time in the LIII treatment, probably due to the rapid accretion that filled up the voids in the mesh, thus trapping hydrogen gas underneath the cathode. The overall trend in pH with lower values during the first five weeks is probably caused by early autumn down-mixing of surface waters. The experiment ran from September 2008 to March 2009, and the temperature had its annual maxima (16.8°C) in September with less pronounced stratification of water masses during the period. Despite the thermo regulation, the temperature could not be kept constant, and the temperature fluctuated during the experiment, as seen in the lower graph.

Growth Measurements

Photographs of the coral fragments were taken at the point of start and end of the experiment, with the pieces placed on a gray perforated panel to provide a reference for measurements (see Figure 10.2). Measurements were made in the free software ImageJ (version 1.42a, Wayne Rasband, National Institutes of Health, USA). Several measurements on each coral fragment were made, and mean growth and maximum observed growth were recalculated to growth in mm yr^{-1} for analysis.

Some of the corals had broken calices after the handling during collection, and these had either died or healed to different degrees. The mostly one-sided growth of healing was measured and noted separately in the protocol. Furthermore, the degree of healing was assigned a symbol; one plus sign (+) for a moderate healing ($0.8 \leq x < 2.0$ mm), two plus signs (++) for a good healing (≥ 2.0 mm), and a minus sign (−) for broken calices that did not heal. A zero (0) denotes that there were no broken calices.

The numbers of new buds were counted, and partial or complete mortality—that is, the number of dead polyps (not counting those that were already dead at the beginning) divided by the total number of polyps on each fragment—was noted.

Statistics

The frequency distributions of the number of new buds and the calices displaying a growth of $0.8 \leq x < 2.0$ mm yr^{-1} or ≥ 2.0 mm yr^{-1} were analyzed by Chi-square (χ^2) tests. A one-way ANOVA was performed on growth rate data, and Pearson Correlations were performed to test whether the different water-flow rates had an effect on the variables' mean and maximum growth or the frequencies of new buds and calices with a growth of ≥ 2.0 mm yr^{-1}.

All statistical analyses were performed with the statistical software SPSS 17.0 (2008).

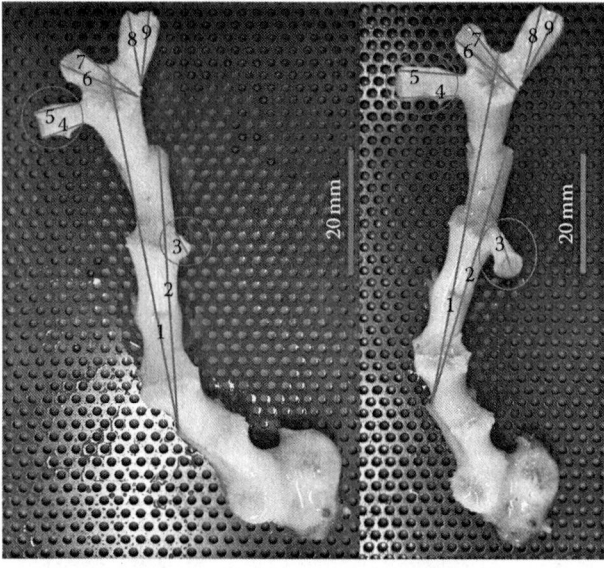

Figure 10.2 An example of images taken at the start (left) and end (right) of the experiment. Measurements were performed in the free software ImageJ (version 1.42a). This particular coral piece was reared in the lowest applied current density (LI: 2.0V, ≤ 0.06 A m^{-2}). A new bud has developed from a small protrusion into a long calice, and the upper left calice (numbered 4 and 5) has grown noticeable.

RESULTS

Mineral Accretions

The accreted mineral layer varied in thickness between the treatments and with distance to the anode. Higher current density produced a thicker layer, and subsequently, the end of the cathode that was closer to the anode received a thicker layer of precipitated minerals (Table 10.2). On the LI cathodes, the measured layer was 1.53 ± 0.74 mm (mean $\pm$ SD) closest to the anode and 0.99 ± 0.60 mm on the end furthest away. The LII treatment had 5.07 ± 0.85 mm and 2.48 ± 0.62 mm

Table 10.2 Start and End Values, Respectively, of Current (A) and the Calculated Current Density (A m^{-2}) per Metal Surface Area for the LI, LII, and LIII Treatments

	Treatment					
	LI		LII		LIII	
	Start	End	Start	End	Start	End
Ampere	0.01	Immesurable	0.12	0.05	0.35	0.21
Current density (A m^{-2})	0.06	—	0.75	0.31	2.19	1.31
	AS		LI		LII	
	Near Anode	Far End	Near Anode	Far End	Near Anode	Far End
Mineral accretions (mm)	7.90	1.98	1.53	0.99	5.07	2.48
±SD	±2.03	±0.30	±0.74	±0.60	±0.85	±0.62

The thickness of mineral accretions in AS, LI, and LII are the sum of the layers on the upper and under sides of the cathodes. The mineral layer on LIII crumbled when dismounting the corals and could not be properly measured; however, layers were up to 10 mm thick on one side only.

Table 10.3 Chi-Square Output and Test Statistics

	Frequency New Buds			Frequency $\geq$ 2.0 mm yr^{-1}		
Treatment	Obs. N	Exp. N	Residual	Obs. N	Exp. N	Residual
C	8	9.7	−1.7	10	6.5	3.5
AS	7	9.7	−2.7	4	6.5	−2.5
GE	13	9.7	3.3	9	6.5	2.5
LI	17	9.7	7.3	11	6.5	4.5
LII	10	9.7	0.3	1	6.5	−5.5
LIII	3	9.7	−6.7	4	6.5	−2.5
Total	58			39		

Test Statistics

	New Buds	$\geq$2.0 mm yr^{-1}
Chi-square	12.345	12.538
df	5	5
Asymp. Sig.	0.030	0.028
Exact Sig.	0.030	0.029
Point probability	0.002	0.005

There were significant differences between treatments in the frequency distribution of new buds as well as in calices with a growth of $\geq$ 2.0 mm yr^{-1}. No cells have expected frequencies less than 5. The differences in the frequency distribution of calices displaying a growth of $0.8 \leq x < 2.0$ mm yr^{-1} was also tested and found to be nonsignificant (not shown): $\chi 2crit = 11.07$ at the .05 level.

thick layers on the close and far end relative to the anode, respectively. Both the treatments had produced a hard crust. The AS treatment had layers between 7.90 ± 2.03 and 1.98 ± 0.30 mm. Noteworthy, the rapid deposition of minerals on the AS treatment produced a very porous layer initially, which subsequently was strengthened, ending up a hard crust. The efforts to counteract corrosion on the AS cathodes failed. The LIII treatment gave a rapid accretion of porous minerals that crumbled while dismounting the corals and thus could not be properly measured; however, layers were up to 10 mm thick on one side only. There were no visible mineral accretions on the GE cathodes. The mineral layer on LI barely covered the cathode, whereas full accretion was accomplished on the LII cathodes. Coral fragments were firmly attached to the cathode surfaces by the mineral accretions.

The amperage decreased rapidly from the initial values during the first week, and after the initial drop, the amperage leveled out and decreased only slightly over the remaining period (Table 10.2). The mineral accretion rates were equal within the treatments, indicating an equal distribution of electrical currents over the four replicate electrode pairs.

Coral Response to the Treatments

There were significant differences between the treatments in the frequency distribution of new buds developing during the experimental period ($\chi^2_{(df\ 5)} = 12.3$, exact $p = 0.03$, and $\chi^2_{crit} = 11.07$ at the 0.05 level). There were 17 new buds in the LI treatment, 13 new buds in the GE treatment, 10 in the LII treatment, and the controls had 8 new buds (see Tables 10.3 and 10.4 and Figure 10.3a). Pairwise Chi-square tests between the controls and the different treatments revealed no significant effects. Testing the substrate per se (AS) against the treatments with an electrode potential (GE, LI, LII, and LIII) also turned out nonsignificant; however, the difference between AS versus LI was close to significant ($\chi^2_{(df\ 1)} = 4.17$, $\chi^2_{(crit)} = 3.84$, and exact $p = 0.06$). The largest difference was found between LI and LIII ($\chi^2_{(df\ 1)} = 9.80$, $\chi^2_{(crit)} = 3.84$, and exact $p = 0.003$).

The number of calices displaying a growth of ≥ 2.0 mm yr^{-1} (see Figure 10.3c) differed significantly between the treatments and were higher in the controls, LI, and GE with 10, 11, and 9 calices, respectively ($\chi^2_{(df\ 5)} = 12.5$, and exact $p = 0.03$). Pairwise tests were nonsignificant, except controls versus LII ($\chi^2_{(df\ 1)} = 7.36$, $\chi^2_{(crit)} = 3.84$, and exact $p = 0.01$). The number of buds displaying a growth rate of $0.8 \geq x < 2.0$ mm yr^{-1} was similar in all treatments except GE and LIII, which had lower numbers, albeit nonsignificant (Figure 10.3b).

The one-way ANOVA performed on growth-rate data turned out nonsignificant, and only diagrams with means and standard error bars are presented to show the trends in the effects on growth (Figure 10.3d through 10.3f). The observed mean growth was slightly higher in the lowest applied current density treatment (LI) than in the controls, that is, 1.27 ± 1.22 mm yr^{-1} and 1.20 ± 0.97 mm yr^{-1}, respectively (mean $\pm$ SD), while all other treatments had lower growth rates (see Table 10.4 and Figure 10.3d). The same pattern was seen in the observed maximum growth where corals in the LI treatment had an average maximum growth of 3.26 ± 2.66 mm yr^{-1} compared to 3.05 ± 1.73 mm yr^{-1} in the controls (Figure 10.3e).

All Pearson Correlations testing whether there were any effects on growth rates or frequencies due to different water flow rates (mL s^{-1}) in the tanks were nonsignificant (Table 10.5). The r^2 value of the maximum growth indicates a lower-range medium-effect size (Kinnear and Gray 2008), but since there was a nonsignificant correlation, this is interpreted as no effect of water-flow rate.

Considering the general health status, it seemed as all corals fared as well; the corals in the controls were extending their polyps to a slightly higher degree, whereas the polyps on the GEs were slightly less extended and vivid, albeit no significant results were found (Figure 10.3f). Looking at the health status derived from checks in one direction only (tanks 1 through 24), it seemed like the corals on the zero-voltage treatment (AS) fared less well; however, this pattern disappeared when pooling the checks from both directions, as presented in Figure 10.3f.

Table 10.4 Table Summarizing the Observed Responses of the Corals to the Six Treatments

Treat		Mean Growth (mm yr^{-1})	Maximum Growth (mm yr^{-1})	Heal Growth (mm)	Degree of Healing	Health Status (Points)	No. of New Buds	No. ≥0.8 mm	No. ≥2.0 mm	Mortalities
C	Mean	1.195	3.051	1.292	8(0)	3.47	8	28	10	Zero mortality
	Minimum	0.005	0.528	0	5(+)	2.88				
	Maximum	3.940	6.290	6.031	3(++)	3.94				
	SD	0.967	1.730	1.937	0(−)	0.41				
AS	Mean	0.843	2.827	0.599	7(0)	3.07	7	32	4	2 Cases of partial mortality (0.14–0.43)
	Minimum	0.156	0.865	0	4(+)	0.25				
	Maximum	1.803	6.426	2.798	2(++)	3.94				
	SD	0.536	1.704	0.939	3(−)	0.86				
GE	Mean	0.863	2.466	1.158	3(0)	2.83	13	17	9	4 Cases of partial mortality (0.29–0.67)
	Minimum	0	0.076	0	8(+)	0.50				
	Maximum	3.301	10.156	7.337	2(++)	3.94				
	SD	0.898	2.458	1.861	3(−)	0.99				
LI	Mean	1.269	3.258	0.583	9(0)	3.10	17	26	11	Zero mortality
	Minimum	0	0.802	0	3(+)	2.31				
	Maximum	3.879	10.228	3.709	2(++)	3.75				
	SD	1.220	2.655	1.083	2(−)	0.43				
LII	Mean	0.579	2.333	0.157	12(0)	3.07	10	27	1	1 Case of partial mortality (0.11)
	Minimum	0.040	1.192	0	3(+)	1.94				
	Maximum	1.914	4.724	1.187	0(++)	4.00				
	SD	0.480	1.066	0.330	1(−)	0.49				

(Continued)

Table 10.4 Table Summarizing the Observed Responses of the Corals to the Six Treatments (Continued)

Treat		Mean Growth (mm yr⁻¹)	Maximum Growth (mm yr⁻¹)	Heal Growth (mm)	Degree of Healing	Health Status (Points)	No. of New Buds	No. ≥0.8 mm	No. ≥2.0 mm	Mortalities
LIII	Mean	0.539	2.172	0.111	11(0)	3.05	3	15	4	8 Cases of partial mortality (0.14–0.29)
	Minimum	0	0	0	4(+)	0				2 Complete mortalities
	Maximum	2.058	6.142	0.967	0(++)	3.88				
	SD	0.640	1.868	0.305	0(−)	0.93				

Mean growth (mm yr⁻¹); maximum observed growth (mm yr⁻¹); heal growth (mm) is the observed one-sided healing growth of broken calices; degree of healing is the frequency of calices displaying the degree of healing within the classification given in brackets, that is, (+) moderate healing ($0.8 \leq x < 2.0$ mm yr⁻¹), (++) good healing (≥ 2.0 mm), (−) no healing, and (0) denotes that there were no broken calices; health status, that is, corals were given a rank based on the degree of extension of the polyps, that is, 0 = fully withdrawn, 1 = partially withdrawn, 2 = extended but slack, 3 = intermediate, 4 = extended and vivid (stiff), and 5 = active tentacles; No. of new buds is the total sum of newly developed calices in each treatment; No. ≥ 0.8 and ≥ 2.0 are the frequencies of calices that displayed a growth rate of $0.8 \leq x < 2.0$ mm yr⁻¹ or ≥ 2.0 mm yr⁻¹, respectively; and mortalities are the observed partial (number of dead polyps/total number of polyps) or complete mortalities. Values for the mean, minimum, maximum, and standard deviation as headlined in the second column is given for the mean, maximum, heal growth, and for health status. C = controls; AS = preprecipitated aragonite substrate (no voltage applied); GEs = galvanic elements (c. 0.4 V, FeIZn); LI = level 1 (2.0 V and 0.01 A); LII = level 2 (2.5 V and 0.12 A); LIII = level 3 (3.0 V and 0.35 A).

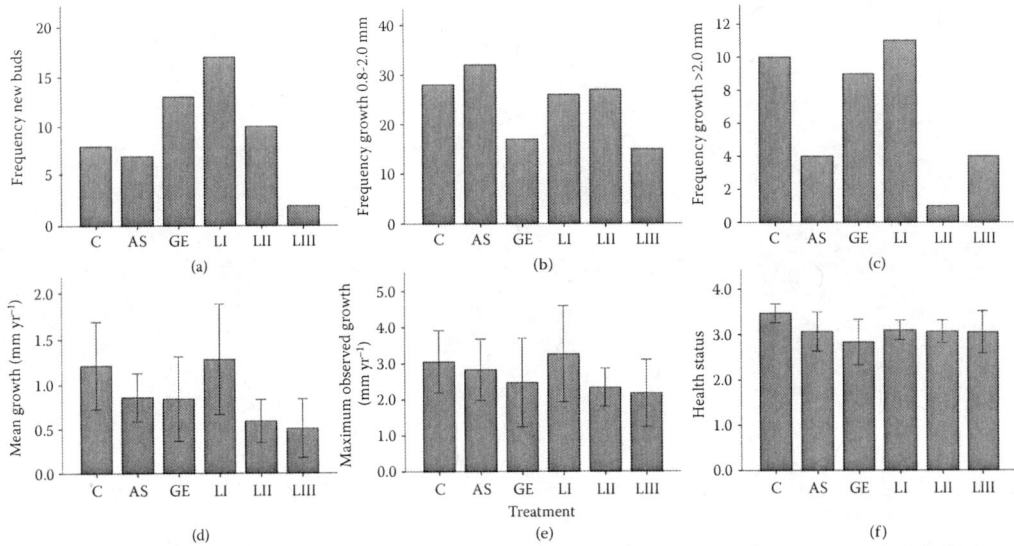

Figure 10.3 In the figure, all new calices that developed during the experimental period is summarized for each treatment; here a peak occurs in the LI treatment, curtailing laterally over the low current density treatments (GE and LII), whereas the highest current density (LIII) produced very few new calices, and the noncharged treatments (C and AS) had moderate budding frequencies. Figures 10.3b and c show the frequencies of calices displaying a growth of $0.8 \leq x < 2.0$ mm yr^{-1} and ≥ 2.0 mm yr^{-1}, respectively. There were significant differences between treatments in the frequency of new buds (Figure 10.3a; $\chi^2_{(df\,5)} = 12.3$; exact $p = 0.03$; point prob. = 0.002; $\chi^2_{crit} = 11.07$ at the .05 level) and in the frequency of calices with growth of ≥ 2.0 mm yr^{-1} (Figure 10.3c; $\chi^2_{(df\,5)} = 12.5$; exact $p = 0.03$; point prob. = 0.005). The differences were not significant in the frequency of calices in the lower range of growth rate (b). Figure 10.3d shows the trends in observed mean growth, and Figure 10.3e shows the maximum observed growth with slightly higher growth rates, albeit nonsignificant, in the lowest current density treatment (LI) compared to the controls for both variables, while all other treatments had lower growth rates. Figure 10.3f shows that the behavior (degree of extension) of the polyps was not affected by the different treatments. Data in Figures 10.3a through 10.3c are presented as frequency distributions (total sums), whereas Figures 10.3d through 10.3f are presented as mean values (four replicate tanks with four coral pieces in each treatment) with standard error bars. C = controls; AS = aragonite substrate, zero voltage; GEs = galvanic elements (Fe|Zn); LI = 2.0 V, 0.01 A; LII = 2.5 V, 0.12 A; and LIII = 3.0 V, 0.35 A.

In the summary table (Table 10.4), there are some additional information presented; heal growth (mm) together with degree of healing (ranks) and mortalities. The healing of broken calices was highest in the controls with an average one-sided growth of healed calices of 1.29 ± 1.94 mm followed by GE, AS, and LI. The controls displayed a high degree of healing, whereas AS, GE, and LI displayed cases of good healing simultaneous with some cases of no healing. Because the number of broken calices differed among treatments, the degree of healing is not entirely comparable; for example, the higher current density treatments (LII and LIII) had very few broken calices to begin with.

There were zero mortalities in the controls and LI, and a few cases of partial mortality in AS and LII (see Table 10.4). In the GE treatment, there were four cases of partial mortality ranging from 29% to 67%; two of these are explained by an infestation by bacteria in one of the tanks. The bacteria were not analyzed, but red crystalline metabolites on the surface of the cathode indicated chemoautotrophic iron-oxidizing bacteria. The highest voltage treatment (LIII) produced a high number of partial mortality—eight cases, ranging from 14% to 29%—and furthermore, two complete mortalities. These could all be explained by the treatment itself, as the rate of mineral accretion was so high that it covered the calices close to the cathode surface and entirely overgrew two small coral

Table 10.5 Results of the Pearson Correlations Show That There Were No Significant Effects of Water Flow Rates on the Variables Mean and Maximum Observed Growth or the Frequencies of New Buds or Calices with a Growth of ≥2.0 mm yr^{-1}

Water-Flow Rate (mL sec^{-1}) versus		
Mean growth (mm yr^{-1})	Pearson correlation (r^2)	−0.004
	Sig. (2-tailed)	0.967
	N	96
Maximum growth (mm yr^{-1})	Pearson correlation (r^2)	0.020
	Sig. (2-tailed)	0.843
	N	96
Frequency of new buds	Pearson correlation (r^2)	0.007
	Sig. (2-tailed)	0.949
	N	96
Frequency ≥ 2 mm yr^{-1}	Pearson correlation (r^2)	0.101
	Sig. (2-tailed)	0.330
	N	96

fragments. In addition, some of the corals in the LIII treatment had skeletons so fragile that when the experiment was terminated, they fell apart while being dismounted from the cathodes.

DISCUSSION

This study is the first to test the method of mineral accretion on cold-water corals. The most striking result of this experiment is the number of new buds that developed in the low current density treatments (Table 10.4 and Figure 10.3a). Although no significant effects were found in the pairwise comparisons between the controls and the different treatments, there was a significant difference ($p = 0.03$) between the treatments in the overall Chi-square test and testing the substrate per se against the charged treatments; the difference between AS versus LI was found to be close to significant ($p = 0.06$). The overall difference between the treatments could largely be attributed to the difference between the LI and LIII (lowest vs. highest applied current density treatment and $p = 0.003$). Looking at the frequency distributions of new buds (Figure 10.3a), the increase lies over the low current density treatments (GE, LI, and LII), and this result is congruent with the previous studies, which is discussed later. The highest current density treatment (LIII) appears to be overcharged and detrimental to the corals, with very few new buds developing as a result.

Also in the frequency distribution of calices with a growth of ≥ 2.0 mm yr^{-1}, there were significant differences between the treatments, although the controls displayed an equally good growth at the lower current-densities (LI and GE), and this effect could therefore not be linked to trickle currents (Figure 10.3c). The pairwise comparisons revealed a significantly lower frequency of calices with a growth of ≥ 2.0 mm yr^{-1} in the intermediate current density treatment (LII), and, rather than a positive effect of applied direct current, there seems to be mainly a negative impact in AS, LII, and LIII. Also in AS and LIII, the frequencies were lower, albeit nonsignificant. The low frequencies of calices with a growth of ≥ 2.0 mm yr^{-1} in AS and LII is somewhat compensated by the higher frequencies of calices with the lower range of growth rate ($0.8 \geq x < 2.0$ mm yr^{-1}).

The growth rates observed in this study were below or in the lower range of the reported rates for the species, 4–25 mm yr^{-1} (Freiwald et al. 2004). Only a few calices on the GEs and the lowest applied current density (LI) showed the growth rates of around 10 mm yr^{-1}, as seen in Table 10.4, where maximum observed growth is presented. Possible causes for this are poor nutritional values of the chosen food (*Artemia*) or effects of stress. The trend in the growth rates seen both in mean and maximum observed growth (Figure 10.3d and e) was a slight increase in the growth rate in LI as compared to the controls, whereas all other treatments had lower rates, the differences being more pronounced in mean growth.

Zero mortality was observed in the lowest applied current density treatment (LI) as well as in the controls, showing that a trickle current has no negative impact on coral transplants if the level is optimized, while overcharging could be detrimental, as seen in the higher current density treatments (LII and LIII). The partial mortalities and overall low performance of the corals in the zero-voltage treatment (AS) and GE could, however, be due to leaking metal ions. Considering the AS treatment, the failure of counteracting corrosion on the spots that had the mineral layer scraped off while mounting the coral fragments could have led to detrimental levels of metal oxides. Rust has been seen to be avoided by settling organisms, leaving bare patches on oxidized metal surfaces (Fitzhardinge and Bailey-Brock 1989). Leaking ions could also explain the low performance of corals on the GEs, as the zinc anodes were oxidized. Interestingly, the partial mortalities caused by the iron-oxidizing bacteria found in one of the GE tanks did not concur with the impaired growth. Noteworthy, three of the corals growing in this specific tank had the highest growth rates observed within the GE treatment. The bacterial growth was restricted to one side of the cathode and thus left the two corals unaffected by direct bacterial overgrowth. Despite some partial mortality, the growth rate of the living calices in one of the affected corals was relatively high. Chemical reactions mediated by the bacteria could have mitigated the effect of leaking zinc ions, thus giving the positive response in growth in this specific tank.

The measured pH differed slightly between the treatments, with highest pH observed in the LII treatment. The pH was initially high also in LIII but dropped after the 12th week (Figure 10.1). This drop could be driven by the fast accretion that filled the voids in the mesh, thus trapping hydrogen gas underneath the cathodes, with equilibrium reactions between the gas and water interface decreasing pH as a result. The fragility of coral skeletons in the LIII treatment could be explained by the lower pH toward the end of the experiment. An additional explanation of the observed fragility could be that the corals incorporate more porous brucite minerals rather than aragonite at high current densities.

The observed drop in pH for all the treatments during the first weeks was probably caused by an early autumn down-mixing of surface waters, as an intense low pressure occurred in August, with winds of storm strength (mean wind velocity 24 m s^{-1}) accompanied by heavy rains (SMHI Report No. 2009–6) that could cause salinity to drop during the following period and affect pH. The water temperature at 30 m depth had its annual maxima in September (temporarily warmer than the surface waters), and the stratification of water masses was less pronounced during the period.

Comparisons with Previous Studies

The stimulatory effect in bud production seen in this study is congruent with the results of the study by Sabater and Yap (2004), where treated nubbins of the branching coral *Porites cylindrica* had significantly higher densities of corallites. Although the results of the present study were nonsignificant in the pairwise Chi-square test between the controls and LI, the concordance with the results of Sabater's and Yap's study strengthens the present results. The stimulated budding is an interesting effect that could prove valuable in transplantation programs, as coral transplants that bud off richly initially will have more growing tips and thus a more rapid outgrowth into a complex matrix of coral branches. The morphology of *P. cylindrica* is very different from that of *L. pertusa*, and perhaps the latter species is better served by an effect of this nature because budding is directly affecting branching. The mechanism for this, however, can only be speculated. Although Sabater and Yap (2004) ascribes this effect to the increased mineral ion concentration, an alternative explanation could be an increase in ATP production at an optimal current density.

As shown by Cheng et al. (1982) in a clinical in vitro study on the effects of electric currents on skin tissue samples (from rats), currents ranging from 50 to 1000 µA had positive effects on ATP generation, with a threefold to fivefold increase in ATP levels. ATP concentrations leveled when applying currents exceeding 1000 µA and were reduced at currents of 5000 µA. This current-density-dependent ATP stimulatory effect could explain why positive effects are restricted to low current density treatments, while higher levels have a negative impact despite the theoretically higher availability of calcium ions with increasing current densities. ATP is needed for the active transport of calcium ions into the calcifying compartment, as well as for removing the hydrogen ions from the same to maintain the alkalinity necessary for precipitation of aragonite within the compartment (Allemand et al. 2004; McConnaughey and Whelan 1997; Tambutté et al. 1996). It is also known that feeding increases calcification rates (Houlbrèque et al. 2003 and 2004), and that respiration rates and ATP concentrations are elevated in the growing tips of coral branches, thus supporting the higher calcification rates (Fang et al. 1989; Gladfelter et al. 1989). Corals use mainly metabolic CO_2 for calcification (Furla et al. 2000), explaining the positive effect of feeding on calcification rates. ATP is thus the limiting factor for calcification in saturated environments. High food availability can support a denser coral colony; could the corals be fooled to branching by artificially elevated ATP levels generated by the electrode potential?

The two lower current densities used in the present study were low (0.06–0.75 A m^{-2}) in comparison to the previously used densities, whereas the highest level (2.19 A m^{-2}), which in this experiment proved to be overcharged, were in the range of previously used levels. The levels used in the referred papers were 2.8 A m^{-2} (Borell et al. 2009); 1.67, 2.38, 3.43, and 2.86 A m^{-2} (Romatzki 2008); 2.8 A m^{-2} (van Treeck and Schuhmacher 1997); and 4.18 A, that is, 9.5 A m^{-2} (Sabater and Yap 2004). The latter authors did not provide a recalculation of the electrical current to amperes per unit metal surface area, and therefore the available information was used to calculate the current density, with possible errors. Noteworthy, Sabater and Yap used solar panels to provide their electrodes with electricity, and thus the given current density is the peak level at noon. van Treek and Schuhmacher (1997) used a constant direct current but turned the electricity off for six hours daily, around noon, to protect the DC source from overheating.

It is clearly seen in both the present study as well as in the study by Romatzki (2008) that the coral response is current-density dependent. In the latter study, three different experiments were conducted on two coral species (*Acropora pulchra* and *Acropora yongei*), and in the first experiment where two current density levels were compared, slightly increased growth rates on the lower current density treatment (1.67 A m^{-2}) were observed, while the higher current density (2.38 A m^{-2}) had a negative impact. In the second and third experiments, the author used 3.43 A m^{-2} and 2.86 A m^{-2}, respectively. Here, without any explanation as to why the higher current densities were used, the author found negative impacts of treatment, with significantly lower growth rates on cathodes compared to controls.

The study conducted by Sabater and Yap (2004) showed significantly higher density of corallites and longitudinal and girth growth in treated corals than in controls during the first four of the six months of mineral accretion. Growth rates were presented as cm per two months, and the difference between the treated and control nubbins was roughly 0.1 cm (controls and treated around 0.5 cm and 0.65 cm, respectively). The corals were reared on a variable current density following a sinusoidal function (0 to 9.5 A m^{-2}). The relatively limited period of peak current density during noon did not seem to impede growth. Although significant results were found, they were not of the magnitude observed in restoration programs (two to six times greater), thus none of the studies (the present study included) on the effects of mineral accretion published so far has confirmed the high growth rates from the restoration programs. The increased density of corallites found by Sabater and Yap (2004) has nevertheless been confirmed in this study.

Both Sabater and Yap (2004) and van Treeck and Schuhmacher (1997) reported high survival rates of corals on cathodes. The latter workers reported high survival rates of coral transplants of a range of species grown on cathodes, except for *Pocillopora damicornis* that were less robust; however, the species is sensitive to fragmentation and transplantation stress (Yap et al. 1992). Hence, the transplantation routine and not the treatment were likely to cause the increased mortality in this species. In the study by Borell et al. (2009), *A. yongei* displayed a pronounced mortality on the cathodes, whereas the mortality of *A. pulchra* was distributed over all the treatments: cathodes, electric field, and controls. The constantly high current density reported by Borell et al. compared to the variable current density reported in the study by Sabater and Yap (2004) and the daily six hours resting period reported in the study by van Treeck and Schuhmacher (1997) could explain the high mortality despite lower current density. In the study by Romatzki (2008), the results regarding survival were ambiguous in the three experiments. Survival was high in the first experiment in both current density treatments (1.67 and 3.38 A m^{-2}) and overall treatments (controls included), whereas survival was lower in some of the treatments in the second experiment, with no apparent pattern. The controls sometimes fared less well than treated corals. The highest survival in the third experiment was observed in corals exposed to an electrical field without direct contact with the cathodes, and the author concluded that the direct contact with the cathodes diminishes growth; however, as suggested by the results of the present study, this was likely due to overcharging.

Notably, many of the cathodic structures used in previous rehabilitation programs have been high, elevating coral transplants several meters up in the water column. This poses a risk for confounding beneficial effects if growth rate comparisons are made with the surrounding natural colonies growing closer to the seafloor. Studies of comparisons between controls at the same height as the corals growing on cathodes have been done, and results will soon be published. With the elevated design, it is possible that the increased water current velocity in the elevated position, with limited exposure to sedimentation compared to the natural habitat, has a beneficial effect that could to some extent explain the increased growth rates. This could also be the case for the coral colonies found on oil rigs (Gass and Roberts 2006). Instead of being caused by trickle currents applied to protect the metal from corrosion, the increased growth rates could be due to the different hydrodynamic regime and absence of predation, sedimentation, competition, and bioeroders. The risk for confounding is apparent, and, although the effect of increased budding that is confirmed in two studies could be a true effect of the treatment, the increased growth rates could be an artifact effect due to the elevated position. Combining the two will give densely branched and fast-growing corals, which indeed is a splendid prospect for a rehabilitation program; however, one should be sure to assign the effects to the right cause.

SUMMARY

Summarizing the results of this study, it is apparent that there is a positive effect solely of the lowest current density treatment (LI), while both higher current densities (LII and LIII) and the substrate per se and GEs produced poor results. The optimal current density in the present study was found to be as low as 0.06 A m^{-2} or less. These results provide new insight in the optimal current density to use in these installations and indicate that cathodes have been severely overcharged in previously published studies.

The zero mortality and the overall performance of the corals in the lowest applied current density (LI) brings the authors to the conclusion that mineral accretion is a suitable method for the target species *L. pertusa*. Although there was no significant gain in the growth rate compared to the controls, the increased budding and firm attachment of coral transplants offer sufficient benefits, and the method is considered worth testing in a field study.

There is obviously a contradiction in the optimal current density considering coral survival and growth versus mineral accretion rates; that is, the optimal level for the corals does not produce a full accretion of the cathodes. Alternating the current density during the day could be a solution to this incompatibility, giving a few hours of higher levels to achieve a thicker layer of minerals. As described in the study by Sabater and Yap (2004), a variable current density gave significantly higher growth rates despite the high peak level of current density at noon.

Considering the rate of mineral accretion, it was slower than seen in tropical environments, as expected, and achieving a sufficiently thick accretion on the cathodes will probably take two years or longer under the local conditions. This could be because aragonite is more soluble in cold waters.

Our results suggest that the electrical stimulation and possible increase in ATP production are the main beneficial effects rather than increased mineral ion concentrations, although it needs further testing.

ACKNOWLEDGMENTS

This study was financed by the Swedish Environmental Protection Agency and the Centre of Underwater Documentation (CUD, T. Lundälv). We are thankful to our technician Roger Johansson for recommending testing the tank constructions on one tank before rebuilding all, and for advising how to do all the connections, as there were a lot of cables and connections involved. Lisbeth Jonsson and Tomas Lundälv are worthy of many thanks for providing the corals, and we are also grateful to Ann Larsson and Martin Ogemark for giving some valuable advice in rearing corals and *Artemia*. Finally, we would like to thank Kerstin Johannesson, Per Jonsson, and Lisbeth Jonsson for proofreading.

The experiment was planned and carried out by SS, in collaboration with TG, who provided scientific advice on the method of mineral accretion and the anode material.

REFERENCES

Allemand D., Ferrier-Pagès C., Furla P., Houlbrèque F., Puverel S., Reynaud S., Tambutté É., Tambutté S., Zoccola D. (2004) Biomineralisation in reef-building corals: From molecular mechanisms to environmental control. *CR Palevol* 3:453–467.

Borell E., Romatzki S., Ferse S. (2010) Differential physiological responses of two congeneric scleractinian corals to mineral accretion and an electric field. *Coral Reefs* 29:191–200.

Broberg E. (2006) Genetic structure of reefs of the deep-sea coral *Lophelia pertusa* in N-E Skagerrak. Master Thesis, Department of Marine Ecology, Tjärnö. University of Göteborg, Sweden.

Brooke S., Koenig C. C., Shepard A.N. (2006) *Oculina* banks restoration project; description and preliminary results. Proceedings of the 57th Gulf and Caribbean Fisheries Institute:607–620.

Cheng N., van Hoof H., Bockx E., Hoogmartens M. J., Mulier J. C., de Ducker F. J., Sansen W. M., de Loecker W. (1982) The effects of electric currents on ATP generation, protein synthesis, and membrane transport in rat skin. *Clin Orthop Relat R* 171:264–272.

Clark S., Edwards A. J. (1994) Use of artificial reef structures to rehabilitate reef flats degraded by coral mining in the Maldives. *B Mar Sci* 55:724–744.

Costello M. J., McCrea M., Freiwald A., Lundälv T., Jonsson L., Bett B. J., van Weering T. C. E., de Haas H., Roberts M. J., Allen D. (2005) Role of cold-water *Lophelia pertusa* coral reefs as fish habitat in the NE Atlantic. In: Roberts and Freiwald (ed) *Cold-water corals and ecosystems* Springer-Verlag Berlin Heidelberg, p 771–805.

Dahl M. (2006) Genetic structure of *Lophelia pertusa* in North East Skagerrak. Master Thesis, Department of Marine Ecology, Tjärnö. University of Göteborg, Sweden.

Eisinger M. (2005) Contributions to ecological and economic aspects of coral transplantation on electrochemically produced substrates as a method for rehabilitation of degraded coral reefs. PhD Thesis, University of Duisburg-Essen, Germany.

Ericson J., Ljunghager F. (2006) Spridningsmönster hos *Lophelia pertusa* i norra Skagerrak (In Swedish). Title translation: Dispersal patterns in *Lophelia pertusa* in the N Skagerrak. Students project at the Department of Marine Ecology, Tjärnö. University of Göteborg, Sweden.

Fang L.-S., Chen Y.-W., Chen C.-S. (1989) Why does the white tip of stony coral grow so fast without zooxanthellae? *Mar Biol* 103:359–363.

Fitzhardinge R. C., Bailey-Brock J. H. (1989) Colonization of artificial reef materials by corals and other sessile organisms. *B Mar Sci* 44:567–579.

Fosså J. H., Mortensen P. B., Furevik D. M. (2002) The deep-water coral *Lophelia pertusa* in Norwegian waters: Distribution and fishery impacts. *Hydrobiologia* 471:12.

Freiwald A., Fosså J. H., Grehan A., Koslow T., Roberts M. J. (2004) *Cold-water coral reefs: Out of sight—no longer out of mind.* UNEP-WCMC, Cambridge, UK.

Furla P., Galgani I., Durand I., Allemand D. (2000) Sources and mechanisms of inorganic carbon transport for coral calcification and photosynthesis. *J Exp Biol* 203:3445–3457.

Gage J., Roberts M. J. (2003) The *Lophelia* reef-associated fauna. EU FP5, project ACES (Atlantic Coral Ecosystem Studies) Deliverable 16:70.

Gass S. E., Roberts J. M. (2006) The occurrence of the cold-water coral *Lophelia pertusa* (Scleractinia) on oil and gas platforms in the North Sea: Colony growth, recruitment and environmental controls on distribution. *Mar Pollut Bull* 52:549–559.

Gladfelter E. H., Michel G., Sanfelici A. (1989) Metabolic gradients along a branch of the reef coral *Acropora palmata*. *B Mar Sci* 44:1166–1173.

Goreau T. J., Cervino J. M., Pollina R. (2004) Increased zooxanthellae numbers and mitotic index in electrically stimulated corals. *Symbiosis* 37:107–120.

Henry L.-A., Roberts J. M. (2007) Biodiversity and ecological composition of macrobenthos on cold-water coral mounds and adjacent off-mound habitat in the bathyal Porcupine Seabight, NE Atlantic. *Deep-Sea Res Pt I* 54:654–672.

Hilbertz W. H., Goreau T. J. (1996) Method of enhancing the growth of aquatic organisms, and structures created thereby, U. S. patent number 5,543,034, p 14.

Houlbrèque F., Tambutté É., Allemand D., Ferrier-Pagès C. (2004) Interactions between zooplankton feeding, photosynthesis and skeletal growth in the scleractinian coral *Stylophora pistillata*. *J Exp Biol* 207:1461–1469.

Houlbrèque F., Tambutté É., Ferrier-Pagès C. (2003) Effect of zooplankton availability on the rates of photosynthesis, and tissue and skeletal growth in the scleractinian coral *Stylopora pistillata*. *J Exp Mar Biol Ecol* 296:145–166.

Husebø Å., Nøttestad L., Fosså J. H., Furevik D. M., Jørgensen S. B. (2002) Distribution and abundance of fish in deep-sea coral habitats. *Hydrobiologia* 471:91–99.

Jensen A., Frederiksen R. (1992) The fauna associated with the bank-forming deepwater coral *Lophelia pertusa* (Scleractinia) on the Faroe Shelf. *Sarsia* 77:53–69.

Jonsson L., Lundälv T. (2006) Associated fauna on deep-water coral (*Lophelia pertusa*) reefs in a fjord area; the importance of reef size and habitat type (submitted) In Jonsson 2006: Ecology of three coastal cold-water cnidarians, in particular the scleractinian *Lophelia pertusa*, PhD Thesis. Department of Marine Ecology, Tjärnö. University of Göteborg, Sweden, p 19.

Jonsson L. G., Nilsson P. G., Floruta F., Lundälv T. (2004) Distributional patterns of macro- and megafauna associated with a reef of the cold-water coral *Lophelia pertusa* on the Swedish west coast. *Mar Ecol-Prog Ser* 284:163–171.

Kinnear P. R., Gray C. D. (2008) *SPSS 15 made simple*. Psychology Press, New York.

Kloth L. C. (2005) Electrical stimulation for wound healing: A review of evidence from in vitro studies, animal experiments, and clinical tests. *Int J Low Extrem Wounds* 4:23–44.

Koenig C. C., Coleman F. C., Grimes G. R. F., Scanlon K. M., Gledhill C. T., Grace M. (2000) Protection of fish spawning habitat for the conservation of warm-temperate reef-fish fisheries of shelf-edge reefs of Florida. *B Mar Sci* 66:593–616.

McConnaughey T. A., Whelan J. F. (1997) Calcification generates protons for nutrient and bicarbonate uptake. *Earth Sci Rev* 42:95–117.

Mortensen P. B., Hovland M., Brattegard T., Farestveit R. (1995) Deep water bioherms of the scleractinian coral *Lophelia pertusa* (L.) at 64°N on the Norwegian shelf: Structure and associated megafauna. *Sarsia* 80:145–158.

Reed J. K. (2002) Comparison of deep-water coral reefs and lithoherms off southeastern USA. *Hydrobiologia* 471:57–69.

Romatzki S. (2008) Determining the influence of an electrical field on the performance of *Acropora* transplants: Comparison of various transplantation structures. In Romatzki 2008: Reproduction strategies of stony corals (Scleractinia) in an equatorial, Indonesian coral reef. Contributions for the reefrestoration (PhD Thesis) 64–90.

Romatzki S. (2009) Determining the influence of an electrical field on the performance of *Acropora* transplants: Comparison of various transplantation structures. *J Exp Mar Biol Ecol*.

Ross S. W., Nizinski M. S. (2007) State of deep coral ecosystems in the US Southeast region: Cape Hatteras to southeastern Florida. P. 233–270 in *The State of Deep Coral Ecosystems*, NOAA, Washington, DC.

Sabater M. G., Yap H. T. (2002) Growth and survival of coral transplants with and without electrochemical deposition of $CaCO_3$. *J Exp Mar Biol Ecol* 272:131–146.

Sabater M. G., Yap H. T. (2004) Long-term effects of induced mineral accretion on growth, survival and corallite properties of *Porites cylindrica* Dana. *J Exp Mar Biol Ecol* 311:355–374.

Schuhmacher H., van Treeck P., Eisinger M., Paster M. (2000) Transplantation of coral fragments from ship groundings on electrochemically formed reef structures. Proceedings 9th International Coral Reef Symposium, Bali, Indonesia 2:983–990.

Tambutté É., Allemand D., Mueller E., Jaubert J. (1996) A compartmental approach to the mechanism of calcification in hermatypic corals. *J Exp Biol* 199:1029–1041.

van Treeck P., Schuhmacher H. (1997) Initial survival of coral nubbins transplanted by a new coral transplantation technology—options for reef rehabilitation. *Mar Ecol-Prog Ser* 150:287–292.

Yap H. T., Aliño P. M., Gomez E. D. (1992) Trends in growth and mortality of three coral species (Anthozoa: Scleractinia), including effects of transplantation. *Mar Ecol-Prog Ser* 83:91–101.

CHAPTER 11

Utilization of Low-Voltage Electricity to Stimulate Cultivation of Pearl Oysters *Pinctada maxima* (Jameson)

Prawita Tasya Karissa, Sukardi, Susilo Budi Priyono, N. Gustaf
F. Mamangkey, and Joseph James Uel Taylor

CONTENTS

Introduction 131
Materials and Methods 132
    Material 132
    Method 132
        Experimental Design 132
        Time and Location 132
        Low-Voltage Electricity Method Installation 133
        Juvenile Stocking 133
        Shell Length Measurements 134
        Weight Measurement 134
        Growth Calculation 134
        Survival Rate 135
        Aquatic Environment Parameters 135
        Data Processing and Statistical Analytics 135
Results 135
Discussion 137
Conclusions 138
Acknowledgments 138
References 138

INTRODUCTION

Pinctada maxima is an oyster that produces highly valuable silver and gold South Sea pearls (Strack 2006; Southgate 2007). Juvenile of *P. maxima* usually grows faster than the adult (Chellam 1978). At least 18 months are needed to achieve sufficient size of oysters for pearl nuclei insertion.

Efforts were taken to accelerate high survival and growth of *P. maxima* juveniles to achieve shorter production time of nuclei-insertion-ready oysters.

As in most mollusks, *P. maxima* grows with increased shell size and body weight. Shell growth is controlled by the mantle organ, which is located between the inner shell and outer epithelium of inner organs or visceral mass. Outer epithelial cells will produce calcium carbonate crystals ($CaCO_3$) in the form of calcite and aragonite (Bubel 1984; Weiner et al. 1984; Simkiss and Wilbur 1989; Lowenstam and Weiner 1989). These cells also form organic protein called conchiolin as the adhesive for lime crystals (Weiner 1984; Suzuki et.al. 2004). The combination of calcite, aragonite, and conchiolin forms shells that protect the mollusk's soft bodies (Bubel 1984).

The composition of coral exoskeleton is similar to *P. maxima* shells containing calcium carbonate ($CaCO_3$). Coral growth can be accelerated by running a low-voltage electrical current through negative and positive terminals inside the seawater, using the low-voltage Biorock method (Dwija 2003; Arifin et al. Chapter 6). The Biorock method, also known as mineral accretion, was developed to be the latest generation of artificial reefs method that is very safe and most effective. Bivalves show rapid growth when they are located near Biorock artificial reef. The purpose of this experiment is to determine the potential benefits of low-voltage electricity on the early growth stage of the pearl oyster *P. maxima* (Jameson).

MATERIALS AND METHODS

Material

Pearl oyster juveniles of *P. maxima*, aged 6 and 10 months, were used. Juveniles of the same age came from the same spawning group at the hatchery department of PT. Cendana Indo Pearl in Penyabangan, Buleleng, Bali. Juveniles were selected based on the length of the shell by passing through a 4 cm hole of a plastic bucket lid for 6 months old and a 6 cm hole lid for 10 months old.

The materials used for installing low-voltage electricity were 6 V, 11 A wet battery as power supply, 1 mm diameter iron wire as cathode, titanium alloy as anode, 10 mm diameter cable, epoxy glue (made from resin and hardener), and 5 cm rubber tube to protect the anode and cable connection.

The tools used in this research were caliper, electric scales, stove oven, light bulb, incinerator (furnace), water sampling, GFF filter paper, thermometers, hand refractometer, pH meter, Erlenmeyer bottle, voltage meter, sounder, pocket net, multitester, and battery charger.

Method

Experimental Design

The method used in this experiment was a 3 × 2 factorial design with four replications. Each replication consists of 10 pearl oysters. There were two treatment factors: electrical treatment and age treatment. Electrical treatments consist of without electricity, electricity without accretion, and electricity with accretion (accretion/cement already exists in one-month cathode). Age treatments consist of 6- and 10-month-old juveniles.

Time and Location

The experiment was conducted in August–October 2006 at PT. Cendana Indo Pearls, Buleleng, Bali, with pearl oysters in panels hanging from long horizontal surface lines. Experiments were carried out by hanging the PN24 rearing panels at 3 m depth from a raft/guarding house located about 500 m from the coastline.

Low-Voltage Electricity Method Installation

The low-voltage electrical treatment consists of four main components: the voltage source/voltage converter, cathode at the negative terminal, anode at the positive terminal, and cable. The cathode is a double-twisted steel wire inserted into the oyster rearing panel to have a direct contact with the juvenile oyster shells. The top edge, middle, and the lower end of double panels were connected by cathode twisted to the cable.

The anode (titanium alloy) was connected to the cable with a strong, electrically insulated seal resistant to water/air. The connection was enclosed in a rubber hose filled with resin and hardener, so water/air could not penetrate. The cathode was connected to the negative terminal, whereas the anode was connected to the positive terminal from the batteries, which had been wired in parallel and placed in a box to protect them from waves and protect the cable connections.

A total of four panels with cathodes were attached to the raft and electrified for one month until the cathode was covered by an accretion of $CaCO_3$. Two wet batteries were used as power supply each day, which last for four hours/day. In the case of electricity with Biorock, the oysters were attached to substrate that had been electrified and built up a layer of minerals. In the case of electricity without Biorock, the oysters were attached to fresh metal substrate on which no minerals had yet grown. Because the mineral growth acts to some degree as an insulator, the electricity without Biorock should have received more electrical current than the electricity with Biorock.

After one month, all panels were hung as shown in Figure 11.1.

Juvenile Stocking

Oysters are selected based on the age and size by passing the oysters through a hole of a plastic bucket lid with diameter of 4 cm for 6-month-old juveniles and 6 cm for 10-month-old juveniles. The shells of the oysters were cleaned from fouling organisms, the excess water was absorbed using cloth or tissue paper, and the oysters were weighed and labeled. Next, the oysters were placed on the rearing panels with a density of 10 juveniles per panel and acclimatized two times. First acclimatization was performed in a tub of running water that contains eight panels and was placed in the room for one night. Second acclimatization was held on the next day at the juvenile rearing location in the sea without low electrical voltage for 24 hours.

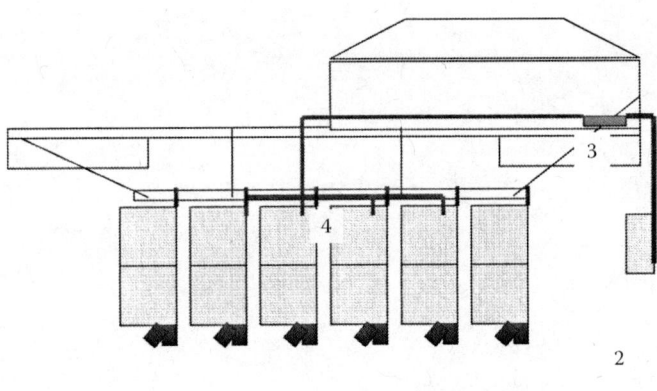

Figure 11.1 The pearl oyster juveniles reared with low-voltage method. Components of low-voltage method setup are cathode (steel wire), anode (titanium alloy), power supply (wet batteries wired in parallel), and cable.

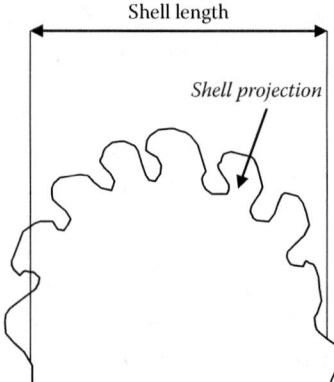

Figure 11.2 Shell length measurement method. Measurement was excluding projections that are common in young shells.

Shell Length Measurements

Length measurements were performed at months zero, one, and two using a 0.01 mm accuracy caliper. Juvenile shell length was measured as shown in Figure 11.2.

Weight Measurement

The shells of the oysters were cleaned from fouling organisms, excess water on the oyster was absorbed using cloth or tissue paper, and then the oysters were weighed and labeled.

Growth Calculation

Absolute growth (Moyle and Cech 1982) in shell length was determined using the following equation from the study by Effendie (1997):

$$\Delta L = L_t - L_0$$

where
ΔL: Absolute growth in the average individual length (mm)
L_t: The average length at time t (mm)
L_0: The average length at the beginning of the experiment (mm)

The absolute growth for wet weight was determined using the following equation from the study by Effendie (1997):

$$\Delta W = W_t - W_0$$

where
ΔW: Absolute growth in the average individual weight (g)
W_t: The average individual weight at the end of the experiment (g)
W_0: The average individual weight at the beginning of the experiment (g)

Survival Rate

Juvenile mortality counts were made every two weeks and at the end of the research. Survival rate was calculated using the following equation (Effendie 1997):

$$S = (N_t/N_0) \times 100\%$$

where
 S: Survival rate (%)
 N_t: Number of living organisms at time t
 N_0: Number of living organisms at the beginning of the experiment

Aquatic Environment Parameters

The observed environment parameters measured every two weeks were water temperature, water current, water clarity (Secchi disk depth), pH, and salinity.

Data Processing and Statistical Analytics

The mean of survival and growth were tested for homogeneity level using the Levene statistic, and continued using analysis of variance (ANOVA) and least significance difference (LSD). All tests were performed by software SPSS 13.

RESULTS

Table 11.1 shows the results of ANOVA of final survival rates, and the experimental results of final survival rate can be found in Table 11.2.

Table 11.2 showed that the final survival rate of *P. maxima* juveniles reared by several electricity methods ranged between 67.5% and 90%. The final survival rate of *P. maxima* juveniles reared with electricity without accretion (90.0%) and electricity with accretion (87.5%) were significantly higher than the juvenile survival rate reared without electricity (67.5%).

Figure 11.3 shows the juvenile survival rates observed during the experiment; there was mortality in each treatment, except for six-month-old juveniles reared by electricity without accretion.

Table 11.1 Results of Analysis of Variance for Survival Rate Using Two Treatment Factors: Electricity and Age. Electricity Treatments Consist of without Electricity, Electricity without Accretion, and Electricity with Accretion. Age Treatments Consist of 6-Month-Old and 10-Month-Old Juveniles

Source of Variety	Sum of Squares (SS)	df	Mean Square (MS)	F	Significance
Corrected model	34–8333	5	6.967	2.697	.055
Intercept	1600.667	1	1600.667	619.613	.000
Electricity	24.333	2	12.167	4.710	.023
Age	4.167	1	4.167	1.613	.220
Electricity × age	6.333	2	3.167	—	.317
Error	46.500	18	2.583	—	—
Total	1682.000	24	—	—	—
Corrected total	81.333	23	—	—	—

Note: $R^2 = .428$ ($R^2 = .269$).

The lowest survival were 10-month-old juveniles reared without electricity. Solid lines refer to 6-month-old oysters, and dashed lines refer to 10-month-old oysters.

Table 11.3 shows that juveniles aged 6 and 10 months showed different growth to the electricity treatment. Juvenile aged six months generally had highest shell length on electricity without accretion treatment. Next were the juveniles reared without electricity, and the last were the juveniles reared with electricity and accretion. The highest wet weight was found on electricity without accretion, followed by without electricity, and last electricity with accretion, although they were not significantly different.

Table 11.2 Final Survival Rate (%) of *P. maxima* Juvenile Pearl Oyster in Each Treatment

Treatment	6 Months	10 Months	Means
Without electricity	72.5[a]	62.5[a]	67.5*
Electricity without accretion	100.0[a]	85.0[a]	90.0**
Electricity with accretion	85.0[a]	90.0[a]	87.5**
Means	85.8[1]	77.5[1]	—

Note: Means followed by same letters, numbers, or sign do not significantly differ (95%)

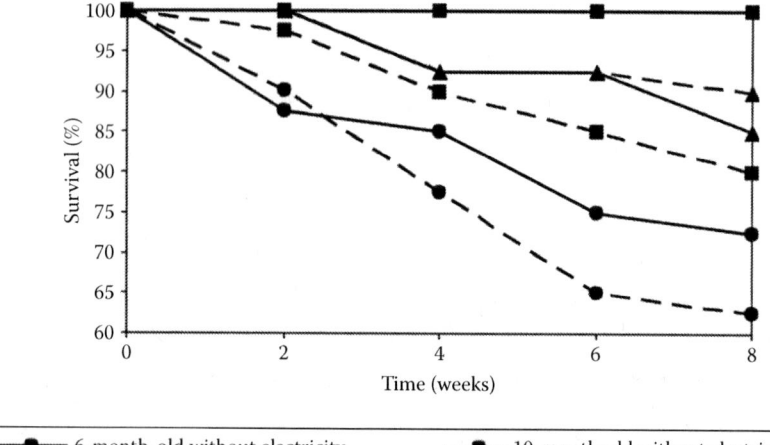

Figure 11.3 Survival rate of *P. maxima* in each treatment.

Table 11.3 Absolute Growth Rate of Pearl Oyster Juvenile *P. maxima*

Treatments	Absolute Shell Length (mm)	Absolute Wet Weight (g)
6-months juvenile		
Without electricity	11.38[ab]	4.60[a]
Electricity without accretion	12.52[b]	4.92[a]
Electricity with accretion	9.00[a]	3.65[a]
10-months juvenile		
Without electricity	9.40[a]	7.08[a]
Electricity without accretion	9.66[a]	9.38[a]
Electricity with accretion	13.09[b]	8.50[a]

Means followed with same letters were not significantly different (95%).

Table 11.4 Environmental Parameters

Parameter	Range Number	Optimum	Notes
Temperature (°C)	27.89–29.50	29.50[a]	Acceptable
Salinity (%)	30.67–33.00	30.00–34.00[a]	Acceptable
Water current velocity (m/s)	0.05–0.11	1.28–1.80[a]	Unacceptable
Water clarity (m)	13.87–18.00	>20.00 m[b]	Unacceptable
Suspended density (g/L)	0.008–0.016	0.010[a]	Acceptable
pH (unit)	7.63–7.83	6.50–8.50[a]	Acceptable
Water depth (m)	30	20–25[b]	Acceptable
Pocket net depth (m)	3.0	2.5[a]	
Bottom substrate condition	Rough sand	Rough sand, coral	Acceptable
Pollutant	None	None[c]	Acceptable

Note: Water quality optimum standard shown according to Anonim (1991), Taylor (1999), and Gervis and Sims (1992).

In contrast with 10-month-old juveniles, electricity with accretion treatment had the highest length increment. Juveniles reared without electricity and juveniles reared with electricity with accretion have similar values. Electricity with accretion also had the highest weight growth. The absolute wet weight and shell growth had the same trend, with electricity without accretion highest, followed by without electricity, and last, electricity with accretion.

Environmental quality parameters are shown in Table 11.4. The table shows that there were few parameters that were substandard for pearl oyster growth, namely low water-current velocity and water clarity, although, in general, the water quality was acceptable for pearl oyster maintenance. High suspended material density (plankton, sediment, organic matter, etc.) indicated that there was enough food for *P. maxima* juveniles.

DISCUSSION

Survival was influenced by environment, food, and diseases (Nasr 1982, Pass 1987, Taylor 1999). Environmental data showed that water current velocity was below desired values. This could lead to reduced food supplies (plankton) and also feces accumulation that are consumed by the *P. maxima* juveniles (Wilson 1987 cit. Taylor 1999). Crabs were common predators of oyster juveniles, and *P. maxima* uses its strong shell for self-defense against predators. Shells are strong not only because they are formed by hard aragonite crystals ($CaCO_3$) but also because of the macromolecular biological matrix that functions to bond the aragonite crystals (Addadi and Weiner 1997).

The Biorock low-voltage method works by influencing the environment and by increasing the pH, $CaCO_3$ contents, and electron flow within several millimeters of the cathode (Global Coral Reef Alliance 2007). Cathode electrochemical processes produce electrons. Electron flow influences the electron transport chain to produce Adenosine triphosphate (ATP) biochemical energy in juveniles (Goreau, Chapter 19). This energy may be used by juveniles to compensate for the consequences of the byssus thread cut used by pearl oyster culture and for lack of food, and could aid formation of the macromolecular shell biological matrix and $CaCO_3$ formation and increase the power of shell closure as self-defense mechanism by *P. maxima* juveniles against crab predation.

Oyster death can be caused by fragility of juvenile shells, making them susceptible to predatory attack by crabs. Shell fragility can be caused by low seawater current velocity causing accumulated organics and reduced food (plankton), and also frequent byssus cuts that will result in less energy to form the shell (Wilson 1987 cit. Taylor 1999, Mariani et al 2002).

Absolute weight was not significantly different when varied low-voltage methods were applied to different aged juveniles. Although the highest absolute weight (Alagarswami et al 1989, Pouvreau 2000, Taylor 1999) was produced by electricity without accretion in two months of experiment, it is possible that if the experiment time was extended with a longer time of electricity supply, then the experimental result might be significantly different. Similar experiments should be done with longer rearing time (six months minimum) using longer electricity supply, with 12 hours during daylight. Experiments using a broader range of ages, including adult oyster, juvenile, spat, and larvae, should be done to understand the appropriate age of each oyster phase to apply low-voltage method and optimize the practical management benefits of the method.

These results might indicate that electricity without accretion has higher electron availability to stimulate ATP production. Electricity without accretion can increase the absolute length increase of six-month-old *P. maxima* juveniles. The absolute weight growth also had significant increases higher than other treatments. Two volts voltage without accretion had higher growth than electricity with accretion. Higher voltages produced higher electron currents and electrochemical reactions, and probably influenced the electron chain transport to produce ATP. These conditions are appropriate for six-month-old juveniles, which have greater proportional growth rates than larger oysters. Higher pH in seawater around the cathode helped juveniles to produce shell (calcification).

Electricity with accretion can increase absolute elongation of *P. maxima* juveniles aged 10 months. Accretion that occurred on the cathode in the electricity with accretion treatment may act as an electrical insulator and reduce the current. No measurements were made of electrical currents or pH change on mantle performance in producing shell, but 1.1–1.3 V electricity did precipitate $CaCO_3$. Minerals precipitated on accretion included aragonite ($CaCO_3$) and brucite ($Mg(OH)_2$), where the proportion depends on the voltage and accretion age. Higher voltage and younger accretion age give more brucite than aragonite (Hilbertz 1992). Gonad development in 10-month-old juveniles may influence energy needs (Pouvreau 2000), so that mineral accretion formed during one month was more appropriate for juvenile *P. maxima* growth, and elevated pH in seawater near the cathode should increase juvenile shell calcification. Lower voltage causes lower electron currents, affecting electron transport chains that produce ATP. Energy produced from ATP is prioritized to gonad development and production of biological matrix macromolecule used for shell formation. Higher temperatures and higher salinity also favor faster accretion (Bachtiar 2003).

CONCLUSIONS

The low-voltage method is very useful for early growth of *P. maxima*, increasing survival and growth. The highest growth was achieved by 6-month-old *P. maxima* juveniles reared with electricity without accretion and 10-month-old *P. maxima* juveniles reared with electricity with accretion method.

ACKNOWLEDGMENTS

Gratitude to Cendana Indo Pearls Company (Atlas South Sea Pearls Ltd.) along with the staffs and employers for the grants and provision of research tools and equipment, also to Tom Goreau from The Global Coral Reef Alliance for inspiration and discussion.

REFERENCES

Addadi, L., and S. Weiner. 1997. A pavement of pearl. *Nature* 389:912–915.

Alagarswami, K., S. Dharmaraj, A. Chellam, and T. S. Velayudhan. 1989. Larval and juvenile rearing of black-lip pearl oyster, *Pinctada margaritifera (Linnaeus)*. *Aquaculture* 76:43–56.

Anonymous. 1991. Pearl Oyster Farming and Pearl Culture, FAO Corporate Document Repository. http://www.fao.org/docrep/field/003/AB726E/AB726E00.HTM. (accessed September 14, 2007).

Bachtiar, R. 2003. Pengamatan Pembentukan Terumbu Buatan dengan Metode Mineral Accretion di Desa Pemuteran, Bali Barat (Observation of Artificial Reef Formation with Mineral Accretion method). Skripsi Sarjana, Jurusan Ilmu dan Teknologi Kelautan FPIK, Institut Pertanian Bogor.

Bubel, A. 1984. Mollusca: Epidermal Cells. In: *Biology of the Integument. 1: Invertebrates*. 400–447 in Bereiter-Hahn, J., A.G. Matolsty, and K.S. Richards (Eds.). Springer-Verlag.

Chellam, A. 1978. Growth of pearl oyster *Pinctada fucata* in the pearl culture farm at Veppalodai. *Indian Journal of Fisheries* 25:77–83.

Dwija, P. N. 2003. Pengamatan Pertumbuhan Karang dan Keragaman Coral Recruitment di Pantai Pemuteran, Kecamatan Gerokgak, Kabupaten Buleleng (Growth and Diversity Observations of Coral Recruitment in Pemuteran Beach, Gerokgak Sub District, Buleleng Regency). Skripsi Sarjana. Fakultas Biologi, Universitas Udayana.

Effendie, H. M. 1997. Biologi Perikanan (Fisheries Biology). Yayasan Pustaka Nusantara.

Gervis, M. H., and N. A. Sims. 1992. The biology and culture of pearl oysters (*Bivalvia: Pteriidae*). ICLARM Stud. Rev. 21. ODA.

Global Coral Reef Alliance. 2007. Summary of Advantages, Disadvantages, and Safety of Biorock® Coral Reef and Fisheries Restoration Technology. http://www.globalcoral.org. (accessed September 14, 2007).

Hilbertz, W. H. 1992. Solar-generated building material from seawater as a sink for carbon. *Ambio* 21:26–129.

Lowenstam, H. A., and S. Weiner.1989. *On Biomineralization*. Oxford, Oxford University Press.

Mariani, S., F. Piccari, and E. De Matthaeis. 2002. Shell morphology in *Cerastoderma* spp (Bivalvia: Cardiidae) and its significance for adaptation to tidal and non tidal coastal habitats. *Journal of the Marine Biological Association of the UK* 82:483–490.

Moyle, P. B., and J. J. Cech. 2000. *Fishes: An Introduction to Ichthyology*. Prentice-Hall, Upper Saddle River, NJ.

Nasr, D. H. 1982. Observations on the Mortality of the Pearl Oyster, *Pinctada margaritifera* in Dongonab Bay, Red Sea. *Aquaculture* 28:271–281.

Pass, D. A. 1987. Investigations into the Causes of Mortality of the Pearl Oyster, *Pinctada maxima* (Jameson), in Western Australia. *Western Australia Department of Fisheries Wildlife Report* 7:1–78.

Pouvreau, S. 2000. Ecophysiological model of growth and reproduction of the black pearl oyster, *Pinctada margaritifera*: potential applications for pearl farming in French Polynesia. *Aquaculture* 186:117–144.

Simkiss, K., and K. M. Wilbur. 1989. *Biomineralization: Cell Biology and Mineral Deposition*. Academic Press, San Diego.

Strack, E. 2006. *Pearls*. Ruhle – Diebener – Verlag, Stuttgart.

Southgate, P. C. 2007. Overview of the Cultured Marine Pearl Industry. In: *Pearl Oyster Health Management: A Manual*. 7–17 in Bondad-Reantaso, M.G., S. E. McGladdery, and F. C. J. Berthe (Eds.). FAO Fisheries Technical Paper No. 503.

Suzuki, M., E. Murayama, H. Inoue, and H. Nagasawa. 2004. Characterization of Prismalin-14, a novel matrix protein from the prismatic layer of the Japanese pearl oyster (*Pinctada fucata*). *The Biochemical Journal* 382:205–213.

Taylor, J. J. U. 1999. Juvenile production and culture of the silver-lip pearl oyster, *Pinctada maxima* (Jameson). PhD diss., James Cook University.

Weiner, S. 1984. Organization of organic matrix components in mineralized tissues. *American Zoologist* 24:945–951.

Weiner, S., W. Traub, and S. B. Parker. 1984. Macromolecules in mollusk shells and their functions in biomineralization. *Philosophical Transactions of the Royal Society of London Series B, Biological Sciences* 304:425–434.

CHAPTER 12

Increased Oyster Growth and Survival Using Biorock Technology

Nikola Berger, Mara G. Haseltine, Joel T. Boehm, and Thomas J. Goreau

CONTENTS

Introduction and History of the Eastern Oyster in NY/NJ Harbor 141
The Study: Background 143
Materials and Methods 143
Results and Discussion 145
    Growth Rates 145
    Growth Comparison 146
    Mortality 147
    Accretion 148
    Chemistry Involved in Accretion Process to Biorock Structure 148
Conclusions 149
References 149

INTRODUCTION AND HISTORY OF THE EASTERN OYSTER IN NY/NJ HARBOR

The eastern oyster (also called the American oyster), *Crassostrea virginica*, can be found from the east coast of Canada to Argentina (Coen, Luckenbach, and Breitburg 1999) and, since the early twentieth century, also on the west coast of the United States (Kurlansky 2006). In the Hudson-Raritan Estuary (HRE), archeological findings suggest that the eastern oyster has made this region its home since the end of the last ice age (Steimle 2005), and shell middens date back 6500 years (Kurlansky 2006). Before the twentieth century, the oyster reefs covered approximately 350 square miles of the HRE, from Sandy Hook, NJ, north as far as Ossining, NY, and especially in the Raritan Bay; the Navesink and Shrewsbury Rivers; the Arthur Kill; Jamaica Bay; and Newark Bay, and served as an economic foundation for much of the region (Ingersoll 1887). The natural beds covered the shoreline and all major islands and shoals (Steimle 2005). But by 1812 many of those beds were depleted, and a commercial oyster industry took over, replenishing large oyster beds with imported seed oysters from Chesapeake Bay and the Long Island Sound.

Oyster reefs provided a large and exceptional habitat for a diverse group of marine life forms. Oysters were a keystone species in the larger Hudson Estuary area, playing a central role in the web of life and linking the benthic and pelagic food webs. The oyster not only sustained a whole industry for centuries but also bolstered the fishing industry by providing extensive fish habitat and refuge and increased water quality. Oysters do not cleanse the water in the sense that they remove harmful pollution permanently, but they trap impurities and expel them in larger particles that sink to the bottom rather than make the water turbid (Comi and Willner 2006). One adult oyster can filter up to 50 gallons of water per day (Comi and Willner 2006), thereby improving water purity and clarity as they remove suspended sediments and microalgae (Newell 1988). Increased water clarity encourages more biological activity at greater depths and results in higher dissolved oxygen levels (Baird 2008). The whole population of oysters in the HRE in the past centuries was counted in billions, filtering the entire estuary in a few days (Coen, Luckenbach, and Breitburg 1999). In addition, about 75% of all commercial fish and shellfish depend on estuaries at some point in their lives (Coen, Luckenbach, and Breitburg 1999), and over 30 fish species were found to be associated with the oyster reefs of the Mid-Atlantic Bight, including juvenile striped bass, tautog, black sea bass, adult black drum, and even the American eel, whose populations have been in decline for decades (Southworth and Mann 1998).

Aiding in oyster reef formation could have large-scale regional benefits. Commercial and recreational fisheries could benefit from the reintroduction of the keystone species of the past, and increased habitat as a result of reef restoration can exhibit a 10-fold increase in species abundance (Coen, Luckenbach, and Breitburg 1999).

THE STUDY: BACKGROUND

Different projects have been initiated to create oyster reefs, with mixed results. The Oyster Growth Study described in this chapter is examining one possible way to help jump-start a reef by growing larger oysters in a shorter amount of time. The method we examined, known as Biorock technology, has been shown to increase coral growth in numerous coral-reef restoration projects. The technology works by using electrically mediated calcium carbonate deposition on submerged metal structures (Figure 12.1). The accreted minerals may be more bioavailable to the oyster on the metal structure and, therefore, possibly promote oyster shell growth. The Biorock technology may also counteract threats to reefs due to increased ocean acidification (Baird 2008). This and other threats are probably due to increased CO_2 partial pressure in the atmosphere. Reefs are also a low cost and self-sustaining armor for vulnerable coasts, which is especially important with an assumed increase in sea-level rise and hurricanes due to global climate change.

One goal of this study was to determine whether this method can be used to aid in oyster reef restoration. Our project also served as an educational tool in collaboration with The River Project (TRP); a nonprofit environmental organization on Pier 40 in Manhattan that houses a wet lab with a natural flow-through estuarine aquarium system. TRP also works closely with students and volunteers from throughout New York City and several high school and undergraduate volunteers aided in data collection and maintenance of this project throughout its duration. Our experimental setup

Figure 12.1 Oyster growth study using Biorock technology.

was similar to one previously conducted by Kaitlin Baird—then a master's student at Columbia University—in the intertidal zone of the East River at College Point, New York. At TRP, we were interested in repeating the study in a more controlled environment, since Baird's results were partially impacted by the large amount of oysters lost due to crab predation and other disturbances such as tidal and other environmental factors, which made data collection a challenge.

MATERIALS AND METHODS

Oysters are called natural bioengineers since they create three-dimensional structures (reefs) as they grow on the backs of older oysters (Dalton, Stringer, and Lock 2003). Oysters cannot naturally sustain themselves without having old shell (cultch) or another hard substrate to attach to. In our study, the source of minerals was artificially produced using the Biorock technology. Oysters, which can grow intertidally, rid themselves of many predators by being able to survive out of the water for long periods of time. This characteristic was a great advantage to us in our study, since we could lift the metal reef structures out of the tanks to attach oysters and subsequently measure them conveniently in the wet lab (see Figure 12.2).

The Oyster Growth Study was conducted over two consecutive years—in the summer/fall seasons of 2007 and 2008—at the facilities of TRP.

We tested if oyster reef formation can be supported through the use of Biorock technology, which possibly promotes faster growth, creating larger and stronger shell as well as reduced mortality in oysters. It was found that shell size is positively correlated with reduced mortality due to the ability to avoid predation (Galtsoff 1964; Arnold et al. 1996). We were able to reduce predation by conducting the study in large tanks using a flow-through system of Hudson river water with screens that stop any organisms larger than a quarter of an inch from getting into the tanks.

The experiment examined the use of Biorock to determine

1. If providing low-voltage current accelerates oyster growth rates
2. If total mortality is reduced in the experimental vs. the control tanks
3. If Biorock technology helps to aid in calcium accretion in an estuarine environment

The mechanism of the Biorock technology, also called Mineral Accretion method, Seament, or Seacrete, is based on creating a difference in pH across the metal reef structure. The cathode is connected to the reef structure, reducing it and thereby attracting positively charged ions like calcium and magnesium. The resulting deposition of $CaCO_3$ through this application of an electric potential may make these important minerals more bioavailable to the oysters due to close proximity (Hilbertz and Goreau 1996; Kurlansky 2006).

Figure 12.2 Adhering oysters to an artificial reef.

The study was conducted in a flow-through system with two identical round fiberglass tanks of approximately 300 gallons each. Tanks 1 and 2 each contained three replicate metal "reefs," housing an equal number of oysters (200 per reef = 600 per tank). Tank 1 used the Biorock technology, and a low-voltage (6–9 V) current was added to the reefs to promote mineral accretion (Experiment), and tank 2 acted as replicate control with no electric current. A random subset of 30 oysters per reef were chosen using a random number generator and marked with nail polish for the duration of the study (see Figure 12.3). The marked 90 oysters per tank were measured twice per month for growth (their "height" as defined by Cardoso et al. 2007) for approximately three months in two consecutive years. Mineral accretion and water chemistry (salinity, temperature, and pH) were also recorded to ensure they were in a normal range, equal to the estuarine water chemistry, and identical in both tank environments.

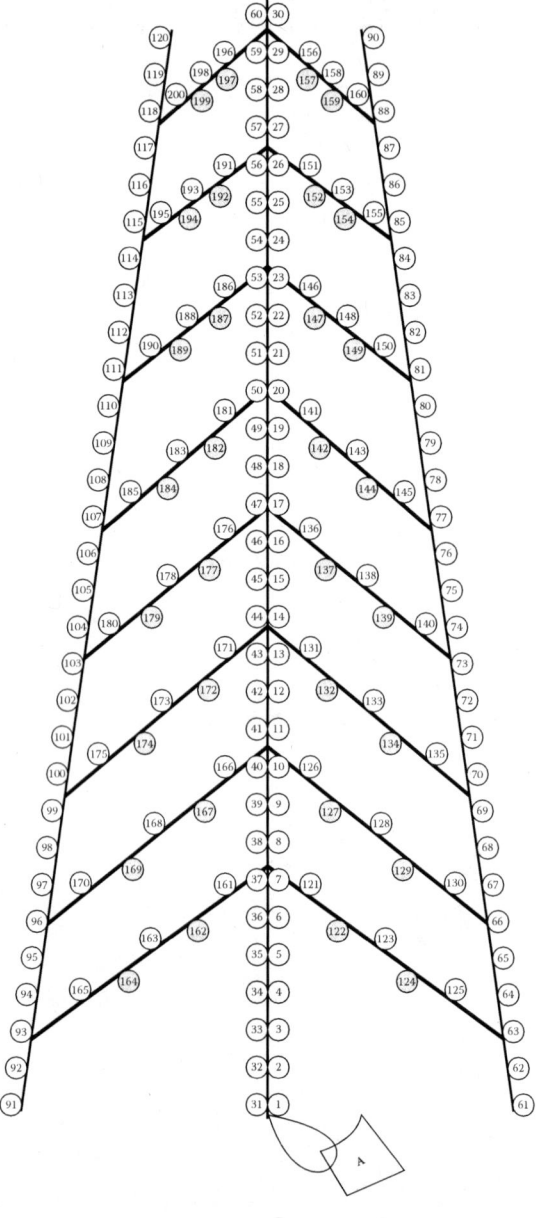

Figure 12.3 Oyster reef layout and oyster numbering system.

RESULTS AND DISCUSSION

The results of the Oyster Growth Study showed that oysters in the experimental tank grew significantly larger and faster than the oysters in the control tanks that were not connected to an electric current. In addition, we found statistically significant differences between the total mortality, with fewer deaths in the experimental than control tanks. The results were consistent for both years we conducted the study.

Growth Rates

Figure 12.4a and b show the average growth rates for all the reefs. The top three lines (circle, square, diamond) follow the growth of the experimental tank oysters, and the bottom three lines (triangles) represent the reefs in the control tank. For both years, we observed that all reefs in the experimental tank grew at a faster rate than the reefs in the control tank and were therefore larger at the end of the study.

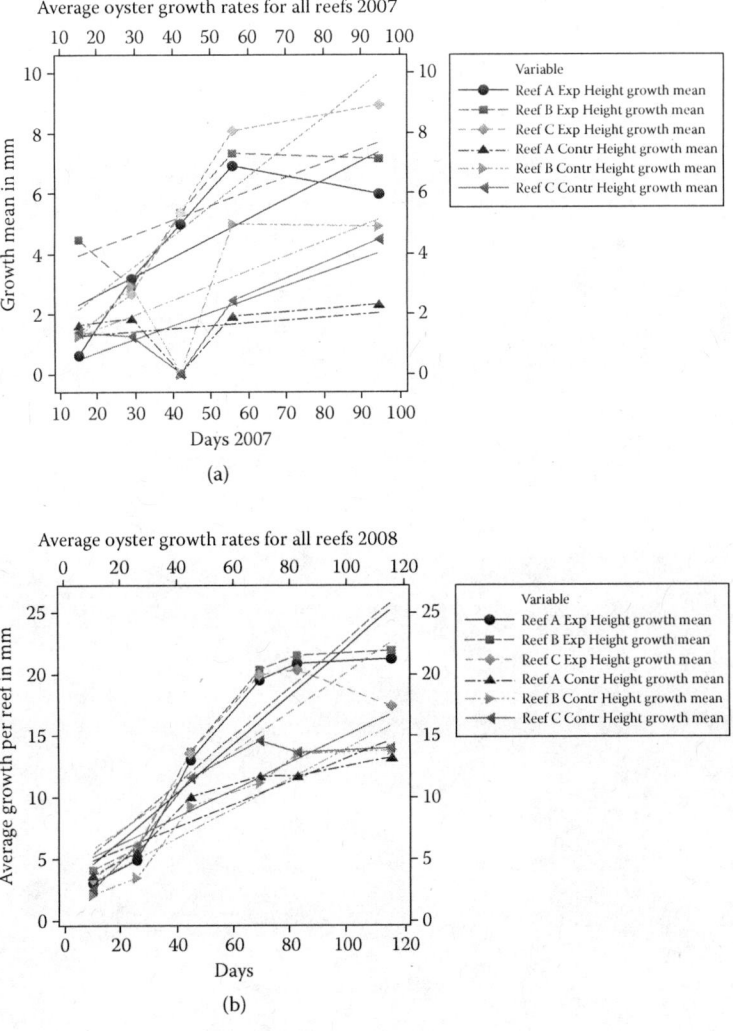

Figure 12.4 Average oyster growth rates in (a) 2007 and (b) 2008.

Growth Comparison

The histograms (Figure 12.5a and b) show how much the oysters grew in each tank from August 15 until November 15, 2007 (Figure 12.5a), and from June 20 until September 15, 2008 (Figure 12.5b). On the left is the control tank (C) showing oyster's final height minus the initial height, and on the right is the experimental (E) tank showing oyster's final height minus initial height. In 2007, the control group grew an average of 2.36 mm in three months and the experimental group grew 6.49 mm (Figure 12.5a). In 2008, the control group grew an average of 13.13 mm in three months, and the experimental group grew 21.28 mm (Figure 12.5b). To statistically validate these increases in growth, we conducted a two-tailed t-test to compare the means of the experimental and control groups in both years to test the null and alternative hypotheses:

H_o: $mu_1 = mu_2$; the sample means are not significantly different
H_a: $mu_1 \neq mu_2$; the samples means are significantly different

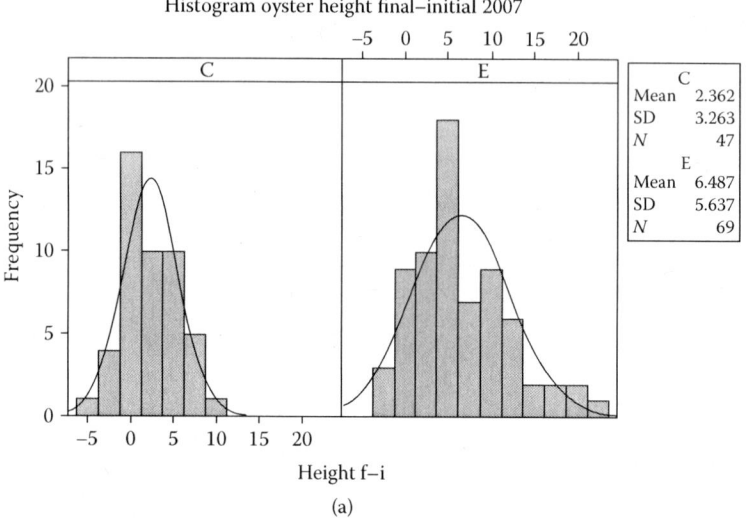

(a)

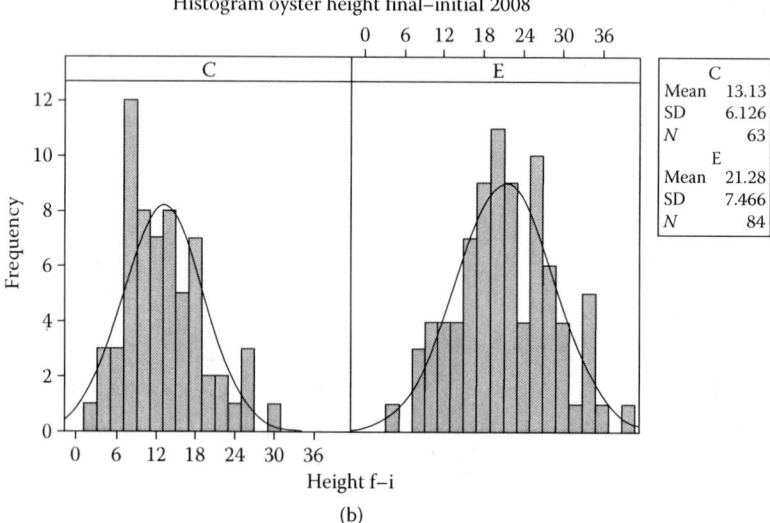

(b)

Figure 12.5 Final oyster heights in (a) 2007 and (b) 2008 in millimeters.

The calculated results for both years can been seen in Table 12.1.

Mortality

We also compared total mortality in both trials. Figure 12.6 depicts a comparison of total mortality for 2007 and 2008. For both years, the experimental tanks showed a lower mortality than the control tanks. We conducted a t-test to determine if the differences were significant between the control and the experimental reefs to test the null and alternative hypotheses:

H_o: $mu_1 = mu_2$; the mean mortality between samples are not significantly different
H_a: $mu_1 \neq mu_2$; the mean mortality between samples are significantly different

The calculated results for both years can be seen in Table 12.2.
For both total mortality and mean growth rates, we found significant differences across trial years.

Table 12.1 Descriptive Statistics of the Growth Data in 2007 and 2008

	N	Mean	SD	Variance	CoefVar	Skewness	Min	Max	Range
2007[a]									
Control	47	2.36	3.26	10.65	138.18	0.37	−4.00	10.5	14.5
Experiment	69	6.49	5.64	31.77	86.89	0.74	−2.70	21.5	24.2
2008[b]									
Control	63	13.129	6.126	37.532	46.66	0.71	2.08	30.75	28.67
Experiment	84	21.278	7.466	55.737	35.09	0.16	4.88	40.69	35.81

[a] Results—T-value: 4.98 at 95% confidence; $p = .000$; $H_0 \neq H_A$ ($p < .05$: the null hypothesis is rejected with >95% confidence).
[b] Results—T-value: 7.26 at 95% confidence; $p = .000$; $H_0 \neq H_A$ ($p < .05$: the null hypothesis is rejected with >95% confidence).

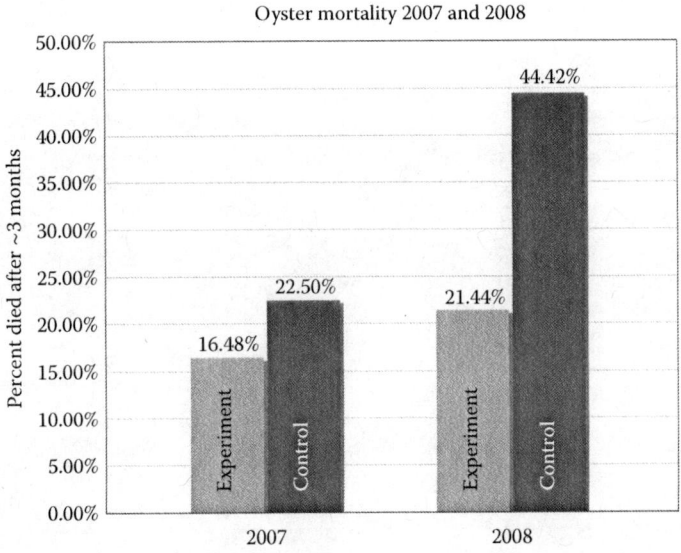

Figure 12.6 Oyster mortality rates in 2007 and 2008.

Table 12.2 Descriptive Statistics of Mortality Rates in 2007 and 2008

	Experiment 2007	Control 2007	Experiment 2008	Control 2008
N (Total—missing)	546	520	555	565
Died	90	117	119	251

T-test for Control 2007 vs. Experiment 2007
Results—T-value: 2.49 at 95% confidence; $p = .01$; $H_0 \neq H_A$ ($p < .05$: the null hypothesis is rejected with >95% confidence).
T-test for Control 2008 vs. Experiment 2008
Results—T-value: 8.42 at 95% confidence; $p = .000$; $H_0 \neq H_A$ ($p < .05$: the null hypothesis is rejected with >95% confidence).

Figure 12.7 Close up of oysters glued to steel "reef" structure. The white patches on the steel are the initial growth of the minerals caused by the Biorock process.

Accretion

We also examined the amount of mineral accretion on the experimental reefs. The diameter of the metal bars of the reef structure used in 2007 and reused in 2008 was about 0.8 cm. Over the course of the study, the metal structure rusted at places but remained about the same size except for the experimental tank's structures. All three reefs in the experimental tank, which were connected to about 9 V of an electric current for the entire study time, showed significant accretion of minerals. Reef A was the first in line and received more of a current than reef B, and reef C received a little less than reef B. The difference was visible in terms of the amounts of accretion of minerals (aragonite and brucite), and reef A seemed to be covered with the most minerals (in terms of overall mineral coating and thickness), reef B with a little less, and reef C with the least. The accretion was thickest on all reefs close to where they were connected to the wire running the current and showed patches of accretion throughout the structures (see Figure 12.7). The most prominent accretion on reef A, close to the wire connection, measured 1.8 cm, which is more than twice the original diameter.

Chemistry Involved in Accretion Process to Biorock Structure

In marine environments, the pH is determined by reactions among dissolved carbon dioxide (CO_2), carbonate ions (CO_3^{2-}), and bicarbonate ions (HCO_3^-):

1. $CO_2 \text{ (g)} + H_2O \text{ (l)} \rightarrow H_2CO_3 \text{ (aq)}$
2. $H_2CO_3 \text{ (aq)} \rightarrow H^+ \text{ (aq)} + HCO_3^- \text{ (aq)}$
3. $HCO_3^- \text{ (aq)} \rightarrow H^+ \text{ (aq)} + CO_3^{2-} \text{ (aq)}$

The increase in pH at the reef structure is caused by the establishment of an electric potential, which promotes the deposition of $CaCO_3$ (Hilbertz 1992):

4. $CO_2 (g) + OH^- (aq) \rightarrow HCO_3^- (aq)$
5. $OH^- (aq) + HCO_3^- (aq) \rightarrow H_2O (l) + CO_3^{2-} (aq)$
6. $CO_3^{2-} (aq) + Ca^{2+} (aq) \rightarrow CaCO_3 (s)$

We found a clear increase in mineral accretion across experimental reefs in both years.

CONCLUSIONS

The initial oyster samples were close in their average size and distributed normally, a factor important as to have a meaningful comparison of the two sample populations at the end of the study. Over the course of approximately three months, a subset of 90 oysters in each of the control and the experimental tank, both containing 600 oysters, were measured from the hinge to the farthest point (height) approximately every two weeks. Oysters in the experimental tank grew statistically significantly faster, 2.75 and 1.62 times faster than oysters in the control tank during the course of both studies (2007 and 2008). Mortality was significantly higher for control oysters than Biorock groups in both years, by 2.08 and 1.36 times (2007 and 2008). Accretion of minerals onto the reef structure was evident and most prominent closest to the attachment point of the wire that ran the electric current. Overall, by increasing growth rate and survival of oysters, Biorock technology may be a viable resource to restore oyster reefs in estuaries where they were once abundant.

REFERENCES

Arnold, W. S., T. M. Bert, D. C. Marelli, H. Cruz-Lopes, and P. A. Gill. 1996. Genotype specific growth of hard clams (genus *Mercenaria*) in a hybrid zone: Variation among habitats. *Marine Biology* 125:129–139.

Baird, K. 2008. Growth and condition of the American oyster (*Crassostrea virginica*) on constructed intertidal reefs: A pilot study using Biorock® accretion technology. Masters Thesis. Columbia University, New York, NY.

Cardoso, J. F. M. F., D. Langlet, J. F. Loff, A. R. Martins, J. I. J. Witte, P. T. Santos, and H. W. van der Veer. 2007. Spatial variability in growth and reproduction of the pacific oyster *Crossostrea gigas* (Thunberg 1973) along the west European coast. *Journal of Sea Research* 57:303–315.

Coen, L. D., M. W. Luckenbach, and D. L. Breitburg. 1999. The role of oyster reefs as essential fish habitat: A review of current knowledge and some new perspectives. In L. R. Benaka (Ed.), *Fish Habitat: Essential Fish Habitat and Rehabilitation*, 438–454. American Fisheries Society, Symposium 22, Bethesda, MD.

Comi M., and A. Willner. 2006. Interview by author. NY/NJ Baykeeper.

Dalton, D., M. Stringer, and M. Lock. 2003. NY/NJ Baykeeper Oyster Restoration Program Report. Baykeeper.

Galtsoff, P. S. 1964. The American Oyster *Crassostrea virginica*. *U.S. Fish and Wildlife Service Fishery Bulletin* 64: 1–480.

Hilbertz, W. 1992. Solar-generated building material from seawater as a sink for carbon. *Ambio* 21:126–129.

Hilbertz, W., and T. J. Goreau. 1996. A method for enhancing the growth of aquatic organisms and structures created thereby. US Patent 08/374993.

Ingersoll, E. 1887. The oyster, scallop, calm, mussel, and abalone industries. In: G. B. Goode (Ed.), *The Fisheries and Fishery Industry of the United States*, Vol. 2, 507–622. U.S. Government Printing Office, Washington, DC.

Kurlansky, M. 2006. *The Big Oyster: History on a Half Shell*. Ballantine Books, New York.

Newell, R. I. E. 1988. Ecological changes in Chesapeake Bay: Are they the result of overharvesting the American oyster, *Crassostrea virginica*? In: M. P. Lynch and E. C. Krome (Eds.), *Understanding the Estuary: Advances in Chesapeake Bay Research*, 536–546. Chesapeake Research Consortium, Publication 129 CBP/TRS 24/88, Gloucester Point, VA.

Southworth, M., and R. Mann. 1998. Oyster reef broodstock enhancement in the Great Wicomico River, Virginia. *The Journal of Shellfish Research* 17:1101–1114.

Steimle, F. W. Jr. 2005. Natural oyster reef habitat in the Hudson-Raritan Estuary [Abstract, oral]. Oyster Restoration in the Hudson-Raritan Estuary Conference, West Long Branch, NJ, February 9–10, 2004.

CHAPTER 13

Electrical Stimulation Increases Oyster Growth and Survival in Restoration Projects

Jason Shorr, James Cervino, Carmen Lin, Rand Weeks, and Thomas J. Goreau

CONTENTS

Introduction: Oyster Biology 151
Causes and Consequences of Oyster Decline 152
Oyster Restoration Strategies 153
Biorock Oyster Restoration 153
Site of Project 154
Materials and Methods 154
Results 155
    2010 Growth Rate with and without Electrical Stimulation 155
    2011 Growth Rate with and without Electrical Stimulation 156
    Winter 2010–2011 Survival Data 156
Discussion 158
Conclusions 158
References 159

INTRODUCTION: OYSTER BIOLOGY

Oyster reefs have suffered massive declines worldwide. New York City once had more than 900 km^2 of oyster reefs, but now has essentially none left because of overharvesting. Because oyster filtration is the most effective way to remove sediments, organic matter, and bacteria from coastal waters, New York City's Green Infrastructure Program is now encouraging oyster reef restoration to improve water quality and avoid the need for vastly more expensive water and sewage treatment plants.

The American oyster (*Crassostrea virginica*) is considered a keystone species, because it builds reefs, provides shelter for other estuarine organisms, improves water quality by filtering particulate matter, and protects shorelines by reducing bank erosion. A keystone species is defined as the one that has a disproportionate impact (relative to its numbers or biomass) on the organization of a biological community. Loss of a keystone species may have far-reaching consequences for the community (Primack 2008).

Oysters are not able to move, and their distribution depends on where the larvae set. The oyster larvae find a solid place to set themselves two to three weeks after spawning. The oyster attaches itself with a drop of liquid cement from its foot and is now called spat (newly attached oysters). Adult oysters filter large amounts of brackish water (water that has a salinity between fresh water and salt water) and remove flagellates and phytoplankton (Stanley and Sellers 1986). Large populations of oysters and other suspension-feeding bivalves filter plankton out of the water so efficiently that they control blooms of phytoplankton and prevent symptoms of eutrophication (Jackson 2001). Oysters most effectively filter particles in the 3–4 μm size range (Stanley and Sellers 1986).

CAUSES AND CONSEQUENCES OF OYSTER DECLINE

There are many reasons for the decline of American oyster reefs (Eastern Biological Review Team 2007). Many scientists attribute the decline of oysters to overharvesting (Newell 1988). Oysters are harvested in a few ways: handpicking of clumps from reefs at low tide, harvesting by hand and by use of tongs from boats, and dragging and dredging from boats (Stanley and Sellers 1986). These harvesting methods degraded the oyster habitat and depleted the oyster beds to a level that inhibited their net reproduction. Chesapeake Bay was once rich with oyster reefs. The oyster reefs before 1870 would have filtered the entire water column every three days. Even though aboriginal and early colonial people harvested these oysters over many millennia, the real problems did not start until the 1870s. This is when mechanical harvesting with dredges was invented, and deep channel reefs were affected (Kirby 2004; Mackenzie 2007). Now, the oysters filter the water column of Chesapeake Bay every 325 days (Newell 1988).

Overfishing could be one of the causes of eutrophication and outbreaks of disease. Eutrophication of Chesapeake Bay got worse in the 1930s. This was about two centuries after land clearing for agriculture, which greatly increased runoff of sediment and nutrients into the estuary. The large quantity of oysters had been able to remove the increase in phytoplankton, but mechanical harvesting progressively destroyed most of the oyster beds from the 1870s to the 1920s (Rothschild et al. 1994; Kirby 2004; Mackenzie 2007). The oysters would have been major consumers of the spring phytoplankton bloom (Newell 1988). The result of eliminating the water filtration by oysters has rippled throughout the ecosystem. When oysters were abundant, the water was clear, light reached the bottom, and algae grew there, providing the basis of the food chain that led to abundant blue crabs, which were also a major traditional fishery in Chesapeake Bay. With the loss of filtration and increased eutrophication from sewage from humans, cattle, pigs, and chickens, as well as agricultural fertilizer runoff, the waters are now turbid and so little light reaches the bottom that the benthic algae have disappeared, causing near total collapse of the crab fishery.

Ocean acidification as a result of CO_2 from fossil fuels is another problem facing all calcareous marine organisms (Gazeau et al. 2007). Carbon dioxide lowers ocean pH and will continue to lower it as humans continue to emit large amounts of CO_2 into the atmosphere. The pH of the oceans has already declined by 0.1 units compared to preindustrial values and is expected to decrease by another 0.4 units by the end of the century. The Intergovernmental Panel on Climate Change predicts atmospheric partial pressure of CO_2 to range from 490 to 1250 ppmv by 2100. Ocean acidification leads to a shift in inorganic carbon balance toward higher CO_2 and lower carbonate ion (CO_3^{2-}) concentrations. This carbonate ion is one of the main building blocks in calcium carbonate ($CaCO_3$).

$$CO_3^{2-} + Ca^{2+} \rightarrow CaCO_3$$

Any change in carbonate ion concentration can hinder calcareous organisms from precipitating calcium carbonate (Gazeau et al. 2007). The calcification rates of oysters and mussels have been shown to decrease with decreasing pH (Gazeau et al. 2007). The larval phases of oysters are especially vulnerable to acidification due to their small size and high surface area. As a result of increasing acidity of seawater in the northeast Pacific, there has been a near total failure of oyster spat settlement in Oregon, Washington, and British Columbia in recent years (Barton et al. 2012).

OYSTER RESTORATION STRATEGIES

Rothschild et al. (1994) proposed a very comprehensive recovery plan for the Eastern American oyster in the Chesapeake Bay in Maryland. They proposed a four-point strategy involving fishery management, repletion, habitat replacement, and broodstock sanctuaries. The first fishery management should be regulated by a scientific size-specified fishery mortality, recognizing that oyster larvae or spat sometimes settle on larger oysters that should not be fished out. The repletion strategy, placing oyster shells on already existing substrate, should be modified to areas that are known to increase oyster growth and survival. The habitat replacement strategy should be implemented to create additional oyster habitat. This new substrate will allow spat to attach and grow new oyster reefs. One of the main problems is that the oysters have no habitat on which to attach. The broodstock sanctuaries would implement a no-fishing zone where engineered reefs were created and where known areas of high spat settlements are known. This broodstock sanctuary idea will help reinforce the first three strategies (Rothschild et al. 1994). If this plan were used throughout all estuaries, then perhaps there would be an increase in all bivalve populations, and cleaner waters would result.

There is a widespread recognition of the need for oyster restoration on a large scale to restore ecosystems and water quality (Beck et al. 2011; Leonard and Macfarlane 2011). Large sums of federal and state funding have been spent to restore the lost oyster reefs of Northeastern US estuaries, not only to revive the oyster industry but also to improve water quality in places where oysters should not be eaten because of toxic contamination. Much of this has been based on building artificial reefs of old oyster shells, fossil shells, concrete, and other exotic materials and attaching oyster spat to them or hoping oysters will naturally settle on them. By and large these have been failures due to poor water quality (Schulte et al. 2009).

It is clear that to restore oyster populations and water quality, a quantum leap in technology is needed, using methods that greatly increase oyster settlement, growth, survival, and resistance to environmental stress. This chapter describes the first field tests of such new approaches that do so.

BIOROCK OYSTER RESTORATION

Restoring oyster beds or reefs is important to protect shorelines from erosion, restore water quality, and supply the human demand for harvest. The first stage of restoring oyster reefs is to supply a hard layer where the oyster larvae can attach (SCORE). A unique strategy to do so is to grow limestone minerals from seawater, the natural material that makes up oyster shell and the preferred substrate for larvae to settle and become spat. The Biorock process does this precisely by utilizing low-voltage electrical currents.

The oyster must pump protons and calcium ions across its membrane to generate a high pH that enables the secretion of shell material. Biorock technology uses a low electrical current that results in electrolysis and thus the precipitation of calcium carbonate and magnesium hydroxide directly to the Biorock structure. Oyster species are thought to grow at accelerated rates because they are utilizing the availability of dissolved calcium and carbonate ions, to increase efficiency of metabolic processes (Goreau and Hilbertz 2005). Using energy supplied by photovoltaic panels, oysters should be able to increase their growth rates and have higher survival.

Electrodeposition of minerals in seawater is known as "mineral accretion" or "Biorock" technology. This technology can be used as building materials for a wide variety of purposes, including artificial reefs (Hilbertz 1979). Using this technology, calcium carbonate, magnesium hydroxide, and hydrogen are created at the cathode, and oxygen and chlorine are created at the anode. At the cathode, water is broken down to create hydrogen gas and hydroxyl ions so the cathode is reducing and alkaline (Hilbertz 1992). At the anode, water is broken down to create oxygen and hydrogen ions, making it oxidizing and acidic (Hilbertz 1992). Calcium carbonate is created and deposited at the cathode when an electric current circuit is created.

The metal structure can be bent or welded to any size or shape, which can create different levels of structure where oysters can attach. This will also increase their filtering efficiency and have positive effects on the water quality (Goreau and Hilbertz 2005). The resulting accretion has similar chemical and physical properties to reef limestone, which will give marine organisms an artificial reef to grow and live on, in, and around. The objective of this project is to increase the size and habitat of the American oyster in the East River at the College Point site by testing cutting-edge Biorock technology for field restoration.

SITE OF PROJECT

The East River was historically one of the best areas for harvest of American oysters. In the 1800s, this part of the East River between Queens and the Bronx was a prime location for harvesting oysters. The East river connects Upper New York Bay in the south with Long Island Sound in the north. The source of freshwater is from the Hudson River and brought to the East River via the Harlem River. The study location was at McNeil Park, Block 3914-Lot No.1.

The site is located within the intertidal zone and consists of mixed mud, sand, and industrial rubble. *C. virginica* are scarce at this site, presumably because of poor environmental conditions. The site lies next to a former Superfund toxic waste site that was formerly a Navy shipyard used for building ships during the Second World War. After the war, the shipyard was closed, and all the electroplating wastes were left on site in old, rusting, steel drums. The site was then used as an illegal waste dump for about 40 years. The remains of the demolished buildings from the 1958 New York World's Fair, and other industrial debris were dumped here during this period.

While salt marsh, mussels, oysters, crabs, and horseshoe crabs have been observed to recover farther away from the former waste site, our projects were placed right next to the edge of the waste site, in an area affected by leaching drainage from the former dump, where no natural ecosystem recovery had occurred. It is thought that this lack of recovery may be due to the effects of the large number of trace metals, polychlorobiphenyls, polycyclic aromatic hydrocarbons, and other toxic materials leaching from the site. In other words, this was a worst-case restoration situation.

The experiments were initiated in September 2009 and started to precipitate calcium carbonate immediately. Measurements and photographs were taken periodically to see how fast they grew. In some places, more than 3 mm of mineral accretion had grown on top of the steel. The first stage of the experiment was to make sure the electric current was working and to see how much mineral accretion would grow over a few months.

MATERIALS AND METHODS

About 600 oysters were used in the spring for growth experiments in the 2010 growing season, another batch in the 2011 growing season, and a third larger-sized batch for survival measurements overwinter 2010–2011. The oysters were donated by Frank M. Flowers and Sons, a major oyster hatchery in Oyster Bay, Long Island, with some of the cleanest water in Long Island Sound.

Solar photovoltaic panels were used to grow artificial reefs where oysters and other benthic organisms could attach and grow. Two of the three solar panels were Sun Electronics model number HS-A-205-fa2, which supplied 205 W, 18.4 V, and 11.15 A at peak power. The third panel was Kyocera model number KC130TM, which supplied 130 W, 17.6 V, and 7.39 A at peak power. The solar panels were used in a direct power mode, with no storage batteries. Because the structures were located in the intertidal zone, they received power only when the sun was shining and the tide was high, probably no more than about a quarter of the time.

At the College Point site, we had four different steel helix-shaped oyster reef structures. These are vertical structures that reach from near the lower intertidal to above the high-tide mark. Oyster bags were attached to the lower portions. The open helical structure allows the free flow of water through the bags from all directions. Three helices acted as the cathodes of solar panels, where the mineral accretion takes place. One of these helixes had no electric current passing through it and served as our control. Close to each cathode there was an anode encased in PVC pipe for protection. The seawater acts as a conductor between the anode and cathode. We attached bags of oysters onto the metal helices.

We measured oyster size periodically to compare growth rates of oysters on the electrified helices and the control helix with no electric power during the 2010 and 2011 growing seasons. The bags were emptied into a tray, sorting live from dead oysters, and photographed with a scale. The dead oysters were removed and the live ones returned to the bag. Measurements were made from the images using the Photoshop measuring tool.

Another set of measurements were made to assess overwinter survival. Measurements were taken on September 5, 2010, and then again at the end of winter. *Crassotrea virginica* were placed in bags and hung from the metal helices on September 5, 2010. They were checked periodically to make sure that they were fixed to the helices and to clear out any sediment clogging holes in the bags. They were then removed from the structures.

2010 Growth Rate with and without Electrical Stimulation

Oysters were attached in late fall 2009 and measured the next year. The oysters in the control structure getting no power decreased in size, and had died by midsummer 2010. The structures getting low current showed very little growth, but the structure getting the highest power showed strong growth (Figure 13.1). However, some of the electrical cables connecting the panels to the solar panels were broken several times during the year by heavy storm-wave action, and we could not be certain how long they had been without power. Because of these problems, the structures were rewired with stronger cables the following year.

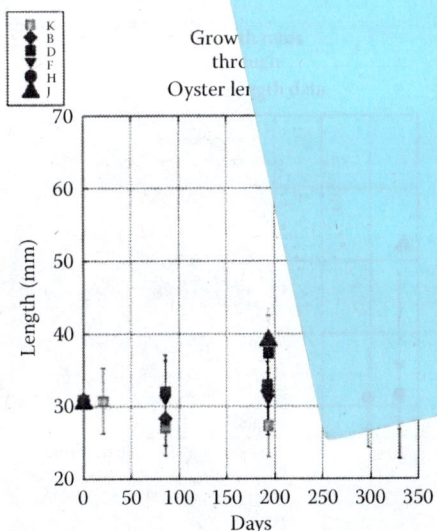

Figure 13.1 Growth of oysters during fall 2010 as a function of time and electrical power. Square symbols represent the controls, circles indicate low current, downward triangles indicate medium current, and upward triangles indicate high current.

2011 Growth Rate with and without Electrical Stimulation

In the 2011 growing season, there were no problems with cable breaks, and power was reliably maintained. The control oysters showed only slight growth of 4.86 mm, the medium-power oysters grew 5.82 times faster, and the higher-power oysters grew 9.30 times faster than the controls (Figure 13.2; Tables 13.1, 13.2, and 13.3).

Winter 2010–2011 Survival Data

Zero day oyster measurements (mm) in random sample of 75/250 oysters. Measurements were made with a digital caliper on September 5, 2010; on April 27, 2011, measurements were made from photographs. Data are presented in Figures 13.3 and 13.4, and Table 13.4.

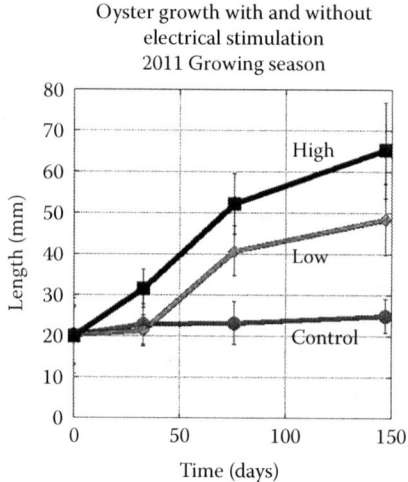

Figure 13.2 Oyster growth in 2011 as a function of time and electrical power.

Table 13.1 Helix 1 (Control, No Current) Average Oyster Length (mm)

	Helix 1	
Date (mm/dd/2011)	Average Length (mm)	SD
7/10	20.13	±6.85
8/12	22.89	±4.92
9/24	23.20	±5.19
12/04	24.99	±4.09

Note: SD, standard deviation.

Table 13.2 Helix 2 (Medium Current) Average Oyster Length (mm)

	Helix 2	
Date (mm/dd/2011)	Average Length (mm)	SD
7/10	20.17	±7.26
8/12	21.42	±3.80
9/24	40.74	±6.14
12/04	48.46	±8.70

Note: SD, standard deviation.

Table 13.3 Helix 3 (High Current) Average Oyster Length (mm)

Helix 3 (24 V and 4 A)		
Date (mm/dd/2011)	Average Length (mm)	SD (±mm)
7/10	20.04	±9.16
8/12	31.55	±4.64
9/24	52.24	±7.35
12/04	65.24	±11.63

Note: SD, standard deviation.

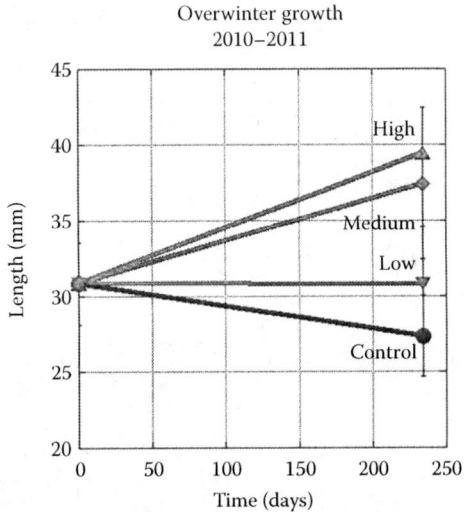

Figure 13.3 Oyster growth over the winter 2010–2011 season. Control oysters (circle) shrank, low power oysters (downward triangle) did not grow, and medium power (diamond) and high power oysters (upward triangle) grew.

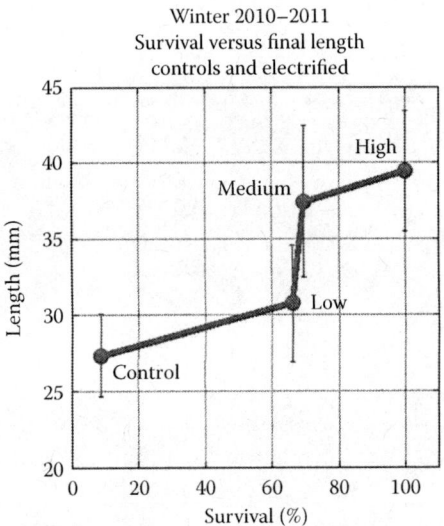

Figure 13.4 Oyster length versus survival over the winter.

Table 13.4 Winter Survival and Growth Data

Start	September 5, 2010 Measurements			
	Mean	SD	Alive (%)	Dead (%)
	30.852	2.712	100.00	0.00
	April 27, 2011 Measurements			
Controls	27.366	2.713	8.54	91.46
Low	30.793	3.815	66.00	34.00
Medium	47.423	5.020	69.23	30.77
High	39.443	3.978	100.00	0.00

Over the winter period, the controls showed more than 91% mortality and a clear decrease in average length. In contrast, the electrified oysters all had high survival, from 66 to 100%. The low-power oysters showed no growth, but those getting higher power had dramatic overwinter growth, even though the winter would normally be the dormant season.

The electrically treated oysters all had high survival, with no mortality at all in the high current treatment. The higher the electrical current, the faster the oyster growth and the greater the survival, indicating that the electrical current greatly increased oyster health and ability to resist winter physical stresses.

DISCUSSION

The results from the overwinter survival study showed that the Biorock treatment had markedly increased growth rates and survival. Young oysters grown in electrical fields under various conditions had very high overwinter survival (mean 79.8%, range 66.0–100.0%) and had shiny white shells. In contrast, control oysters in the same habitat had almost complete mortality (survival 8.5%), and the shells had a red-yellow color and appeared eroded. Spat placed in the fall were 30.9 ± 2.71 mm long. Electric oysters had clear growth over the normally dormant winter season (mean length 37.4 ± 5.0 and 39.4 ± 4.0 mm in two separate experiments). In contrast, control oysters shrank in length over the winter to 27.4 ± 2.71 mm. The shrinkage of the controls was presumably due to acid water, since carbon dioxide is more soluble in cold water. The electricity appears to reverse the effects of acidity and physical stress.

The growth rate experiment also showed positive results for the Biorock experimental plots. There was an increase in size over time with more electricity. Figures 13.1 and 13.2 show that the oysters with the most power grew the largest compared to controls. Survival rate for helix 1 was about 43%. Survival rate for helix 2 was about 67%, and the survival rate for helix 3 was about 75%. Mortality increased as electricity decreased. The cause of oyster death is uncertain but can possibly be a result of water pollution, weather conditions, diseases, or predation.

CONCLUSIONS

Results obtained from the experiments that were performed at the College Point Biorock site showed that the electrical field had positive effects, increasing both growth rate and survival of *C. virginica* oysters. While all oyster groups that received electrical stimulation responded with increased growth and survival, the oysters given the highest amount of electrical stimulation had the fastest growth and highest survival rates. These findings indicate that electrical currents have induced positive growth in *C. virginica* oysters, even under the most severe pollution conditions when no growth would be expected.

Estuary sites that have been polluted by runoff and sewer drainage present ideal locations for future Biorock projects, particularly if these locations once flourished with oyster and other benthic populations. The benefits of these artificial oyster nurseries would be to create a habitat for oyster growth where they otherwise would not be able to survive in natural conditions. Increasing the oyster population in heavily polluted estuaries would create a natural filter and improve water quality in these areas.

"Restoring an abundant population of filter feeding oysters to the Chesapeake Bay and its tributaries might really help lend new meaning to the phrase *clean up the bay*" (Newell 1988). Using this idea and applying it to the Hudson River and New York harbor, we can hopefully make a difference in the water quality surrounding the city. Because our site is a worst-case scenario, in which control oysters steadily shrink and die, if electrical stimulation can increase growth and survival under such severe conditions, then better results would be expected in cleaner waters, and it should be possible to restore vanished oyster populations even in places where no other restoration method would work. Large-scale application of this method should be started wherever oyster populations need to be restored. In particular, the method may be of particular use in places like the Pacific Northwest, where oyster growth is failing due to ocean acidification (Barton et al. 2012).

REFERENCES

Barton, A., B. Hales, G. G. Waldbusser, C. Langdon, and R. A. Feely. 2012. The Pacific oyster, *Crassostrea gigas*, shows negative correlation to naturally elevated carbon dioxide levels: Implications for near-term ocean acidification effects. *Limnology and Oceanography* 57:698–710.

Beck, M. W., R. D. Brumbaugh, L. Airoldi, A. Carranza, L. D. Coen, C. Crawford, O. Defeo, et al. 2011. Oyster reefs at risk and recommendations for conservation, restoration, and management. *Biosciences* 61:107–116.

Eastern Oyster Biological Review Team. 2007. Status review of the eastern oyster (*Crassostrea virginica*). Report to the National Marine Fisheries Service. Northeast Regional Office. February 16, 2007. NOAA Tech. Memo. NMFS F/SPO-88, p. 105.

Gazeau, F., C. Quiblier, J. M. Jansen, J-P. Gattuso, J. J. Middelburg, and C. H. R. Heip. 2007. Impact of elevated CO_2 on shellfish calcification. *Geophysical Research Letters* 34: L07603.

Goreau, T. J., and W. Hilbertz. 2005. Marine ecosystem restoration: Costs and benefits for coral reefs. *World Resource Review* 17:375–409.

Hilbertz, W. H. 1979. Electrodeposition of minerals in sea water: Experiments and application. *IEEE Journal on Oceanic Engineering* Vol. OE-4, No. 3, July 1979.

Hilbertz, W. 1992. Solar-generated building material from seawater as a sink for carbon. *Ambio* 21:126–129.

Kirby, M. X. 2004. Fishing down the coast: Historical expansion and collapse of oyster fisheries along continental margins. *PNAS* 101:13096–13099.

Leonard, D. L., and S. Macfarlane. 2011. *Best Management Practices for Shellfish Restoration*, ISSC, Columbia, SC.

Mackenzie, C. L. 2007. Causes underlying the historical decline in eastern oyster (*Crassostrea virginica* GMELIN, 1791) landings. *Journal of Shellfish Research* 26:927–938.

Newell, R. I. E. 1988. Ecological changes in the Chesapeake Bay: Are they the result of overharvesting the American oyster? in M.P. Lynch, E.C. Krome, & 37 editors. Understanding the estuary: Advances in Chesapeake Bay research. Chesapeake Bay Research Consortium Publication 129, pp. 537–646.

Primack, R. 2008. *A Primer of Conservation Biology*, Fourth Edition. Sinauer Associates, Sunderland, MA.

Rothschild, B. J., J. S. Ault, P. Goulletquer, and M. Heral. 1994. Decline in the Chesapeake Bay oyster population: a century of habitat destruction and overfishing. *Marine Ecology Progress Series* 111:29–39.

Schulte, D. M., R. P. Burke, and R. N. Lipcius. 2009. Unprecedented restoration of a native oyster metapopulation. *Science* 325:1124–1128.

Stanley, J. G., and M. A. Sellers. 1986. Species profiles: Life histories and environmental requirements of coastal fishes and invertebrates (mid-Atlantic): American oyster. *U.S. Fish and Wildlife Service Biological Report* 82:1–25.

CHAPTER 14

Restoration of Seagrass Mats (*Posidonia oceanica*) with Electrical Stimulation

Raffaele Vaccarella and Thomas J. Goreau

CONTENTS

Introduction 161
Materials and Methods 162
Results 163
    Giovinazzo 163
    Torre Guaceto 165
    Otranto Electroduct 165
Conclusions 166
References 167

INTRODUCTION

Seagrasses provide unique biodiverse communities in shallow water, essential habitat for juvenile fishes, and provide shore protection against erosion by stabilizing offshore sediments (Brasier 1975). Seagrass clones can grow for thousands of years (Arnaud-Haond et al. 2012), but they are in serious decline worldwide (Short et al. 2011) and in the Mediterranean (Procaccini et al. 2003).

Part of this decline is due to direct damage by dredging and anchors, but the major part results from land-based sources of pollution, mainly by both nutrients from inadequately treated sewage, which causes seagrasses to be overgrown and smothered by weedy algae (Lapointe et al. 1994), and erosion of sediments from land after deforestation, which causes sediment buildup in the water that kills seagrasses by decreasing the light they need for photosynthesis and by direct smothering.

Restoration of seagrass communities has long been seen as a priority in many coastal areas, because loss of seagrass beds plays a major role in loss of coastal fisheries habitat—especially for juvenile fish populations—and in accelerating coastal erosion. But in general, most restoration have proven to be expensive failures, because when seagrasses are transplanted into areas where previously existing seagrass beds had been killed by deteriorating water quality, transplanted seagrasses usually die for the same reasons as the previous populations. There are few locations where water quality deterioration has been permanently reversed and improved so that transplanted seagrasses are able to survive.

Because of the slow growth or poor survival with conventional methods of seagrass restoration in most locations (Curiel et al. 1994; Faccioli 1996; ICRAM 2001; Boudouresque et al. 2006; Uhrin et al. 2009; Paling et al. 2009; van Katwijk et al. 2009), improved methods are needed that increase seagrass growth rates, attachment, and survival in order to successfully restore this important habitat. Our results with restoration of *Posidonia oceanica* in the Adriatic Seawaters of the Province of Bari, Puglia, Italy, show that such technology is now available.

MATERIALS AND METHODS

Seagrass transplantation projects using electrified mesh substrates were carried out from June to September 2008 at two very different locations, Giovinazzo and Torre Guaceto. The depth, habitat, and power sources were different at each site (Table 14.1). At each site, three or four pieces of metal mesh, 50 cm on each side, were nailed to hard substrate or laid over soft substrate. Mesh spacing was 4 cm, and seagrass plants with roots were planted in the spaces, attached by ties to the mesh to secure them against wave surge until established.

At Giovinazzo (Figure 14.1), the shallow (1.5 m) station was located in sand patches created by erosion between nearby *Posidonia oceanica* beds. The deeper station was located on dead *Posidonia* root mats that had been overgrown by *Nanozostera noltii*, also close (about 1 m away) to living *Posidonia* beds. At Giovinazzo, these two seagrasses occur in patches mixed with a third species, *Cymodocea nodosa*. The deepest site, at Torre Guaceto (Figure 14.2), was free of seagrasses because the seagrass could not attach to the limestone hardground due to lack of sufficient sediment and high wave action. The bottom was largely covered with algae.

Detailed descriptions of these sites, including bathymetry, sedimentology, and biology, are included in Vaccarella and Ciccolella (2008) and in Vaccarella and Goreau (2008).

Table 14.1 Comparison between Sites

	Giovinazzo	Torre Guaceto
Depth (m)	1.5 m, 4 m	8 m
Wave exposure	Protected	Exposed
Bottom type	Sand, seagrass	Rocky hardground, algae
Power source	Transformer 24 hours	Solar panel daylight only
Surroundings	Urban	Nature reserve

Figure 14.1 Giovinazzo locations marked with X symbols. The site is urban and protected. The 1.5 m site is inshore, which had 50 m of cable from the power supply, and the 4 m site is farther out with higher wave exposure and 120 m of cable. (From Google Earth.)

RESTORATION OF SEAGRASS MATS (*POSIDONIA OCEANICA*) WITH ELECTRICAL STIMULATION

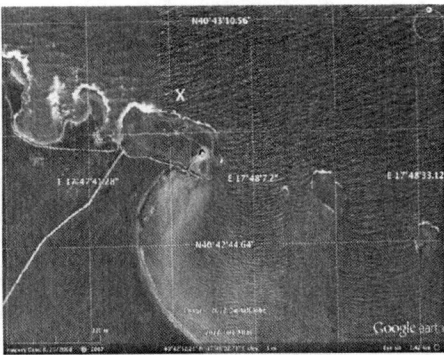

Figure 14.2 Torre Guaceto marine-protected area site. The experiment was at a depth of 8 m in front of an exposed limestone cliff. The solar panel was located on top of the cliff, and the cable length was 190 m. (From Google Earth.)

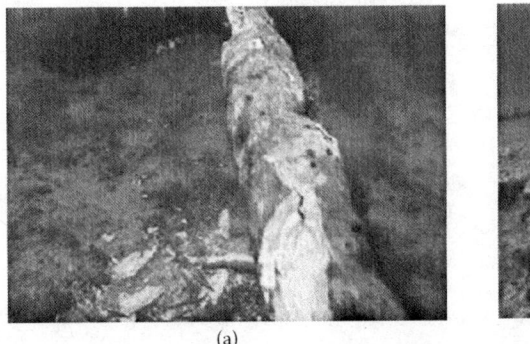 

(a) (b)

Figure 14.3 Otranto electroduct (a) with mineral growth covering and (b) with minerals broken off, showing copper cable without corrosion.

In addition, comparison was made with an underwater high-voltage direct current transmission line, the Otranto electroduct (Figure 14.3). This was an uninsulated copper cable at a depth of 40 m, and more than 8 cm thickness of soft minerals, almost entirely brucite, grew over it in 3 years.

RESULTS

Giovinazzo

Minerals grew very rapidly on the meshes because of the high power from the transformer. Time series color photos of the results at all sites are included in the appended CD. The transplanted *Posidonia* began growing well, but by the end of the experiment the thickness of mineral growth completely filled in the spaces between the meshes and smothered most of the seagrass plants (Figure 14.4). The shallower station meshes, which had shorter cables than the deeper ones, drew more current and grew and filled in faster, taking about two months to fill in, while the deeper ones took about four months (Figure 14.5). Samples of material were analyzed by X-ray diffraction and found to be 91% brucite (magnesium hydroxide), 7.5% aragonite (calcium carbonate), and the rest composed of sand and mud cemented by the mineral growth (Figure 14.6).

In September, the power was terminated and the meshes left in the water for two years. By that point, the soft minerals had turned hard and cemented themselves to the substrate, making their

Figure 14.4 Biorock mesh at 1.2 m depth at the end of the experiment. (Photograph by Raffaele Vaccarella.)

Figure 14.5 Biorock mesh at 4 m depth at the end of the experiment. (Photograph by Raffaele Vaccarella.)

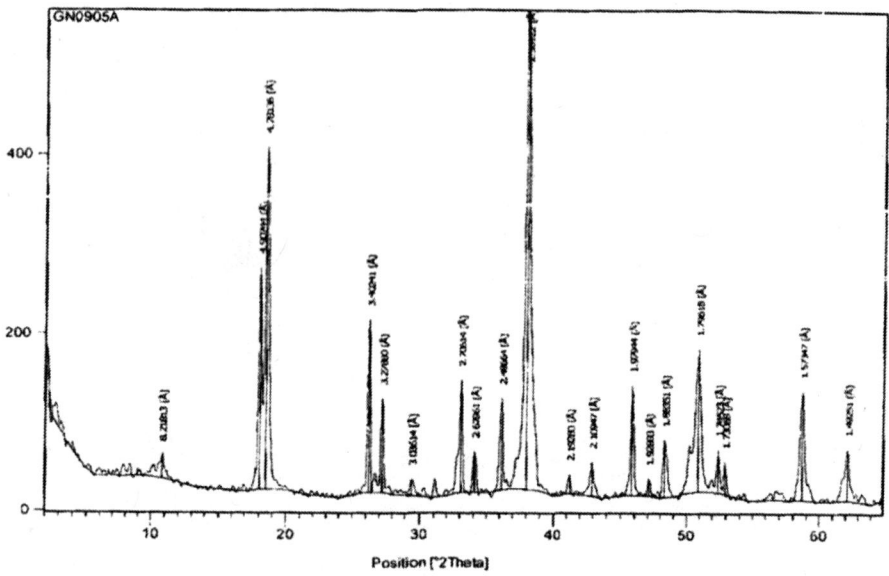

Figure 14.6 X-ray diffractogram petrographic analysis of fresh material from Giovinazzo Biorock project. The major peaks are due to brucite, magnesium hydroxide, with less than 10% aragonite, calcium carbonate. (Image by Raffaele Vaccarella.)

retrieval very hard. They were extensively covered with marine life, including algae (*Caulerpa racemosa* and *Halopteris filicina*), barnacles (*Balanus perforatus*), oysters (*Anomia ephippium*), chitons, and serpulids. An *Octopus vulgaris* had created a den under one of the meshes. Color photos of these organisms are included in the CD in the back of the book.

Torre Guaceto

At Torre Guaceto marine-protected area, the power was supplied by a solar panel in a direct mode, with no battery, so it received power only when it was sunny. It was clear that the power was very much less than at Giovinazzo, the minerals grew much more slowly and harder, and they achieved a thickness of only 1.5 mm (Figures 14.7 and 14.8). The mesh cemented itself to the hard bottom. The *Posidonia* transplants grew very well, and after only three months all the plants added new young and intermediate leaves, and about 80% of them showed clear root growth, from 6.2 to 79.7 mm (Figure 14.9a and b).

Organisms found on and around the Biorock meshes are shown in Table 14.2.

Otranto Electroduct

The copper cable remained free of corrosion despite being in the sea for years. The organic or inert material aggregated by the deposits of brucite was overgrown by vegetal and/or calcareous residues of *Posidonia oceanica, Cladocora caespitosa, Filigrana implexa, Bittium reticulatum, Laevicardium oblungum, Myriapora truncata, Pentapora fascialis, Retepora septentrionalis,*

Figure 14.7 Two Biorock meshes at 8 m at the end of the experiment. (Photograph by Raffaele Vaccarella.)

Figure 14.8 Closeup showing how mesh has cemented itself to hard bottom. (Photograph by Raffaele Vaccarella.)

Figure 14.9 (a) Fishes were attracted to seagrass grown in an area where otherwise no seagrass would grow. (b) The roots showed prolific growth. (Photograph by Raffaele Vaccarella.)

Table 14.2 List of Species Associated with Biorock at Torre Guaceto Marine-Protected Area

Algae	Annelids	Crustaceans	Fish	Phanerogams	Molluscs	Echinoderms
Caulerpa racemosa	Nereis sanguinea	Xantho incisus	Diplodus annularis	Posidonia oceanica	Gibbula sp.	Ophioderma longicaudum
Udotea petiolata	Lisidice ninetta	Brachynotus sexdentatus	Diplodus vulgaris	Nanozostera noltii	Hexaplex trunculus	Echinaster sepositus
Halimeda tuna	Pontogenia chrysocoma		Gobius bucchichi		Octopus vulgaris	Paracentrotus lividus
Peyssonnelia squamaria			Symphodus mediterraneus			Holothuria tubulosa
			Coris julis			

spines, and plates of echinoidea. Apart from the biogenic components, there were also aggregated mineral components, sand, and mud.

Among the organisms detected on the magnesium hydroxide deposits, the alga *Codium tomentosum*; the porifera *Axinella polypoides*; the annelids *Filigrana implexa, Sabella penicillus*, and *Sabella spallanzani*; the bryozoan *Pentapora fascialis*; and the echinoderm *Echinaster sepositus* were also observed. The organisms that covered the deposits of magnesium hydroxide showed healthy development, and the fish swam without any problem near the electroduct together with molluscs (*Tonna galea*).

CONCLUSIONS

Biorock meshes grown at a low charging rate were able to produce healthy and rapid growth of *Posidonia oceanica* on hardground where they normally could not attach. Meshes that were overcharged grew too fast and overgrow the seagrass. In all cases, the minerals hardened, cemented themselves to the substrate, and attracted colonization by a wide variety of local marine life.

These results suggest that the Biorock method can be used to restore biodiverse seagrass habitat in a wide variety of settings, including those where inappropriate substrate or high wave energy would normally prevent their establishment and growth. Further experimentation is needed to optimize the method and apply it to restoration of damaged seagrass habitats.

REFERENCES

Arnaud-Haond, S., C. M. Duarte, E. Diaz-Almela, N. Marba, T. Sintes, and E. A. Serrao. 2012. Implications of extreme life spans in clonal organisms: Millenary clones in meadows of the threatened seagrass *Posidonia oceanica*. *PLoS ONE* 7:1–9, e30454.

Brasier, M. D. 1975. An outline history of seagrass communities. *Palaeontology* 18:681–702.

Boudouresque, C. F., G. Bernard, P. Bonhomme, E. Charbonnel, G. Diviacco, A. Meinesz, G. Pergent, C. Pergent–Martini, S. Rutton, and L. Tunesi. 2006. Preservation and conservation of plants of *Posidonia oceanica*, (Accord RAMOGE: France-Italie, 1976) pubbl. No. ISBN 2-905540-30-3; RAMOGE. pp. 1–202.

Curiel, D., A. Rismondo, A. Solazzi, M. Marzocchi, and M. Scattolin. 1994. Assessment of the health of populations of the marine phanerograms in the Venice Lagoon and experiments replanting *Cymodocea nodosa, Zostera marina, and Zostera noltii*. *Mediterranean Marine Biology* 1:407–408.

Faccioli, F., 1996. The morphological restoration of the Venice Lagoon. *Quaterly, New Venice Consortium, Italy, supplement* 3–4:1–24.

ICRAM. 2001. Analytical methods reference: Monitoring program for control of the marine coastal environment (2001–2003). Ministry of Environment and Environmental Protection, Sea Defense Service, Rome.

Lapointe, B. E., D. A. Tomasko, and W. R. Matzie. 1994. Eutrophication and trophic state classification of seagrass communities in the Florida Keys. *Bulletin of Marine Sciences* 54:696–717.

Paling, E. I., M. Fonseca, M. M. van Katwijk, and M. van Keulen. 2009. Seagrass restoration. In G. M. E. Perillo, E. Wolanski, D. R. Cahoon, and M. M. Brinson (Eds.). *Coastal Wetlands: An Integrated Ecosystem Approach*. 687–713. Elsevier. Amsterdam.

Procaccini, G., M. C. Buia, M. C. Gambi, M. Perez, G. Pergent, C. Pergent-Martini, and J. Romero. 2003. The Seagrasses of the Western Mediterranean. In E. P. Green, and F. T. Short (Eds.). *World Atlas of Seagrasses*. 48–58. UNEP. Berkeley, CA.

Short, F. T., B. Polidoro, S. R. Livingstone, K. E. Carpenter, S. Bandeira, J. S. Bujang, H. P. Calumpong, et al. 2011. Extinction risk assessment of the world's seagrass species. *Biological Conservation* 144:1961–1971.

Uhrin, A. V., M. O. Hall, M. F. Merello, and M. S. Fonseca. 2009. Survival and expansion of mechanically planted seagrass sods. *Restoration Ecology* 17 (3): 359–368.

Vaccarella, R., and A. Ciccolella. 2008. Morphobathymetry and ecological communities at two sites in Apulia located in the coastal areas of Giovinazzo and Torre Guaceto. 61–76. In PO.PRU.RA: Prog. Interreg IIIA; Grecia-Italia, 2000–2006. Provincia di Bari.

Vaccarella, R., and T. J. Goreau. 2008. Applications of electrodeposition towards restoration of mats of *Posidonia oceanica*. 93–105. In PO.PRU.RA: Prog. Interreg IIIA; Grecia-Italia, 2000–2006. Provincia di Bari.

van Katwijk, M. M., A. R. Bos, V. N. de Jonge, L. S. A. M. Hanssen, D. C. R. Hermus, and D. J. de Jong. 2009. Guidelines for seagrass restoration: Importance of habitat selection and donor population, spreading of risks, and ecosystem engineering effects. *Marine Pollution Bulletin* 58:179–188.

CHAPTER 15

Electrical Fields Increase Salt Marsh Survival and Growth and Speed Restoration in Adverse Conditions

James Cervino, Dajana Gjoza, Carmen Lin, Rand Weeks, and Thomas J. Goreau

CONTENTS

Introduction 169
Site History 170
Spartina Ecology and Physiology 170
Pollution Effects 171
Methods 173
Results 175
Extreme Stress Site 176
Discussion 177
Conclusions 177
References 178

INTRODUCTION

Salt marshes have been severely impacted by clearance for land development and are being eroded from the ocean side by global sea level rise (Adam 2002; Silliman et al. 2009). The ecosystem is threatened in many places because, as sea level rises, coastal development prevents its migration inland. Essential habitat for birds, fishes, shellfish, the American eel (*Anguilla rostrata*) (ASMFC 2006), and many other organisms, and the ability of salt marsh mussels, and oysters to filter water and improve water quality have been greatly reduced. Loss of salt marshes leads to erosion. Jamaica Bay, New York City's largest salt marsh, is steadily retreating as sea level rises (Hartig et al. 2002). The ability of salt marsh to keep pace with sea-level rise depends on the density and growth rate of the intertidal salt marsh grass, *Spartina alterniflora* (Morris et al. 2002; Morris 2007; Kirwan et al. 2010).

There have been many efforts to plant salt-marsh grass to trap suspended sediments and reduce coastal erosion. These are generally successful where plants can be protected against erosion from storm waves before they are fully established, but fail in many places where storms erode them before they are deeply rooted. Methods that accelerate their growth rates, densities, root growth, overwinter survival, and ability to grow deeper in the intertidal, seaward of the normal lower limit, would greatly aid those efforts. Here, we demonstrate for the first time a new method that enhances growth rates and survivorship.

SITE HISTORY

College Point is located in Queens, New York, on Long Island Sound next to Flushing Bay, near where the East River connects to New York Harbor. The bay was an important source of food for the Matinecock and Lenape Native Americans and early European settlers, providing abundant fish, shellfish, and waterfowl until the early nineteenth century (Waldman 1999; Kurlansky 2006). *Spartina* salt marshes provide food and shelter for many species, including oysters, mussels, crabs, fish, and birds, maintaining biodiversity. *Spartina* spp. inhabit the same ecosystem as economically valued mussels (*Mytilus edulis*) and American oyster, *Crassostrea virginica*.

After the civil war, the area surrounding the bay became a waterfront resort for the wealthy and then for industrial development, including shipyards. By 1900, it had become a degraded and polluted ecosystem with few living organisms in the coastal zones. In the early 1950s and 1960s, it was a dumping ground for incoming ships to dispose of wastes as they entered New York Harbor. Pollution caused major die-off of marine plants and shellfish along most New York coastlines. *S. alterniflora* is native to New York City; however, it has become rare except in Jamaica Bay. With water quality restoration efforts, in particular the construction of a sewage treatment plant in Flushing in 1977, *Spartina* are now starting to recover in College Point and Flushing Bay (http://www.dec.ny.gov/lands/5489.html).

The New York State Department of Environmental Conservation Tidal Wetlands Program was established to protect wetlands under the Tidal Wetlands Act (Article 25 of the Environmental Conservation Law). Marine intertidal wetlands trend analyses have been conducted by the New York State Department of Environmental Conservation for the past 20 years, with an emphasis toward mitigation and restoration of intertidal coasts in New York City. We chose this intertidal wetland location for its variety of pollutant runoff. One end of the intertidal zone is a NYC Park (Herman McNeill Park), and the other was an illegal landfill that has recently been converted into a residential development. This development is itself adjacent to a State Superfund site that has yet to be remediated. It was declared a Superfund site by the Department of Environmental Conservation after pollutants were found (http://www.nytimes.com/2011/10/23/realestate/college-point-queens-living-in-attention-shore-lovers.html?pagewanted = all).

McNeill Park remains in good environmental condition, but the landfill under the new residential development has been shown to leach dangerous HAZMAT chemicals like polychlorinated biphenyls (PCBs), polycyclic aromatic hydrocarbons (PAHs), hydrocarbons, and solid waste trace metals such as mercury, cadmium, and lead. While *Spartina* has begun to regenerate in front of McNeill Park, there has been no natural recovery in front of the former dump site.

Although preliminary (builder and state) cleanup was somewhat successful, our research group conducted chemical tests in soil samples along the coast of the development that still show "fingerprint" signatures of high concentrations of hazardous materials. No *Spartina* grows naturally along the severely contaminated shoreline of the development. The Resource Conservation and Recovery Act (RCRA), enacted in 1976, is the principal federal law in the United States that governs the disposal of solid wastes and hazardous wastes.

SPARTINA ECOLOGY AND PHYSIOLOGY

Spartina and other intertidal marine plants are photosynthetic autotrophs that convert carbon dioxide into organic compounds, most notably sugars, using adenosine triphosphate energy made from sunlight. *S. alterniflora* is a perennial grass commonly found in intertidal wetlands and salt marshes: it dominates coastal salt marshes along low-wave-energy coasts in mid and high latitudes. Its distribution is affected by inundation periods, controlled by elevation in the intertidal zone, that affect the salt, temperature, and nutrient balance (Brown et al. 2006; Kamel et al. 2009), and by wave energy.

Intertidal marine plants benefit the ecosystem because of their ecological interactions with many marine organisms such as mussels, oysters, fish, crabs, and shore birds (Chung et al. 2004). Mussels help hold *Spartina* plants in place during tidal surges and intense wave energy. Marine plants also play a significant role in the coastal carbon and nitrogen cycles. Reintroducing mussels and oysters simultaneously prevents oxygen deficiency in the rooting zone, which occurs in polluted degraded intertidal sites that are not restored (Drew 1997). Loss of shellfish leads to poor sediment retention after rain or tidal fluctuations, causing depressed growth and yield.

We have tested whether low-voltage direct current electrical stimulus can protect and help accelerate restoration of *Spartina* in severely stressed urban intertidal zones.

POLLUTION EFFECTS

This wetland restoration experimental site is next to a former landfill and US Navy shipyard (Figures 15.1 and 15.2) and has experienced high amounts of pollution from PCBs, PAHs, hydrocarbons, and other RCRA hazardous metals from previous industrial use and illegal dumping (http://www.queenstribune.com/deadline/TideTurnsToxicOnCollegePoi.html; http://www.queenstribune.com/feature/WhatLiesBeneathScientistsa.html).

Sediments at the site where our experimental *Spartina* plots were planted at College Point have been polluted by heavy metals such as copper, zinc, mercury, and lead from the seeps draining the former dump (Figure 15.3). *S. alterniflora* has a low tolerance for highly concentrated toxins in the sediment, such as sulfur, PCBs, and copper (Burke et al. 2000; Mateos-Naranjo et al. 2008).

Due to past losses of salt marsh, it is important to increase *Spartina* biomass to restore biodiversity and improve water quality through the filtering action of salt marsh plants and shellfish. Increasing *Spartina* biomass will increase mussels and other shellfish, which will increase fish and birds, helping restore a healthy marine ecosystem. The roots of *Spartina* are also deep enough to

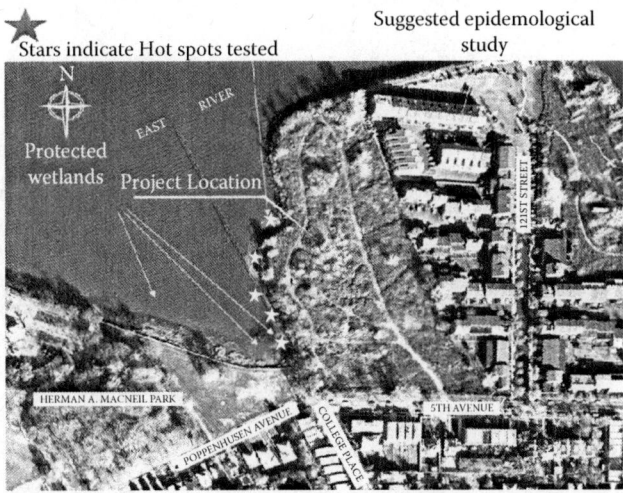

Figure 15.1 Google Earth view of site at low tide. The stars mark sites of seeps draining water into the sea from the former illegal waste dump. Note the area of the former dump that is shown forested in this image has since been cleared and converted into a housing development.

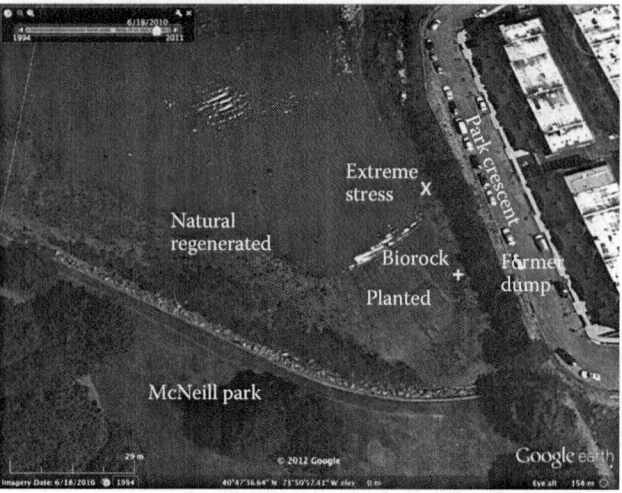

Figure 15.2 Google Earth closeup view of the site at low tide showing planted areas and natural regenerated areas with regard to the former dump site, which had already been covered by a housing development at the time this image was taken on June 18, 2010.

Figure 15.3 One of the groundwater seeps draining the former dump into the sea. The material is almost entirely made up of construction debris. The reddish color is iron stains. The iron dissolves in the anoxic groundwater of the former dump and is precipitated as rust when it is exposed to oxygen.

maintain costal shorelines by preventing sediment erosion in the winter. Marine plants, specifically *Spartina*, are also linked to controlling low levels of metal pollution.

Mateos-Naranjo et al. (2008) designed an experiment to test the effect that copper has on photosynthesis and growth of *Spartina densiflora*. *Spartina* resisted high concentrations because its roots acted as a "barrier" to copper uptake. The leaf concentration hardly increased when the external concentration exceeded 9 mM of copper during these experiments, but copper and other metals were found to be highly toxic to plants in amounts exceeding 9 mM. Copper reduces plant growth by reducing photosynthetic and respiratory activities of most plants. Exposure of high and lethal concentrations of copper and other metals to plants may lead to chlorosis. Fortunately, *Spartina* has high tolerance to heavy metals due to its ability to control excessive ion transport into leaves and sequester them in tissues or cellular compartments; therefore, the species has not completely died off, even in severely contaminated areas. Additionally, there was a rise in nitrogen concentration at the highest copper concentration tested, 9 mM. Copper levels over 9 mM can inhibit the uptake efficiency of water, calcium, magnesium, and phosphorus needed for photosynthesis and growth, due to chloroplast membrane damage, leading to a decline in *Spartina* growth. Our planting efforts were in the areas near the dump where there has been no natural regeneration, although there has been farther away.

We hypothesized that *Spartina* under electrical stimulus has higher growth rate and ability to function under stressful conditions, based on previous experiments with corals and oysters. If so, implementation of this technology in severely distressed urban environments may promote restoration of marine vegetation, higher biodiversity, reintroduction of shellfish, help improve coastal water quality, and enhance protection against shoreline erosion while buffering runoff damage to coastal zones.

METHODS

This experiment was conducted on the southern shore of the Western end of Long Island Sound in College Point, New York, near 5th Avenue and 119th Street. The *Spartina* reestablishment is a 7.62 × 7.62 m² quadrat with 15 cm spacing between each plant. Each quadrat contained approximately 30 plants. Three different *S. alterniflora* quadrats were grown (Figures 15.4, 15.5, and 15.6).

Young *Spartina* plants averaging 20.3 cm tall from Pinelands Nursery were planted at a density of approximately 12–15 plants/m². Quadrats 2 and 3 were much closer to the polluted landfill adjacent to the bay than the control sites, and so would have been expected to show slower growth.

For planting, a shovel was inserted into the sediment to the full depth of the blade to create a V-shaped opening. With the shovel still in place, an ounce of slow-release Miracle Gro® fertilizer was added. Marsh grass requires a small amount of fertilizer at the time of planting to provide nutrients throughout the first growing season. The young plant was inserted into the hole and the sediment around it was compacted by hand and foot.

Quadrat 1 is *S. alterniflora* that had naturally regenerated since the 1970s and served as the control. Quadrat 2, with low electrical current, is *S. alterniflora* planted in June 2010. The *S. alterniflora* in Quadrat 3, planted in June 2010, was grown under higher electrical current. The current was supplied from photovoltaic modules, which supplied electricity to metal grids at ground level with *Spartina* growing in 6 inch (15 cm) square spaces.

Figure 15.4 The planted sites in the spring (May 21, 2011). The higher-current plot is in the foreground, the lower-current plot in the mid-ground, and the control site in the background.

Figure 15.5 The planted sites in the summer (August 15, 2010). The control site is in front, the charged sites behind it, and the solar panel is visible in the background.

Figure 15.6 The planted sites in the fall (October 31, 2010) are in the foreground and the naturally regenerated sites in the background.

The heights of 20 random *S. alterniflora* plants from each quadrat were measured by tape measure every week at low tide in July 2010 to September 2010, and then again from May 2011 to September 2011. To measure the height, the plants were delicately stretched, to make the leaves stand straight and vertically, and measured from the top of the highest leaf to the sediment surface.

Helical steel loops were attached to each quad to monitor carbonate deposition. If calcium carbonate coated the loop, this indicated that carbonate was being deposited from seawater. For all quadrants, the plants' mean height and standard deviation were calculated and plotted using Excel. A best fit line was calculated. Growth rates were calculated using $G = \frac{\Delta H}{\Delta T}$ where G = growth rate, ΔH = change in mean height, and ΔT = time change.

We used common concrete reinforcement metal mesh of 1/8″ steel wire welded together to form a 6-inch-square grid pattern. Three 1 m² metal wire grids were secured to the sediment surface. The higher-current mesh grid was connected directly to a solar panel, and the second grid received a much lower current via an indirect connection via the first grid, while the third was left without electrical stimulus, as a control.

Another set of charged and control sites, called the extreme stress site, was planted below the extreme lower limit of *Spartina* growth and in an area affected by drainage from a groundwater spring draining the former waste dump (Figures 15.7 and 15.8). These plots were followed for three years.

Figure 15.7 The extreme stress site is planted lower in the intertidal than the lower limit of the salt marsh and is further stressed by being in the outflow of the groundwater seep draining the former waste dump as shown in Figure 15.3, which causes the trickle of water seen at left. In the foreground is the Biorock salt-marsh grass, in the mid-ground are the control plantings, which are doing poorly and die every winter, at background left are the planted quadrats, and at background right is the naturally regenerated salt marsh.

Figure 15.8 *Spartina* growing in electrified grid at the extreme stress site. Note that the metal grid, which was severely rusted at the start, has had all rusting stop and is coated with a white mineral layer as a result of the electrical current.

RESULTS

Height averages for each quadrat were calculated in mid-September 2011. The higher-current electrically stimulated *Spartina* had the fastest growth rate (11 cm per week) and the tallest plants. The low current group had the second-highest growth rate (9 cm per week), and the control group had the lowest growth rate (5 cm per week). Figure 15.9 shows the average heights for each of the three quadrats for the first 11 weeks of the experiment.

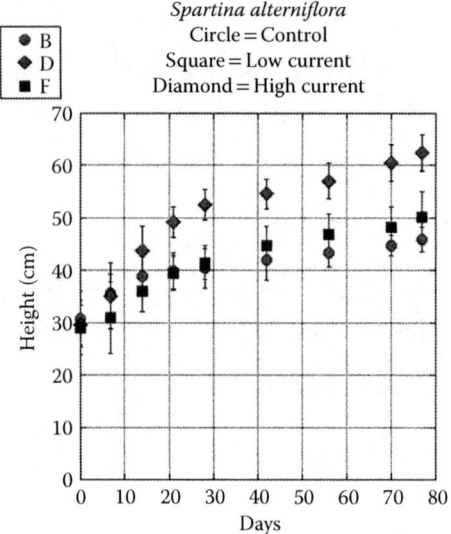

Figure 15.9 Growth of *Spartina* in control and electrified quadrats.

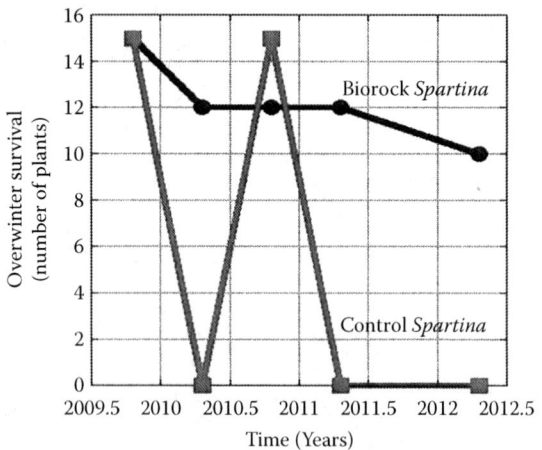

Figure 15.10 Survival of *Spartina* at a severely stressed site. Fifteen *Spartina* were planted in the electrified grid, and 10 of them were still alive three years later. In contrast, *Spartina* planted in nearby control sites on either side without electricity suffered 100% mortality each winter.

EXTREME STRESS SITE

Three grids were set up in front of the metal-rich seep shown in Figure 15.3 to specifically investigate the role of electrical stimulus closest to the landfill in an area where no natural *Spartina* regeneration had taken place. This site is severely stressed for two reasons. First, it is right next to the edge of the landfill, located in a groundwater seep that drains the landfill. Second, this site is lower in the intertidal than the normal lower tolerance limit of *Spartina*. Three plots were established: two controls and one electrified plot.

For three growing seasons, we saw positive growth and regrowth following the winter season in only the electrified plot, with very high overwinter survival. In contrast, the control plots grew much more slowly than the electrified plot and suffered 100% mortality every winter (Figure 15.10). New plants were planted every spring, and they did poorly and died the following winter.

DISCUSSION

Mateos-Naranjo et al. (2008) suggest that metal contaminants in sediment may slow *Spartina* growth rate. The rapid plant growth seen in severely metal-polluted electrified site suggests that electrical stimulus improves *Spartina* growth despite what should be a toxic habitat.

Since *Spartina* has many positive ecological relationships with other species, the use of electrical stimuli to increase *Spartina* growth, biomass, and survival could help restore an important ecosystem whose growth has suffered badly from pollution. Salt marsh ecosystems in Louisiana, Mississippi, Alabama, and Florida damaged by the Gulf oil spill in 2010 could benefit greatly from the results of this experiment. The BP oil spill resulted in damage to many ecosystems, including coastal salt marshes (Lin and Mendelssohn 2012). The positive effect of electrical stimuli on *S. alterniflora* in severely stressed environments may provide a valuable restoration tool for ecosystems impaired by the oil spill.

The procedure used to measure *Spartina* growth could be improved by calculating leaf surface area, leaf and shoot numbers, and the date on which the plant begins to flower, in addition to measuring height. In future experiments, we hope to calculate the leaf surface area, photosynthetic capacity, the number of leaves, the number of stems per clump, and root and rhizome biomass, in addition to height when analyzing the effect of electrical stimulus on *Spartina* growth. It will also be important to see what the effects are on root growth (Valiela et al. 1976), sediment trapping (Mudd et al. 2010), and carbon storage. The faster and denser the plants grow, the better they trap sediment and cause vertical growth of the marsh surface (Mudd et al. 2010) and the better their chance of keeping up with sea level rise.

Coastal wetlands are significant carbon sinks, and it will be important to see how this process affects carbon accumulation underground as roots, rhizomes, and organic carbon (Rabenhorst 1995; Chmura et al. 2003). It will also be important to assess how the number and biological activity of mussels and other animals are affected, because they play key roles in stabilizing the roots against wave erosion, filtering water, trapping sediment, and oxygenating sediments, which benefits root growth (Drew 1997).

CONCLUSIONS

The *Spartina* that received electrical stimulus had faster growth and greater height than the control *Spartina*, the leaves appeared distinctly darker green, and the roots appeared darker and thicker at the holdfast. Electrically charged *Spartina* had high survival under conditions of high metal pollution and deeper in the intertidal than the species could otherwise survive, under which the controls all died. This implies greatly improved root and rhizome growth. Near the end of the experiment, a hard layer of calcium carbonate had formed over the grid. This could promote growth of mussel and oyster shells, increasing biodiversity. Calcium carbonate acts as a buffer to increase the pH of the seawater to ensure survival of more marine organisms, in areas of high-impact CO_2 acidification.

Our work shows that salt marsh can be restored using electrical currents even in severely stressed sites where normal growth is impossible. Because *Spartina* can now be grown seaward of its normal lower limit, salt marshes can be extended seaward, reversing the coastal erosion caused by the rising sea level. The techniques developed here could be applied to Jamaica Bay, the largest salt marsh in New York City, which has lost a significant amount of Zostra (eel grass) and high marsh marsh area and is steadily retreating as a result of global sea-level rise (Hartig et al. 2002). Many coastal wetlands around the world are threatened by sea-level rise, and these methods could assist in maintaining marshes by making them grow faster, a critical parameter in adaptation to sea-level rise. In particular, this could also be done on a large scale in places like Louisiana, where coastal salt marshes are vanishing from sea-level rise and oil pollution (Lin and Mendelssohn 2012) at rates of up to hundreds of meters a year.

REFERENCES

Adam, P. 2002. Saltmarshes in a time of change. *Environmental Conservation* 29:39–61.

Atlantic States Marine Fisheries Commission (ASMFC). 2006. Terms of reference and advisory report to the American eel stock assessment peer review. ASMFC Stock Assessment Report 06-01.

Brown, C. E., S. R. Pezeshiki, and R. D. DeLaune. 2006. The effects of salinity and soil drying on nutrient uptake and growth of *Spartina alterniflora* in a simulated tidal system. *Environmental and Experimental Botany* 58:140–148.

Burke, D. J., J. S. Weis, and P. Weis. 2000. Release of metals by the leaves of the salt marsh grasses *Spartina alterniflora* and *Phragmites australis*. *Estuarine, Coastal and Shelf Science* 51:153–159.

Chmura, G. L., S. C. Anisfield, D. R. Cahoon, and J. C. Lynch. 2003. Global carbon sequestration in tidal, saline, wetland soils. *Global Biogeochemical Cycles* 17:22.1–22.12.

Chung, C., R. Zhuo, and G. Xu. 2004. Creation of *Spartina* plantations for reclaiming Dongtai, China, tidal flats and offshore sands. *Ecological Engineering* 23:135–150.

Drew, M. C. 1997. Oxygen deficiency and root metabolism: Injury and acclimation under hypoxia and anoxia. *Annual Review of Plant Physiology and Plant Molecular Biology* 48:223–250.

Hartig, E. K., V. Gornitz, A. Kolker, F. Mushacke, and D. Fallon. 2002. Anthropogenic and climate change impacts on salt marshes of Jamaica Bay, New York City. *Wetlands* 22:71–89.

Kamel, H., J. Martinez, M. Grandour, A. Albouchi, A. Soltani, and C. Abdelly. 2009. Effect of water stress on growth, osmotic adjustment, cell wall elasticity and water use efficiency in *Spartina alterniflora*. *Environmental and Experimental Botany* 64:321–319.

Kirwan M. L., G. R. Guntenspergen, A. D'Alpaos, J. T. Morris, S. M. Mudd, and D. Temmerman. 2010. Limits on the adaptability of coastal marshes to rising sea level. *Geophysical Research Letters* 37, L23401.

Kurlansky, M. 2006. *The Big Oyster: History on the Half Shell*. New York: Ballantine.

Lin, Q., and I. A. Mendelssohn. 2012. Impacts and recovery of the *Deepwater Horizon* oil spill on vegetation structure and function of coastal salt marshes in the northern Gulf of Mexico. *Environmental Science and Technology* 46:3737–3743.

Mateos-Naranjo E., J. Cambrolle, and M. Figueroa. 2008. Growth and photosynthetic responses to copper stress of an invasive cordgrass, *Spartina densiflora*. *Marine Environmental Research* 66:459–465.

Morris, J. T. 2007. Ecological engineering in intertidal salt marshes. *Hydrobiologia* 577:161–168.

Morris, J. T., P. V. Sundareshwar, C. T. Nietch, B. Kjerfve, and D. R. Cahoon. 2002. Responses of coastal wetlands to rising sea level. *Ecology* 83:2869–2877.

Mudd, S. M., A. D'Alpaos, and J. T. Morris. 2010. How does vegetation affect sedimentation on salt marshes? Investigating particle capture and hydrodynamic controls on biologically mediated sedimentation. *Journal of Geophysical Research* 115, F03029.

Rabenhorst, M. C. 1995. Carbon storage in tidal marsh soils, pp. 93–104. In R. Lal, J. Kimble, E. Levine, and B. A. Stewart (Eds.). *Soils and Global Change*. Boca Raton, FL: CRC Press.

Silliman, B. R., E. Grosholz, and M. D. Bertness (Eds.). 2009. *Human Impacts on Salt Marshes: A Global Perspective*. Berkeley, CA: University of California Press.

Valiela, I., J. M. Teal, and N. Persson. 1976. Production and dynamics of experimentally enriched salt marsh vegetation: Below ground biomass. *Limnology and Oceanography* 21:245–252.

Waldman, J. 1999. *Heartbeats in the Muck: A Dramatic Look at the History, Sea Life, and Environment of New York Harbor*. New York: Lyons Press.

CHAPTER **16**

Postlarval Fish Capture and Culture for Restoring Fisheries

Gilles Lecaillon

CONTENTS

Introduction 179
Overview of the Life Cycle and Nonimpact of Postlarvae Collection 180
Postlarvae Fishing Techniques: The Four Main Methodologies 181
    Crest Nets Set Up on Barrier Reefs 181
    Hoa Nets Set Up between Small Islands on Reef Ridges 181
    Light Trap 182
    Collect by Artificial Reef Eco-Friendly 182
    Postlarval Capture and Culture 183
    Market Opportunities 183
    Alternative Aquarium Fish Farming 183
    Supplementary Aquaculture 184
    Restocking 184
    Restocking as Mitigation for Maritime Impact 185
Postlarval Capture and Culture as a Provider of New Eco-Jobs 186
Conclusions 186
Perspectives 187
References 187

INTRODUCTION

    The management and use of marine fisheries, regardless of the scale at which they occur, involves removing a part of the stocks, which are often limited and finite. This presents a real problem because for certain species, these stocks are increasingly depleted and sometimes even exhausted. Such overexploitation of resources is evident not only for food species, but also for reef species that are popular for the aquarium fish market.
    In general, most fishing techniques take adults, often breeders, thereby diminishing not only current stocks but also future stocks. Open-ocean fishing techniques, whose destructiveness varies depending on the technique used, such as gillnets, and the amount of fish caught, do, however, spare habitat. This is not the case for coral-reef fishery techniques, which, depending on the method employed, can directly affect habitat conditions, such as using cyanide and explosives. Russ and

Alcala (2004) found that 75% of the coral reefs in the Philippines have been damaged from such practices. In other words, not only is there overfishing, but also fish habitats have been destroyed, leaving nature with no way to recover from such disturbances.

Eggs, larvae, and postlarvae of marine fish and crustaceans are usually considered to be a nonexploitable marine resource, as opposed to juveniles and/or adults, which are actively harvested. Recent books shed light on the rarely described stages of those postlarvae (Juncker 2007; Lecaillon et al. 2012; Maamaatuaiahutapu et al. 2006).

The approach of postlarval capture and culture (PCC), however, develops these previously overlooked resources through the capture and rearing of presettlement postlarvae. Adults reared from postlarvae can then be a genuinely sustainable alternative to the capture of wild adult marine animals for a variety of species, including both high-quality food fish and valuable ornamentals. This is not as far-fetched as it sounds, because the postlarvae stage is the final stage of the pelagic larval phase of coastal marine animals, which ends with the colonization and settlement of the postlarvae into their habitat, usually a coastal environment such as coral reef or rocky or sandy shores. Most demersal (bottom-dwelling) reef fish species have pelagic (open-ocean) larvae. However, after metamorphosis to juveniles, these postlarvae suffer catastrophic mortality rates during the settlement process, with more than 95% dying within the first week of their return, mainly due to predation (Doherty et al. 2004; Planes and Lecaillon 2001). By capturing an insignificant proportion of these postlarvae before the mortality phase, we effectively exploit a resource that would otherwise be largely, and naturally, wiped out (Bell et al. 2009; Lecaillon and Lourié 2007).

Collection of postlarvae is certainly not the only solution to the overexploitation of demersal species, but it is nevertheless a path worth exploring, not only for developing an innovative and sustainable type of aquaculture, but also for repopulation efforts, which are just beginning (Delbeek 2006).

OVERVIEW OF THE LIFE CYCLE AND NONIMPACT OF POSTLARVAE COLLECTION

Most coastal fish (not only coral-reef fish, but also demersal fish in temperate zones) have oceanic larval phases at the beginning of their life cycles (Leis 1991; Leis and Carson-Ewart 2000; Sale 1980).

This phase allows them to colonize new habitats, thereby facilitating the species' broad distribution and, consequently, their persistence (Choat and Robertson 1975; Lobel 1978; Victor 1986). Depending on the species, larvae spend from 20 days (Pomacentridae) to more than 100 days (Aulostomidae) in the open ocean (Brothers et al. 1983; Victor 1986; Victor and Wellington 2000; Wellington and Victor 1989). Size of postlarvae range from 6 mm to 80 mm for a large majority of fishes (see Figure 16.1). Fairly passive during most of this phase, they finally become active—they enter their competence phase (Cowen et al. 2000; Doherty and Williams 1988; Fauvelot et al. 2003)—to look for their new habitat. This settlement phase takes place at night, if possible, when there is no moonlight. Still, most of those postlarvae (more than 95%) disappear during the week that follows settlement, mainly due to predation (Doherty et al. 2004; Planes and Lecaillon 2001; Planes et al. 2002).

Postlarvae collection techniques make it possible to recover these animals before this phase of high natural mortality. Given the very large number of postlarvae arriving from the ocean, collecting a small percentage of them has almost no impact (Bell et al. 1999, 2009). These techniques provide access to a previously unexploited marine resource, without impacting stocks or damaging the environment (Lecaillon 2004; Lecaillon and Lourié 2007).

Today, thanks to a range of existing collection tools and the know-how developed by certain private and public agencies, these postlarval fish can be kept alive, weaned, and grown out to become a new marine resource, while respecting the spirit of sustainable development and biodiversity conservation.

Figure 16.1   A 17 mm wrasse postlarva. (Courtesy of A.Frezel/Ecocean©.)

POSTLARVAE FISHING TECHNIQUES: THE FOUR MAIN METHODOLOGIES

The following techniques are described in the Moana Initiative (2007) booklet.

Crest Nets Set Up on Barrier Reefs

This technique consists of setting up a net on the barrier reef that encircles the lagoon, with the open end toward the ocean, to catch the postlarvae surfing over the crest to get into the lagoon. This technique was developed by French and Australian laboratories (Ecole Pratique des Hautes Etudes [EPHE] of Perpignan) (Dufour 1991) (Australian Institute of Marine Science) in collaboration with the WorldFish Center (Hair and Doherty 2003). A private firm based in Moorea, French Polynesia, which is no longer in business, used it.

Setting up the nets requires a significant number of staff to put up, on the reef crest, the poles needed to prop up the net. The equipment wears out quickly, because it is constantly hit by waves. These nets can be used only near amphidromic points (where tidal ranges are very small) and, by definition, in those areas where there are crests/ridges, which considerably reduce the number of countries where the technique can be used. For instance, these nets cannot be used in temperate settings. Finally, the crests sometimes have a lot of *Sargassum*- and *Turbinaria*-type seaweeds that get caught in the collector that can abrade the postlarvae and damage them.

Hoa Nets Set Up between Small Islands on Reef Ridges

Certain islands, particularly coral atolls, have very shallow (2 m) channels on their reef ridges, surrounded by dry land (called *motu* in Polynesian). The channels, called *hoa* in Polynesian, meaning "marine rivers," allow the ocean to fill the lagoon. This technique consists of setting a net across these *hoa* to catch the postlarvae concentrated in the water masses passing between the *motu* and entering the lagoon.

When a site has a *hoa*, setting up the net is simpler than for crest nets. The force of the waves on the gear is smaller, so it is easier to set the net up and take it down. This device traps everything going through the *hoa* and so is very effective. It is sometimes the victim of its high level of effectiveness, because when the postlarvae of a given species are particularly abundant, millions of postlarvae can be caught. But because there are too many of them in the collector, most die from a lack of oxygen. This is not profitable for either the fisher or the environment.

This technique arose from the efforts to optimize the use of crest nets and so has appeared more recently. Of course, one must have a *hoa* to use the method. Unfortunately, these geomorphologic structures are even rarer than reef ridges, so use of this technique is, once again, geographically limited. A private firm based in Bora Bora, French Polynesia, used it mainly for ecotourism restocking.

Light Trap

Many different models of light traps exist, because scientists have used this method for many years. First developed by the Australians (Doherty 1987), then optimized by the French, it consists of a casing surrounding an autonomous underwater lamp. Postlarval fish, attracted by the light, are trapped when they go through the slots into the trap.

This technique is widely used for scientific missions, because it can be set up rapidly (Watson et al. 2002). All one has to do is deploy it; no reef ridges or *hoa* are needed, so it can be used anywhere. Light traps have their limits, however, because the postlarvae have to find the slots (vertical in the French model and horizontal in the Australian model) to be trapped, and this reduces their effectiveness. In addition, certain small pelagic fish (e.g., sardines) are also attracted to the light, and, because of their horizontal swimming style, get trapped, panic, and then die.

Collect by Artificial Reef Eco-Friendly

Collect by Artificial Reef Eco-Friendly (CARE) is a new technique that was recently developed by a French company (patented by Ecocean in 2002; www.ecocean.fr). It uses a lighted artificial reef that takes advantage of the behavior of new recruits to trap them; that is, their attraction to light (phototropism), desire to come in contact with a solid object (thigmotropism), and desire to find shelter from predators. These sensory elements are important for postlarval fish, which have very acute senses during recruitment (Lecchini 2003; Sweatman 1988).

This technique has the advantages of light-trap fishing, while being more efficient and free of the problem of attracting pelagic species. It attracts the postlarvae of reef or demersal species that are in the settlement phase. The postlarvae choose to take shelter in the artificial reef, while unwanted small pelagic fish just swim over the reef.

A new smaller and cheaper recent version (2012) of the CARE trap has been specially designed for ease of use in developing countries and other locations where specialist technical support may not be available:

- The CARE trap can be universally applied in different weather conditions (tide, current, turbidity, etc.) and is a highly efficient tool for collection.
- The collection devices are ergonomic for the user, easy to set up at sea from a regular fishing boat, and highly durable (see Figure 16.2).

Figure 16.2 Madagascar fishermen fishing with CARE traps in Toliara lagoon. (Courtesy of G.Lecaillon/Ecocean©.)

- If the equipment gets damaged, it is easy to repair using locally available materials, thus reducing reliance on imported equipment and minimizing costs.

Postlarval Capture and Culture

A new technique used to collect and rear the marine resource is called PCC. Breeding marine fish is highly expensive, labor-intensive, and technically difficult. However, PCC offers a solution whereby Mother Nature does all the hard work, and where the abundant hatcheries of the sea can provide large quantities of postlarval fish with relative ease.

The PCC technology involves the following steps:

1. Catching live, healthy, and uninjured postlarval fish
    Postlarvae are collected at night in specially designed traps such as CARE traps, and then fishers simply pour them into plastic or foam boxes containing seawater, with oxygen supplied from small air pumps, and bring the postlarval fish to land-based facilities for sorting.
2. Identifying and segregating species that could harm or consume one another
    Sorting can be done without the need of a magnifying glass, as distinct shapes can already be seen with most common species. Postlarval size ranges from 6 mm for a Napoleon wrasse up to 80 mm for larger families such as jacks, goatfish, surgeonfish, squirrelfish, and filefish. Once sorted, postlarval fish are then segregated into separate aquaria to avoid predation. However, fish species with similar characteristics (feeding, growth, etc.) are reared together.
3. Weaning and growing the fish for sale or for restocking
    One of the most important advantages of the PCC technique is that the rearing (or grow-out) facilities are easy to establish using intermediate technologies, thus reducing the need for technologically advanced and expensive laboratories and hatcheries. Initially, the sorted postlarval fish are fed with artemia (brine shrimp), which are hatched daily. Fish-food granules, the sizes of which are fitted to the needs of different species and the size of their mouth openings, rapidly replace the live food. Weaning is simple and can be done quickly for most species. The time taken varies from one day for species like damsels to a few months for scorpion fish. The postlarval fish can be fed up to four times per day.
4. Through know-how developed mainly by the French company Ecocean, postlarval fish, collected with whatever method, can now be farmed so as to produce large quantities of marketable product. Ecocean's PCC technology has been developed through extensive and in-depth scientific research on the life cycle of coastal reef fishes and crustaceans. Ecocean's team has over 14 years of experience of collecting and rearing postlarval fish for prevention and/or mitigation of overexploitation of tropical reefs and temperate seas.

Market Opportunities

Depending on the year and site, part of some postlarval fish and shrimps collected (average of 30%) can go to the ornamental market (e.g., butterfly fish and damselfish), especially because marine aquarium fish fetch a high unit price when there is a market for them. Some (average of 40%) can be used for aquaculture (e.g., rabbit fish, groupers, and snapper), but all can provide income for local fishers, with the added benefit that postlarval collecting and rearing reduces fishing pressure on the already stressed fish stocks in the area.

Alternative Aquarium Fish Farming

Here, opportunities exist mainly in developing countries like the Philippines and Indonesia, whose current levels of exported aquarium fish account for 80% of the world market. The steps for postlarvae collection, grow-out, and shipping are very well known, and several trials have been successfully carried out in the Reunion Island, Comoros, Hawaii, French Polynesia, and the Philippines. Actually, in the Indian Ocean, two entities—MadaFia (Madagascar island) and EcoFish Mauritius (Mauritius island)—are running a business for marine aquarium trade. On an average, about three months of grow-out are needed to get fish of a "small" marketable size (less than one month for Pomacentridae but more than four months for Labridae and Chaetodontidae). Reared juveniles caught as postlarvae stage also become accustomed to human handling and to elevated levels of fish by-products in their tanks. In this way, they become adjusted to the tank environment, and as they acclimatize, they become much less prone to stress. Therefore, they are much easier to transport and to keep in captivity if sold into the ornamental trade.

This new procedure makes it possible to produce immunized and disease-resistant specimens, thereby bringing a certain level of quality to a declining market for a wild product, which has mortality rates of up to 90% between the points of collection and final purchase by aquarium-fish lovers (Schmidt and Kunzmann 2005).

In summary, tank-raised fish from postlarval collection offer many benefits both for conservation and for the marine aquarium trade. They reduce the number of wild-caught fish, which in turn decreases the pressure from marine ornamental collectors on coral-reef habitats. In addition, their acclimatization, freedom from disease, and general hardiness improves life span in marine aquaria and enhances demand for such eco-friendly fish while reducing pressure to restock aquaria due to fish mortality.

Supplementary Aquaculture

Here again, opportunities exist mainly in developing countries. This activity can provide ciguatera-free protein for local consumers as well as a product that could be destined for the Asian live fish market.

It should be noted that most food-fish families have "large" (> 2 cm) postlarvae, which makes it possible to produce meal-portion-sized specimens after about six to eight months of in-cage grow-out. Currently, trials are under way in Madagascar, in collaboration with Institut Halieutique Service Maritime (Tuléar). Initial results for the in-cage grow-out of species of Siganidae and Lethrinidae indicate growth coefficients that were 1.8 times greater than those in land-based tanks.

Another project underway (September 2006–September 2007) in the Philippines, with funding from the US-based National Fish and Wildlife Foundation, is attempting to transfer knowledge about larvae harvest techniques to local communities. This project has the unanimous support of the various fisher, farmer, and decision-maker circles. The farm belongs to a local nongovernmental organization (NGO) Feed the Children, and the operational project is part of a Coastal Resource Management Plan set up by the Municipality of Tubigon. This project also has the goal of repopulating a local marine reserve with some 10,000 juvenile fish.

Some developed countries may also be interested in collecting postlarvae, particularly to study the growth rates of certain target species before attempting to improve particular reproduction phases (e.g., increasing gamete production rates, limiting stress, etc.) Finally, biotechnical companies could be interested in the biomolecules contained in oceanic postlarvae, which have low parasite levels.

Restocking

Although new jobs can be created through aquaculture production for ornamental markets or for consumption, a percentage of the stocks collected should always be returned in the form of restocking in Marine Protected Areas (MPAs) (Lecaillon and Lourié 2007).

Restocking of different coral-reef fish species from postlarval collection is an entirely new concept. Once grown and prepared, juveniles are brought to the restocking location. Once divers have released the fish, marine scientists from local NGOs or academic institutions who also study their behavior monitor and count survival rates. Experiments have shown good results, where released fish were still near restocked habitat after two weeks. Juveniles have wild behavior a few hours after the release.

The impact of restocking can be measured against the baseline data already collected prior to restocking. These data include habitat as well as the condition and abundance of fish and invertebrates. Tagging has been tested, but suffers from several disadvantages (cost, time, stress to fishes during the tagging procedure, etc.). Results are determined through a long-term process of counting fish population before restocking and at regular intervals after restocking of the site, and comparing the results with a local control area with similar habitat and fishery characteristics (to ensure like-to-like comparison).

The successful restocking of juveniles may also lead to synergistic effects on the recruitment of other larval fishes from outside the MPA (due to noise and pheromone attraction). Genetic diversity is often lower in disturbed and degraded areas, with a possible risk of extinction. Postlarvae collected by the CARE trap have higher genetic diversity than the resident adult fish population, as they have not yet undergone major mortality events. Therefore, restocking with the resulting juveniles will increase genetic diversity and ensure population continuity (Frankham et al. 2002).

Ideally, restocking should focus on species that are either lacking within the restocking area or whose characteristics fit the needs of the MPA. For example, if coral-reef cover is low and algae are abundant, algal grazers can be introduced to clear the algae, which in turn should enhance new coral recruitment within the MPA.

The restocking activities can also help promote the local MPAs and open up the area to improve tourism. Restocking (or monitoring) is well suited to this kind of destination. Fisher families may earn additional income by providing transport, accommodation, and food, and guiding visitors through the MPA and the restocking facility. The potential for such activities has yet to be fully explored.

In any restocking case, juvenile release must be done according to microhabitat. The best results are found where habitat structures of suitable size and shape provide shelter for juveniles against predation, such as natural or artificial mangrove roots (Cocheret de la Moriniere et al. 2004; Nagelkerken and Faunce 2007, 2008; Nagelkerken et al. 2010; Verweij et al. 2006) or Biorock structures (Goreau 2010). These microhabitats can be diverse, and Biorock will work perfectly for juvenile restocking from postlarval collection.

Restocking as Mitigation for Maritime Impact

Finally, a process named BioRestore© has just being validated in 2011 as a mitigation measure for coastal projects (http://www.ecocean.fr/en/ecological-restoration-applications.asp). This process will include a postlarval capture step, a specific culture onto small juveniles step, and a restocking step associated with microhabitats (see Figure 16.3). BioRestore® is custom-made for companies,

Figure 16.3 Juveniles of sea bream just released near artificial microhabitat. (Courtesy of G. Saragoni/CNRS©.)

NGOs, cooperatives, municipalities engaged in sustainable coastal development, ecological restoration, mitigation for unavoidable impacts, and coastal infrasctructure projects such as ports, marinas, off-shore energy, seafloor mining and underwater pipelines.

As at totally sustainable process involving local stakeholders, BioRestore® can facilitate community and government acceptance of the project, while at the same time enhancing the knowledge of local marine biodiversity. BioRestore® helps to restore impaired marine ecosystems and boost ecological resilience towards self-sustainability.

POSTLARVAL CAPTURE AND CULTURE AS A PROVIDER OF NEW ECO-JOBS

The United Nations Education, Scientific and Cultural Organization (UNESCO) through its Man and Biosphere program (MAB), has funded a booklet about the eco-jobs that are economic opportunities generated through environmentally oriented capacity-building. The following jobs are among those that can be created through a postlarval collection and rearing project, implemented with and for local people, through the conservation and sustainable use of coastal and marine biodiversity (Moana Initiative 2005):

Postlarvae fishermen (collectors)
Work in the aquaculture facility, including cleaning and maintenance of equipment and various fish-rearing activities (feeding, medication, etc.). Employment opportunities can range from jobs for unskilled laborers to those for highly skilled technicians
Handling, packing, and shipping
Sales and marketing
Management and administration, including accounting and quality assurance
Ecotour guiding for visitors to the facility and the MPA
Provision of visitor transport and facilities

CONCLUSIONS

PCC has a double positive impact: 1-Collection from within a huge nonexploited natural stock enables a reduction of fishing effort on the normally exploited stock. 2-This will save future genitors, allowing production of an increasing number of saleable products from abundant postlarval stock. The high-quality fish resulting from the eco-friendly collection of postlarvae are extremely robust, exhibiting a high rate of survival, because tank-raised fish are better suited to life in fish farms, sea cages, or aquaria. They are accustomed to human handling, acclimatized to wastewater in their tanks, and are weaned on pellets.

… and no known negative impacts.

The nonimpact of PCC technology on the marine environment is widely validated by the scientific community. PCC has been already tested in several countries and regions, such as the Philippines, French Polynesia, Madagascar, Fiji, Comoros, Hawaii, Florida, and the Mediterranean Sea. There is ample proof of concept.

As a consequence, PCC technology has already been labeled "good practice" by the International Coral Reef Initiative (www.icriforum.org). Organizations such as the UNESCO, through its MAB program, have found PCC suitable for building quality economies through the conservation and sustainable use of coastal and marine biodiversity. A recommendation from the Ministry of Environment in France concerning the use of PCC has just been included in the "Grenelle de la mer" last June 2009 (http://www.legrenelle-mer.gouv.fr/IMG/pdf/LIVRE_BLEU_Grenelle_Mer.pdf). PCC is a completely new and fully sustainable technology, because it effectively creates "something from nothing" by exploiting a resource that would otherwise simply be overlooked.

PERSPECTIVES

A PCC-based aquaculture project uses an eco-friendly technology, which is quite simple to set up, and can generate employment and new income for local communities in developing countries. The use of PCC also serves to reduce the pressure exerted on adult reef fish by offering access to a new and un- or underexploited marine resource.

However, PCC's application is not limited to tropical seas and developing countries. It is also suitable for use by advanced nations in temperate seas as an additional economic, conservation, and restocking resource (mitigation banking), especially where conventional conservation strategies have already failed. The heavily overfished Mediterranean and North Sea fisheries are good examples of areas that could benefit from such an approach. A new project called GIREL has just started to mitigate Marseille Harbor impact.

Finally, PCC is an excellent tool to study and understand this little-known part of the life cycle of coastal marine species. Thanks to Ecocean's automation of the postlarvae capture processes, complete with timing devices if needed, PCC can be used in any marine environment for research and monitoring of planktonic biodiversity. A LIFE and EU funded project has just been funded about larval fish biodiversity using CARE devices and PCC technology (Life10/NAT/FR/00200). LIFE is the EU's financial instrument supporting environmental and nature conservation projects throughout the EU. This project named SUBLIMO, is hosted by the University of Perpignan in South of France.

REFERENCES

Bell, J. D., E. Clua, C. A. Hair, R. Galzin, and P. J. Doherty. 2009. The capture and culture of post-larval fish and invertebrates for the marine ornamental trade. *Reviews in Fisheries Science* 17:223–40.

Bell, J. D., P. J. Doherty, and C. A. Hair. 1999. The capture and culture of post-larval coral reef fish: Potential for new artisanal fisheries. *SPC Live Reef Fish Information Bulletin* 6:31–34.

Brothers, E. B., D. McWilliams, and P. F. Sale. 1983. Length of larval life in twelve families of fishes at "One Tree Lagoon", Great Barrier Reef, Australia. *Marine Biology* 76:319–24.

Choat, J. H., and D. R. Robertson. 1975. Protogynous hermaphroditism in fishes of the family Scaridae. 263–283 In: Rheinboth R (Ed.) *Intersexuality in the animal kingdom*. Springer Verlag. Berlin.

Cocheret de la Morinière, E., I. Nagelkerken, H. van der Meij, and G. van der Velde. 2004. What attracts juvenile coral reef fish to mangroves: Habitat complexity or shade? *Marine Biology* 144:139–45.

Cowen, R. K., K. M. M. Lwiza, S. Sponaugle, S. C. B. Paris, and D. B. Olson. 2000. Connectivity of marine populations: Open or closed? *Science* 287:857–59.

Delbeek, J. C. 2006. Advances in Marine Fish Aquaculture, Marine Fish and Reef USA, 2006 Annual:110–16

Doherty, P. J. 1987. Light-trap: Selective but useful devices for quantifying the distributions and abundances of larval fishes. *Bulletin of Marine Science* 41:423–31.

Doherty, P. J., V. Dufour, and R. Galzin. 2004. High mortality during settlement is a population bottleneck for a tropical surgeonfish. *Ecology* 85:2422–28.

Doherty, P. J., and D. McWilliams. 1988. The replenishment of coral reef fish populations. *Oceanography and Marine Biology: Annual Review* 26:487–551.

Dufour, V. 1991. Variations in the abundance of fish larvae in reef environments: effect of light on larval colonization. *CR Académie des Sciences* 313:187–94.

Fauvelot, C., G. Bernardi, and S. Planes. 2003. Reductions in the mitochondrial DNA diversity of coral reef fish provide evidence of popultion bottlenecks resulting from holocene sea level change. *Evolution* 57:1571–83.

Frankham, R., J. D. Ballou, and D. A. Briscoe. 2002. *Introduction to conservation genetics*. Cambridge University Press. Cambridge.

Goreau, T. J. 2010. Coral reef and fisheries habitat restoration in the Coral Triangle: The key to sustainable reef management. 244–53 In: J. Jompa, R. Basuki, Suraji, M. Tesoro, and E. T. Lestari (Eds.) Proceedings of the COREMAP Symposium on Coral Reef Management in the Coral Triangle.

Hair, C., and P. Doherty. 2003. Report on the capture and culture of fish in Solomon Islands. *Bulletin de la CPS* 11:13–18.

Juncker, M. 2007. Young coral reef fish of Wallis Islands and the Central Pacific, identification guide. Environment Department of Wallis & Futuna for the CRISP Programme.

Lecaillon, G. 2004. The "CARE" (Collect by Artificial Reef Eco-friendly) system as a method of producing farmed marine animals for aquarium market: An alternative solution to collection in the wild. *SPC Live Reef Fish Information Bulletin* 12:17–20.

Lecaillon, G., and S. M. Lourié. 2007. Current status of marine post-larval collection: Existing tools, initial results, market opportunities and prospects. *SPC Live Reef Fish Information Bulletin* 17:3–10.

Lecaillon, G., M. Murenu, F. Felix-Hackardt, and P. Lenfant. 2012. Identification Guide to the post-larval fish of the Mediterrenean. Edited by Ecocean for the SEARCH Foundation for Biodiversity.

Lecchini, D. 2003. Ecological characteristics of fishes colonizing artificial reefs in a coral garden at Moorea, French Polynesia. *Bulletin of Marine Science* 73:763–69.

Leis, J. M. 1991. Vertical distribution of fish larvae in the Great Barrrier Reef lagoon, Australia. *Marine Biology* 109:157–66.

Leis, J. M., and B. M. Carson-Ewart. 2000. *The Larvae of Indo-Pacific Coastal Fishes: An Identification Guide to Marine Fish Larvae.* (Fauna Malaysian Handbook 2.) BRILL. Australian Museum.

Lobel, P. S. 1978. Diel, lunar and seasonal periodicity in the reproductive behavior of the pomacanthid fish, *Centropyge potteri*, and some other reef fishes in Hawaii. *Pacific Science* 32:193–207.

Maamaatuaiahutapu, M., G. Remoissenet, and R. Galzin. 2006. Guide to the identification of larval reef fish of French Polynesia. Éditions Téthys.

Moana Initiative. 2005. New eco-jobs from marine post-larval collection, Moana Initiative, MAB/Unesco and ReefCheck. www.moanainitiative.org

Moana Initiative. 2007. PCC, a useful method for conserving value and biodiversity, Foundation for Marine Biodiversity, MAB/Unesco and CRISP project. www.moanainitiative.org

Nagelkerken, I., and C. H. Faunce. 2007. Colonization of artificial mangroves by reef fishes in a marine seascape. *Estuarine Coastal and Shelf Science* 75:417–22.

Nagelkerken, I., and C. H. Faunce. 2008. What makes mangroves attractive to fish? Use of artificial units to test the influence of water depth, cross shelf location, and presence of root structure. *Estuarine Coastal and Shelf Science* 79:559–65.

Nagelkerken, I., A. M. de Schryver, M. C. Verweij, F. Dahdouh-Guebas, G. van der Velde, and N. Koedam. 2010. Differnces in root architecture influence attraction of fishes to mangroves: A field experiment mimicking roots of different length, orientation, and complexity. *Journal of Experimental Marine Biology and Ecology* 396:27–34.

Planes, S., and G. Lecaillon. 2001. Caging experiment to examine mortality, during metamorphosis of coral reef fish larvae. *Coral Reefs* 20:211–18.

Planes, S., G. Lecaillon, P. Lenfant, and M. Meekan. 2002. Genetic and demographic variation in new recruits of *Naso unicornis*. *Journal of Fish Biology* 61:1033–49.

Russ, G. R., and A. C. Alcala. 2004. Marine reserves: Long term protection is required for full recovery of predatory fish populations. *Oecologia* 138:622–27.

Sale, P. F. 1980. The ecology of fishes on coral reefs. *Oceanography and Marine Biology: An Annual Review* 18:367–421.

Schmidt, C., and A. Kunzmann. 2005. Post-harvest mortality in the marine aquarium trade, a case study at an Indonesian export facility. *Live Reef Information Bulletin* 13:3–12.

Sweatman, H. 1988. Field evidence that settling coral reef fish larvae detect resident fishes using dissolved chemical cues. *Journal of Experimental Marine Biology and Ecology* 124:163–74.

Verweij, M. C., I. Nagelkerken, D. de Graaff, M. Peeters, E. J. Bakker, and G. van der Velde. 2006. Structure, food, and shade attract juvenile coral reef fish to mangrove and seagrass habitats: A field experiment. *Marine Ecology Progress Series* 306:257–68.

Victor, B. C. 1986. Duration of the planktonic larval stage of one hundred species of Pacific and Atlantic wrasses (family Labridae). *Marine Biology* 90:317–26.

Victor, B. C., and G. M. Wellington. 2000. Endemism and the pelagic larval duration of reef fishes in the eastern Pacific Ocean. *Marine Ecology Progress Series* 205:241–48.

Watson, M., R. Power, S. Simpson, and J. L. Munro. 2002. Low cost light traps for coral reef fishery research and sustainable ornamental fisheries. *Naga, the ICLARM Quarterly* 25:4–7.

Wellington, G. M., and V. C. Victor. 1989. Planktonic larval duration of one hundred species of Pacific and Atlantic damselfishes (Pomacentridae). *Marine Biology* 101:557–67.

CHAPTER 17

Mariculture Potential of *Gracilaria* Species [Rhodophyta] in Jamaican Nitrate-Enriched Back-Reef Habitats
*Growth, Nutrient Uptake, and Elemental Composition**

Arlen Havenner Macfarlane

CONTENTS

Introduction 190
    Gracilaria Mariculture for Agar Production 190
    Gracilaria Mariculture Potential in Tropical Countries 192
    Gracilaria Mariculture Potential in Jamaican Nitrate-Enriched Back-Reef Habitats 193
    Research Objectives 194
Materials and Methods 195
    Field Sampling of *Gracilaria* Populations 195
    In Situ Culture Sites 195
    In Situ Culture Methods 196
        Rope Culture 196
        Vexar Cage Culture 196
        Net Culture 197
    Nutrient Enrichment and Controlled Grazing Experiments 197
    Epiphyte Observations and Quantification 198
    Grazing Observations and Quantification 199
    Weather Observations 199
    Nutrient Uptake Experiments 199
    Physical and Chemical Parameters 200
    Agar Extractions and Determinations 201
    Tissue Analysis for C, N, and P Contents 202
    Statistical Analysis 202
Results 202
    Strain Selection for Mariculture 202

* This chapter on optimizing growth of economically valuable tropical algae was originally presented as a thesis at the University of Miami in 1991 and is being published here for the first time.

Adaptability to Culture, Growth, and Epiphyte Resistance of Collected *Gracilaria*
Species and Strains..202
Agar Contents of Collected *Gracilaria* Species 210
Between-Habitat Variability 211
Salinity/Nutrient Relationships of Culture Sites...211
Growth, Epiphyte, and Grazing Differences between Sites......................................213
Nutrient Enrichment Experiments 217
G. terete..217
G. domingensis...217
G. crassissima..222
C, N, and P Contents and Atomic Ratios of Cultured and Natural Populations of
G. domingensis...224
Flowing Aquaria Cultures..225
Unenriched Seawater Culture 225
Groundwater-Enriched Seawater Culture 225
Natural Populations: East Discovery Bay 225
Nitrate and Phosphate Uptake Kinetics 226
In Situ Cultures..226
Nitrate Uptake 226
Phosphate Uptake 227
Flow-Through Cultures: *G. domingensis*...227
Nitrate Uptake (Pulse 1)..227
Phosphate Uptake 227
Nitrate and Ammonium Uptake (Pulse 2) 228
Discussion 231
Growth and Productivity of *Gracilaria* in Nitrate-Enriched Back-Reef Habitats....................231
Factors Influencing Growth of *Gracilaria* Cultures in Back-Reef Habitats.............................232
Salinity and Temperature 232
Nitrate Concentration 232
Phosphate Concentration...232
Epiphytes...233
Herbivory..233
Depth of Culture 234
Weather Conditions 235
Assessment of Nutrient Limitation of *In Situ Gracilaria* Cultures and *G. domingensis*
Natural Populations...235
Growth Rates...235
Nutrient Status: *G. domingensis*..236
Nitrate and Phosphate Uptake 237
Conclusions..239
Acknowledgments..240
References..240

INTRODUCTION

Gracilaria Mariculture for Agar Production

Mariculture of *Gracilaria* species for agar production and as a local food resource in several countries has become increasingly important as an alternative to the harvesting of wild, agar-producing seaweeds (agarophytes). Increasing demand for agar by the food-processing, pharmaceutical, microbiological, and biotechnological industries has led to a world shortage of agarophytes and the possibility of permanent ecosystem damage (McLachlan and Bird 1983; Sullivan 1983; McHugh 1984; McLachlan et al. 1986).

Species of *Gelidium* and *Gracilaria* (Rhodophyta) are the principal commercial agar sources (Chapman and Chapman 1980). While *Gelidium* produces agar of higher gel strength and commands a higher price, some *Gracilaria* species can be pretreated with alkali to significantly increase the gel strength (Hansen et al. 1981) and have, in part, replaced agars from species of *Gelidium* (Mclachlan et al. 1986). With *Gelidium* resources diminishing since the early 1960s, use of several *Gracilaria* species surged, and this genus presently accounts for a major portion of the commercial agarophyte biomass. Due to the rapid evolution of vegetative propagation of *Gracilaria* and the adaptability of some species to cultivation, mariculture of this genus has rapidly increased (Hansen et al. 1981).

The development of *Gracilaria* mariculture is particularly appropriate in tropical developing countries where the high diversity of *Gracilaria* species (Taylor 1960; Doty 1980; McLachlan and Bird 1984) constitutes a broader resource base for the production of diverse agars. Several agars have been found from a variety of *Gracilaria* species that are presently unutilized but appear to have useful properties (Duckworth et al. 1971; Doty 1980). Some species have recently been found to contain unusually good-quality agars. These include *G. cylindrica* (Doty and Santos 1983), *G. sjoestedt*ii (Craigie et al. 1984), and *G. crassissima* (Lahaye et al. 1988). These species are all local to the Caribbean. In Jamaica, at least 16 *Gracilaria* species have been identified (Taylor 1960; Chapman 1963) (Table 17.1).

Table 17.1 *Gracilaria* Species Local to Jamaica: Previously Identified (Taylor 1960; Chapman 1963) and Collected during This Study

Species	Location Collected	Dates Collected
G. terete	Llandovery	July 12, 1987
G. armata T		February 9, 1988
G. verrucosa TC		April 17, 1988
		May 7, 1989
	St. Ann's Bay	December 11, 1987
		January 7, 1988
	E. Discovery Bay	October 10, 1988
	Salem	May 7, 1989
G. blodgettii TC[a]	St. Ann's Bay	December 11, 1987
	Llandovery	April 17, 1988
G. cervicornis TC[a]	DBML lagoon	July 14, 1987
		October 10, 1988
		April 19, 1989
	St. Ann's Bay	December 11, 1987
	Mamee Bay	January 7, 1988
	Llandovery	February 9, 1988
G. compressa TC		
G. crassissima TC[a]	Ocho Rios	July 10, 1987
	Bali Hai	March 20, 1988
		April 10, 1988
		October 11, 1988
		February 12, 1989
	E. Discovery Bay	October 10, 1988
		April 15, 1989
	Salem	May 7, 1989

(Continued)

Table 17.1 *Gracilaria* Species Local to Jamaica: Previously Identified (Taylor 1960; Chapman 1963) and Collected during This Study (*Continued*)

Species	Location Collected	Dates Collected
G. curtissiae TC		
G. damaecornis TC[a]	St. Ann's Bay	December 11, 1987
G. domingensis TC[a]	St. Ann's Bay	December 11, 1987
		January 7, 1988
	E. Discovery Bay	October 10, 1988
		December 17, 1988
		February 12, 1989
		April 29, 1989
G. foliifera TC[ab]	St. Ann's Bay	July 8, 1987
		December 11, 1987
		January 7, 1988
	Llandovery	February 9, 1988
G. mammillaris TC[a]	Ocho Ríos	July 10, 1987
	DBML lagoon	July 12, 1987
		October 10, 1988
		February 2, 1989
G. sjoestedtii TC[a]	Rocky point	July 10, 1987
G. caudata C		
G. cornea C		
G. cylindrica C[a]	St. Ann's Bay	December 11, 1987
G. divaricata C		

[a] Species collected between July 1987 and April 1988. Species previously identified as *G. armata* or *G. verrucosa* are now of questionable identification and have been referred to as *G. terete* pending further taxonomic clarification (Smith et al. 1986).

[b] *G. foliifera* var. angustissima, which is distinct from *G. foliifera* var. foliifera, has been renamed *G. tikvahiae* (Mclachlan 1979). Species identified as *G. foliifera* in Jamaica bear little resemblance to *G. tikvahiae* and may be the latter variety.

T, identified by Taylor (1960); C, identified by Chapman (1963).

Gracilaria Mariculture Potential in Tropical Countries

Factors that can limit the productivity of *Gracilaria* include light, temperature, nutrient availability, salinity, epiphytes, grazing, and the inherent maximum growth rate of the cultivated species (Mathieson and North 1982; Hanisak and Ryther 1984). High year-round sunlight and temperatures in tropical countries permit high productivity of *Gracilaria*, a highly productive and photosynthetically efficient genus (Mclachlan and Bird 1983), when adequate levels of nutrients are available and when epiphytes and grazing can be minimized by cultivation methods and selection of resistant strains (Hansen, 1984). Epiphytes on cultured *Gracilaria* reduce light penetration, slow growth, and at high densities may kill the host (Edelstein et al. 1976; Ryther et al. 1979).

Although some tropical developing countries are in various stages of developing *Gracilaria* mariculture (Raju and Thomas 1971; Tseng 1982; Doty 1978; McHugh 1984; Ren-Zhi et al. 1984), Taiwan is the only country where commercial *Gracilaria* cultivation has been established on a large scale. This has become a widespread means of livelihood for farmers since 1971 due to declining harvestable wild populations, price increases of dried seaweed, and improvements in culturing techniques (Hansen et al. 1981). Three *Gracilaria* species are cultivated in shallow tidal ponds formerly used for milkfish production. The ponds are maintained at a salinity range for optimizing growth by frequent water exchanges. Adequate nutrients are supplied by periodic additions of urea or fermented pig or chicken manure. A novel feature of this system is the introduction of milkfish (*Chanos chanos*) to control epiphytes that will otherwise overgrow the cultured *Gracilaria*. These fish will satisfactorily remove all epiphytes by grazing, but will consume *Gracilaria* after all epiphytes are gone. At this point, large fish must be removed and small ones introduced. Grass shrimps and crabs, which also graze epiphytes, are sometimes introduced to provide additional income. *Gracilaria* productivity in Taiwan, which stops from December to March due to cold weather, is around 40 tons dry wt/ha/year and provides much higher profits than milkfish culture (Shang 1976; Chiang 1981).

Several *Gracilaria* species known locally as sea moss (St. Lucia) or Irish moss (Jamaica) are in great demand in several West Indian islands and communities for the preparation of popular drinks or puddings. *Gracilaria* species have a wide distribution and diversity, but generally low abundance in the Caribbean. Limited wild stocks are partly due to overharvesting, lack of management, and hurricane damage of productive beds. In Jamaica, wild populations of Irish moss have been overexploited, and many of the formerly most productive beds have been eliminated.

Experimental cultivation of several *Gracilaria* species [*G. crassissima*, *G. domingensis*, *G. cornea*, and *G. terete*, a species of uncertain identification that may have been previously identified as *G. verrucosa* or *G. armata* (Smith et al. 1986)] leading to viable mariculture has recently been initiated in St. Lucia (Smith et al. 1984, 1986) by adopting procedures used for *Eucheuma* farming in the Philippines (Doty 1980). Vegetative propagation, in which fragments are inserted between braids of polypropylene rope that is then attached to floating bamboo rafts in shallow coastal areas, has had varying degrees of success. Initial problems included the accumulation of sediment and grazing by herbivorous fish. *G. terete* was found to be the most successful candidate for commercial mariculture due to its high growth rate and its remaining free of sediment and epiphytes. Herbivory was minimized by keeping culture lines at least 1 m above the seagrass out of the apparent normal range of herbivorous fish. *G. domingensis* was found to be less suitable for this method of vegetative propagation due to its more fragile morphology, but could be successfully propagated by spore recruitment from natural populations. This work suggests that further development of *Gracilaria* mariculture in the Caribbean will depend on additional research relating to the selection of improved strains for increased growth rates and intrinsic crop value; further characterization of successful culture sites with respect to seasonal salinity, sedimentation, and nutrient fluctuations; the potential benefits of nutrient fertilization; appropriate culture methods suitable for other species with high intrinsic value; and improvement of techniques for spore recruitment and grow-out.

Gracilaria Mariculture Potential in Jamaican Nitrate-Enriched Back-Reef Habitats

Back-reef habitats along the north coast of Jamaica are highly productive ecosystems fed by limestone groundwater springs and rivers containing very high nitrate levels (D'Elia et al. 1981; Goreau et al. 1986). Local strains of *G. terete* cultured in the Discovery Bay back-reef, using techniques similar to those in St. Lucia, doubled in weight in 8–12 days (Macfarlane et al. 1988). These rates are significantly higher than those obtained in St. Lucia, where selected fast-growing strains double in weight in around 15 days (Smith, in preparation).

Of 36 elements analyzed in limestone groundwaters along the Jamaican north coast, nitrogen was the only element found to be unusually elevated (Goreau et al. 1988). Phosphate levels are very low in these groundwaters, which have N:P ratios of around 625:1 (D'Elia et al. 1981). This situation suggests that growth of algae cultured on artificial substrates in back-reef habitats may be subjected to phosphate limitation.

Flowing seawater culture experiments with a fast-growing strain of *G. tikvahiae* showed that continuous enrichment of seawater with nitrogen and phosphorus has led to extreme epiphyte growth, smothering and eventually killing the plants (Lapointe and Ryther 1978). Epiphytes were inhibited by the strategy of pulse feeding, in which the plants were periodically removed from the culture tanks and immersed in a concentrated nutrient solution for as little as six hours. During this treatment, they were able to rapidly assimilate and store internally enough nitrogen to grow at their maximum rate in unenriched seawater for up to two weeks (Ryther et al. 1981). Applied to *in situ Gracilaria* cage cultures in the Florida Keys, this technique resulted in high year-round yields of 35 g dry wt/m^2/day (Lapointe and Hanisak 1984). Pulsed additions of phosphate may therefore be a useful technique for maximizing growth rates and vigor while minimizing the growth of epiphytes on *Gracilaria* cultures in nitrate-enriched back-reef habitats.

In studies on the growth of pulse-fed *in situ G. tikvahiae* cage cultures as a function of concentration and frequency of pulses, it was found that with increasing phosphorus limitation in seawater, a higher-pulse frequency was required to maintain maximum growth rate (Lapointe 1985), implying that *Gracilaria* has a lower storage capacity for phosphorus than for nitrogen. Although much is known about the N storage pools and capacity of *Gracilaria*, little is known about the storage capacity for P (Bird et al. 1982; Lapointe 1985).

Measurement of photosynthesis and growth rates of *in situ G. tikvahiae* cultures in the Florida Keys subsequent to nutrient pulses, in conjunction with tissue C, N, and P levels, have been used to establish year-round phosphate limitation of these cultures (Lapointe 1987). Under these conditions, it was found that at all levels of N enrichment, increasing phosphate enrichment elevated the levels of tissue N. Pulsed phosphate enrichment may thus increase the efficiency of nitrate uptake and utilization by *Gracilaria* cultures in nitrate-enriched back-reef habitats.

Friedlander and Dawes (1985) found that phosphate uptake by *G. tikvahiae* had a rapid linear diffusion phase at high phosphate concentrations and a low saturation phase at lower concentrations. Thus, information on phosphate uptake rates, over a wide range of phosphate concentrations, of *Gracilaria* species will be important to the design of possible fertilization strategies for maximizing mariculture productivity.

Research Objectives

Research carried out in this study was designed to

1. Identify *Gracilaria* species and strains with desirable qualities for mariculture and agar production in nitrate-enriched back-reef environments. Criteria included adaptability to cultivation, growth rate, epiphyte resistance, and agar content.
2. Determine suitable back-reef sites for mariculture on a gradient of high to low groundwater input and test the hypothesis that intermediate locations on this gradient are maximal for the growth of *Gracilaria*. Factors of site variability was investigated, and those that had a major influence on growth of *Gracilaria* cultures included nutrient availability, epiphyte accumulation, and grazing by herbivorous fish.
3. Determine the effects of pulsed phosphate and groundwater enrichment on growth and nutrient status (as C, N, and P composition and ratios) of back-reef and flow-through *Gracilaria* cultures in order to assess possible nutrient limitation and the potential for enhancement of mariculture productivity by phosphate enrichment. Seasonality of nutrient limitation of natural *G. domingensis* populations was investigated as nutrient status relative to that of N- and P-enriched cultures.
4. Determine phosphate and nitrate uptake rates of selected *Gracilaria* species at various phosphate concentrations during pulsed enrichment and possible interactions between simultaneous nitrate and phosphate uptake.

MATERIALS AND METHODS

Field Sampling of *Gracilaria* Populations

A total of 10 *Gracilaria* species (Table 17.1) were collected at various sites along the north and east coasts of Jamaica between July 1987 and April 1988, with nine collected in large enough quantities (usually >100 g) for initial field culture experiments in the Discovery Bay back-reef. These sites include Bali Hai, Discovery Bay, Llandovery, Lover's Beach, St. Ann's Bay, Mamee Bay, and Rocky Point (Figure 17.1). Some species (*G. terete*, *G. cervicornis*, and *G. domingensis*) had a wide range of morphological features, and it remains to be determined whether these varieties should be grouped as one species or separate species or subspecies (Chapman 1961; DeOliveira et al. 1983; Abbot and Norris 1984; Smith et al. 1986). Populations of five species (*G. terete*, *G. cervicornis*, *G. domingensis*, *G. crassissima*, and *G. mammillaris*) were located in the vicinity of Discovery Bay in large enough quantities for repeated sampling and further experimental culture carried out between May 1988 and June 1989. All experimental culture and nutrient-uptake experiments were carried out at the Discovery Bay Marine Laboratory (DBML). The plants were maintained in flowing seawater aquaria (1 m × 1 m × 0.1 m [deep]) in the wet lab or in outside concrete holding tanks (~2 m × 2 m × 0.6 m [deep]) during taxonomic identification, life history stage determinations, and attachment to ropes before deployment for *in situ* culture experiments.

In Situ Culture Sites

Experimental culture sites (designated L2–L4 and BR0–BR2) and sampling locations in the Discovery Bay back-reef are shown in Figure 17.2. Habitat features of the principal culture sites are as follows:

L2: Located in a lagoon close to spring-fed sources of groundwater and somewhat protected from wave action by limestone formations. Depth 1.5–2.0 m with a fine sand and sediment substrate.
L3: Close to L2 but more exposed to wave action and seawater mixing. Depth 1.0–1.5 m with substrate covered with a sparse *Thalassia* bed with open patches of fine sand and sediment.

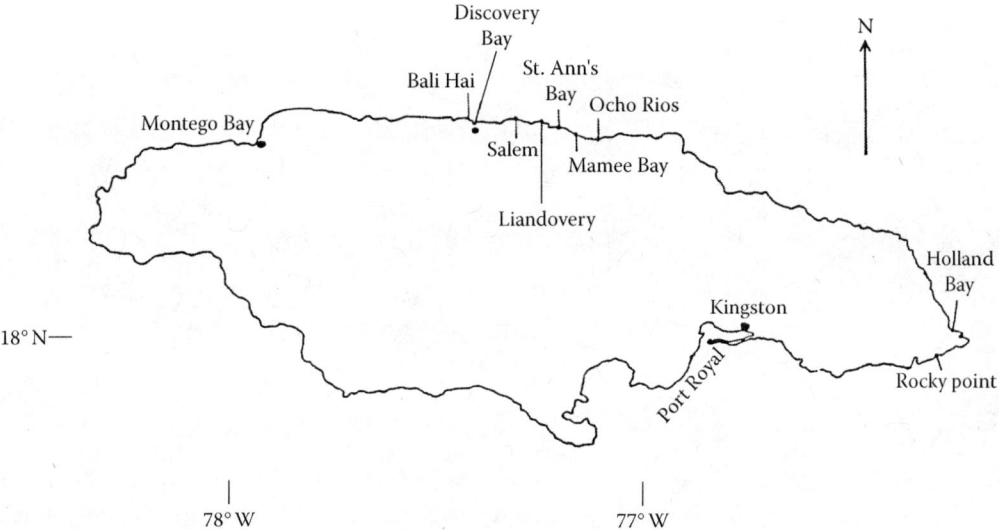

Figure 17.1 *Gracilaria* collection sites along the north and east coast of Jamaica.

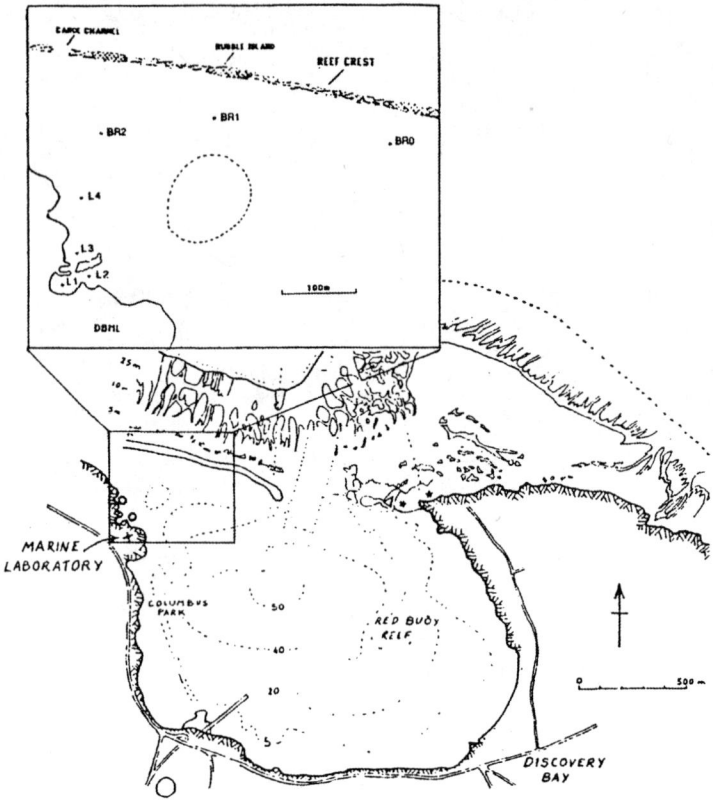

G. domingensis natural populations

△*G. crassissima, G. terete* isolated natural populations

○*G. cervicornis, G. mammillaris* natural populations

Figure 17.2 *Gracilaria* natural populations and culture sites, Discovery Bay (depth contours within the Bay in meters).

L4: Approximately 30 m from shore receiving intermediate groundwater input and relatively exposed to wave action. Depth 2.5–3.0 m with substrate covered with *Thalassia*. Within 5 m of a small patch coral community.

BR1: In the open back reef, ~75 m from the reef crest receiving only low or sporadic groundwater input. Depth 2.0–2.5 m with a clear sandy bottom with some patches of macroalgae and small corals.

In Situ Culture Methods

Several culture methods were used in experiments designed to determine the adaptability of a variety of *Gracilaria* species to mariculture and the growth rates of unfertilized cultures at several locations on a gradient of high to low groundwater input.

Rope Culture

Plants were attached to lengths of polypropylene rope either by tying pieces on with nylon string or by intertwining fronds within the strands of the rope depending on morphology. Seeded ropes were then attached horizontally or vertically to floating rectangular bamboo rafts tethered to rocks at the desired location. Vertical rope cultures were weighted with stones to maintain a vertical orientation and stability with respect to currents. Culture depth ranged from 0.1 to 0.5 m by the position of the plant on a vertical rope or by attaching stones near the two ends of longer horizontal ropes. Rope cultures were easily detached from the rafts for periodic cleaning, weighing, observations, and application of desired treatments. Culture ropes were wet-weighed before the attachment of plants in order to determine the weight of cultures at initial and subsequent weighings that were carried out at intervals of one week to one month, depending on the culture and experiment. Cultures were cleaned and shaken loose of excess water before weighing. See photo below of *Gracilaria terete* growing on rope culture in bamboo raft. (Photograph by Dr. Peter D. Goreau.)

Vexar Cage Culture

Plants were enclosed in cylindrical or oval (in cross section) vexar cages with a 1.3–2.5-cm mesh size also attached to floating bamboo rafts. These cages floated near the surface and the plants were kept within the top 30 cm of the water column. Plants were removed from cages for periodic weighing and harvesting, at which time the cages were thoroughly cleaned of epiphytes with a stiff brush. Cage cultures were more appropriate for species with fragile or highly branching morphology and were provided some protection from grazing, but conditions within cages differed significantly from those on ropes with respect to light and water movement, particularly with the accumulation of dense epiphyte cover on cages.

Net Culture

Plants were enclosed in rectangular pockets (~0.3 m² in surface area) in 2.5-cm nylon gill-net structures horizontally stretched across floating bamboo rafts. This method was tested for use with species having highly branching morphologies and as a more appropriate means, relative to cages, for their *in situ* mass culture. Plants were removed from the pockets for periodic weighing, at which time the nets were cleaned by rinsing in fresh water and sun-drying.

Nutrient Enrichment and Controlled Grazing Experiments

This design was used for experimental cultures of three species (*G. terete, G. domingensis,* and *G. crassissima*) at three locations with high, medium, and low groundwater input, respectively (L2, L4, and BR1), in order to

1. Assess nutrient limitation at culture sites and the effects of phosphate enrichment on growth and vigor (as morphology and epiphyte resistance) of these species (nutrient enrichment experiment).
2. Compare growth rates at different locations and quantify epiphyte and grazing differences with respect to location (controlled grazing experiments).

Vertical rope cultures of individual plants (5–15 g) were suspended from horizontal ropes on rafts within and exterior to large hemicylindrical vexar cages (1.3-cm mesh, 2 m × 0.7 m open surface area) open to the surface, permitting equal irradiance on enclosed and nonenclosed cultures while isolating enclosed cultures from fish grazing. Replicate enclosed and nonenclosed cultures, suspended at two depths of 0.1 and 0.4 m (3 replicates/position = total of 12 replicates), were used for controlled grazing experiments. Twelve to 15 shallow enclosed cultures were used for nutrient enrichment experiments (4–5 treatments, 3 replicates/treatment). At weekly intervals, all cultures were cleaned of epiphytes and weighed, and quantitative and qualitative estimates of epiphyte abundance were made.

At this time, nutrient pulse treatments were applied to the shallow-caged cultures. Treatments consisted of soaking the plants in a range of seawater/groundwater mixtures for 12 hours overnight in 18-L plastic pails that were aerated using aquarium pumps to facilitate nutrient uptake. Treatment media included the following:

1. Control (ambient seawater from the site at which the cultures were maintained)
2. A seawater–groundwater mixture obtained from a lagoon or limestone channel close to freshwater spring inputs with a nitrate concentration in the range of 10–20 µM (NB: µM means micromolar) and a salinity range of 28–33 ppt. This was the control treatment for cultures at a site (L2) in the lagoon near a freshwater spring (abbreviated as +GW).
3. Medium 2 plus low phosphate (~2 µM); N:P = 10 (abbreviated as +LP)
4. Medium 2 plus medium phosphate (~20 µM); N:P = 1 (abbreviated as +MP)
5. Medium 2 plus high phosphate (~200 µM); N:P = 0.1 (abbreviated as +HP)

NaH_2PO_4 was used for all phosphate enrichment solutions. Duplicate water samples of each treatment medium were taken at the start and end of the pulse using a 50-mL syringe fitted with a GF/F filter and frozen for later analysis in order to determine nutrient uptake during pulse treatments. All plants used in these and in the flow-through culture experiments were ascertained to be of the same life-history phase (tetrasporophytes) in order to eliminate possible variability caused by different growth rates or responses of different reproductive phases. These experiments were continued for 4–5 weeks (3–4 pulses) until a clear trend in growth data and/or morphology was observed. Cultures were not periodically harvested back to their original weight because of the difficulty of harvesting new growth tissue from attached cultures in an unbiased manner. At the end of the experimental growth period, all cultures were thoroughly cleaned and rinsed in distilled water to remove debris and organisms and oven-dried at 60°C for 48 hours for later analysis of C, N, and P contents.

The range of phosphate concentrations used in the nutrient pulses were chosen to give approximate N:P molar ratios of 10, 1, and 0.1 for the three phosphate-enriched treatments, respectively. This N:P range appears appropriate for investigating the effects of P enrichment and nutrient uptake of *Gracilaria* in high-nitrate and low-phosphate habitats. This phosphate concentration range was also similar to that used by Friedlander and Dawes (1985) to demonstrate multiphasic phosphate uptake in *G. tikvahiae*.

Growth rates (k) of all cultures were calculated and expressed as % increase or decrease in fresh weight per day using the equation following:

$$k = \frac{100\left[\log_2\left(W_t/W_o\right)\right]}{t}$$

where W_o is the initial biomass and W_t is the biomass at time t (in days) at the end of the interval t between successive weighings (Brinkhuis 1984). From this equation, the doubling time in days can be conveniently calculated as the reciprocal of $k/100$.

Epiphyte Observations and Quantification

The major categories of epiphytes occurring on *in situ* cultures included the following:

1. Diatoms, accumulating as a loose film
2. Attached red and green filamentous species
3. Green parenchymatous species, mainly *Ulva* and *Enteromorpha* spp.
4. Attached red macrophytic species such as *Acanthophora spicifera*, *Hypnea*, and *Laurencia* spp.
5. Loose drifting filamentous or macrophytic species such as *Chaetomorpha* or *Sargassum* spp. The occurrence of epiphytes in these major categories on *Gracilaria* cultures were recorded when the cultures were cleaned and weighed.

During nutrient enrichment and controlled grazing experiments, an attempt was made to quantify the occurrence of the most consistent epiphytes while recording the presence and relative abundance of less = common species. Diatoms on several replicate vertical rope cultures at each location were shaken loose in a fixed volume of seawater. With the diatoms in homogeneous suspension, several subsamples of 100 or 200 mL were filtered through a dry preweighed filter paper. These were dried and weighed to obtain a mean dry diatom weight per subsample, which was used to calculate the total quantity of diatoms shaken loose. Quantities of diatoms as % diatom dry weight of *Gracilaria* dry weight were determined. Attached *Ceramium* spp. were very time-consuming to completely remove by hand and thus difficult to quantify by weight. A visual estimate was made of relative abundance of *Ceramium* and *Enteromorpha* spp. on each replicate culture.

Grazing Observations and Quantification

The occurrence of grazing was evident as clearly cut fronds of *Gracilaria* and the presence of herbivorous fish in the vicinity of the rafts. Relative amounts of cut *Gracilaria* fronds were used as a qualitative estimate of grazing. An attempt was made to quantify grazing intensity at different locations by comparing differences in growth rates between caged and uncaged replicate cultures (described above in the section Nutrient Enrichment and Controlled Grazing Experiments). This measure of grazing intensity is based on a simple, noninteractive plant–herbivore model, with assumptions appropriate to a short-term growth experiment (Caughley 1976). In this model, $g' = g - cH$ (where g' = growth or decline rate of grazed plant population, g = intrinsic growth rate of plant population, c = herbivore grazing rate, and H = herbivore population size). Differences in growth rates between grazed and ungrazed cultures thus provide an estimate of cH or grazing intensity.

Weather Observations

During nutrient enrichment experiments (April–June 1989), daily weather observations, including relative wind intensity and direction, cloud conditions, rain, and water conditions, were recorded.

Nutrient Uptake Experiments

An outdoor flow-through culture system was designed to investigate in more detail the nutrient uptake during pulse treatments similar to those used for *in situ* nutrient enrichment experiments and to compare the effects of these treatments on *Gracilaria* cultures under more controlled conditions. One species (*G. domingensis*), for which more comprehensive data on the effects of nutrient enrichment on *in situ* growth rates and nutrient status were obtained, was used in these experiments.

Five gallon plastic pails were sectioned longitudinally and used as replicate culture chambers with a capacity of 12.35 L each. Low-nutrient and groundwater-enriched seawater were each gravity fed to five culture chambers (total of 10 chambers) from two 3-m³ concrete holding tanks. Gravity-fed water was filtered through a nylon screen with a mesh size of 0.2 mm. One holding tank was fed by the existing DBML seawater intake system and the other by a separate plumbing and intake system supplying groundwater-enriched seawater from a nearby limestone channel. Nutrient levels and physical parameters (temperature and salinity) of the flow-through system are shown in Table 17.2. Flow rates providing exchange rates of ~70 culture volumes/day for each chamber were maintained with adjustable clamps. High exchange rates were maintained in order to prevent overheating of the cultures during the day and to optimize conditions for CO_2 and nutrient exchange (Lapointe and Ryther 1978, 1979). Cultures were aerated for 9–12 hours during the day via a system of perforated plastic tubing fed with compressed air from SCUBA tanks.

Plants used in these experiments were collected from East Point, Discovery Bay, and acclimated in back reef (BR1) cage or net culture for several weeks before the experiments. Three replicate plants (6–20 g fresh weight/plant) were placed in each chamber and were individually identified by attaching small pieces of nylon string of a different color to each plant. In an initial experiment, the plants were acclimated to the flow-through system for one week. At this time it became apparent that high light intensities due to the high reflectivity of the culture chambers were damaging the cultures (observed as pale green color, bleaching, and fragmenting of the plants), and this experiment was terminated. Subsequent measurements with a Li-Cor model LI-185A light meter using a 2-pi quantum sensor demonstrated much higher light intensities in the culture chambers than *in situ* culture conditions (Table 17.3). Light intensity in the culture chambers were subsequently maintained at levels similar to *in situ* culture conditions by shading with a fine vexar mesh screen.

In a second two-week experiment (June 6–20, 1989), the cultures received two pulse treatments: at the start and after one week of flow-through culture. During treatments, the inflow was stopped, but aeration continued and the plants soaked in a range of enriched media (one treatment/chamber), similar to that used for the *in situ* cultures, for eight hours starting during late afternoon. Nutrient depletion during all treatments was followed by removing duplicate water samples every two hours

Table 17.2 Flow-Through Culture System: Physical Parameters and Nutrient Levels during a 2-Week Experiment (June 6–20, 1989)

Treatment	Date	Temp. (°C)	Salinity (ppt)	Dissolved Nutrients (µM)			N:P
				(NH_4^+)	NO_3^-	PO_4^{3-})	
Unenriched	6/8/89	29.0–29.6	36.0				
Seawater	6/16/89			0.85	0.68	0.18	8.5
Culture	6/17/89	29.3–29.6					
Chambers	6/20/89		36.0	2.86	1.59	0.23	19.3
Groundwater	6/8/89	27.9–28.5	28.7				
Enriched	6/16/89			1.65	18.52	0.35	57.6
Seawater	6/17/89	27.7–28.2					
Culture Chambers	6/20/89		28.9	5.44	16.07	0.27	79.7

Table 17.3 Irradiance of Flow-Through and *In Situ* Culture Systems (All Readings Taken between 11.30 AM and 12.30 PM under Full Natural Irradiance)

Culture Location		Irradiance ($\mu E/m^2/s$)
Flow-Through Culture Chamber		
Unscreened		2100–2460
Screened		1200–1500
***In Situ* (BR1) Hemicylindrical Vexar Cage**		
Top	empty	1230
	with *G. terete* cultures	1080
Bottom	empty	1200
	with *G. terete* cultures	795
Open water (BR1, 10–20 cm)		1575
Inside closed vexar cage (BR1, 10–20 cm)		1080

using a 50-mL syringe fitted with a GF/F filter. These samples were immediately frozen for later analysis. During the initial pulse, nutrient-uptake data were obtained for two replicate sets of treatments. The plants were weighed before each pulse, and tissue samples of cultures receiving each treatment were taken before and after the second pulse, rinsed in distilled water, and oven-dried (60°C for 48 hours) for later determination of C, N, and P contents. At the end of the experiment, all cultures were weighed and similarly processed for C, N, and P determinations.

During the second pulse treatment, additions of ammonia (as NH_4Cl) were mistakenly substituted for additions of phosphate; thus, phosphate-uptake data were available only for the initial pulse. Data pertaining to ammonium and nitrate uptake by *G. domingensis* were available from the second pulse.

Physical and Chemical Parameters

Water temperature and salinity were measured at approximately weekly intervals at all culture locations during *in situ* culture experiments. During initial field culture trials (November 1987–May 1988), water samples for nutrient analysis were collected from culture locations at approximately monthly intervals, filtered (GF/F), and frozen for later analysis. During nutrient enrichment experiments (April–June 1989), water samples were collected at weekly intervals from each culture location, filtered, and frozen for later nutrient analysis. Salinities were determined by measuring sample densities (using a hydrometer) and temperatures, and using a density–temperature–salinity nomogram to determine the corresponding salinities. Frozen water samples, including those from nutrient-uptake experiments, were subsequently analyzed for nitrate, ammonia, and phosphate concentrations on a Technicon II autoanalyser. Some water samples were analyzed for nitrate and phosphate at DBML using the methods of Morris and Riley (1963) and Murphy and Riley (1962). For these samples, reliable replicate nitrate concentrations were obtained, but highly variable replicate phosphate concentrations were obtained and were not included in the results.

Agar Extractions and Determinations

All samples used for agar extractions were cleaned of epiphytes, rinsed in freshwater, and oven-dried at 60°C for 48 hours. Bleached samples were sun-dried, with periodic freshwater rinses. Extractions were carried out by heating 4 g dried *Gracilaria* in 100 mL distilled water in covered beakers in a boiling water bath for three to four hours or in a pressure cooker for one hour at 15 psi. The resulting liquid was filtered through cheese or J cloth on a preheated vacuum filter and allowed to cool, and the resulting gel cut in squares and frozen. Frozen gels were later thawed to separate out water and impurities, and the freeze–thaw process repeated for some gels. If significant amounts of agar remained unextracted (indicated by the condition of the plant material and amount of agar obtained from a second extraction), then second or sometimes third extractions were carried out with remaining plant material from previous filtrations. Separated agars were then oven-dried at 60°C for 24 hours and weighed in order to calculate % agar content by weight.

Pretreatments with NaOH were carried out in order to determine their effects on extracted agar quantity by heating separate samples in 1% or 2% NaOH solutions in a 70°C water bath for one hour. The plant material was then rinsed and neutralized with 1% HCl followed by several additional rinses before extractions carried out as above.

Tissue Analysis for C, N, and P Contents

Carbon and nitrogen contents were determined using a Carlo Erba 1106 Elemental Analyzer, and total phosphorus was determined using a modified persulfate digestion method (Menzel and Corwin 1965) followed by phosphate analysis using the method of Murphy and Riley (1962).

Statistical Analysis

Single classification analysis of variance (ANOVA) followed by a priori tests of differences among means (Sokal and Rohlf 1969) were used to assess the effects of nutrient enrichment on growth rates and tissue C, N, and P levels of *in situ* and flow-through cultures, as well as the effects of culture location, grazing, and depth of culture on growth rates of *in situ* cultures. Temporal growth rate differences of *in situ* cultures and the potential effects of variable weather conditions were also assessed using ANOVA.

RESULTS

Strain Selection for Mariculture

Adaptability to Culture, Growth, and Epiphyte Resistance of Collected Gracilaria Species and Strains

G. terete

This species had terete, dichotomously branched fronds, 1–2 mm in diameter, that grew up to lengths of 30 cm or more. Fronds were flexible and resilient and highly amenable to rope culture by intertwining within the strands of rope. This was the most abundant species sampled and for which most initial growth data were obtained.

During initial growth experiments with horizontal rope cultures at L2 and BR1 (July 14–August 17, 1987), high growth rates, with doubling times of 7.2–8.9 days, were obtained during the first week of culture (Figure 17.3 and Table 17.4). In similar growth experiments carried out at three locations (L2, L3, and BR1; November 14–December 20, 1987) using plants that had grown untended at BR1 for two months, lower growth rates (with doubling times of 10.6–17.9 days) were obtained. This appeared to be related to the reproductive condition of the plants that commonly had small sporeling clumps attached to the parent thalli and tended to fragment more easily. The occurrence of *Ceramium* sp. was first noticed on the cultures at BR1 during this experiment. These cultures were subsequently increasingly overgrown by *Ceramium* with associated lower growth and fragmentation.

Similar culture experiments with freshly collected *G. terete* plants carried out at BR1 at different times of the year (February 2–March 29, 1988; April 22–May 14, 1988; May 10–June 23, 1989) resulted in high growth rates similar to those obtained with the original cultures (Table 17.4),

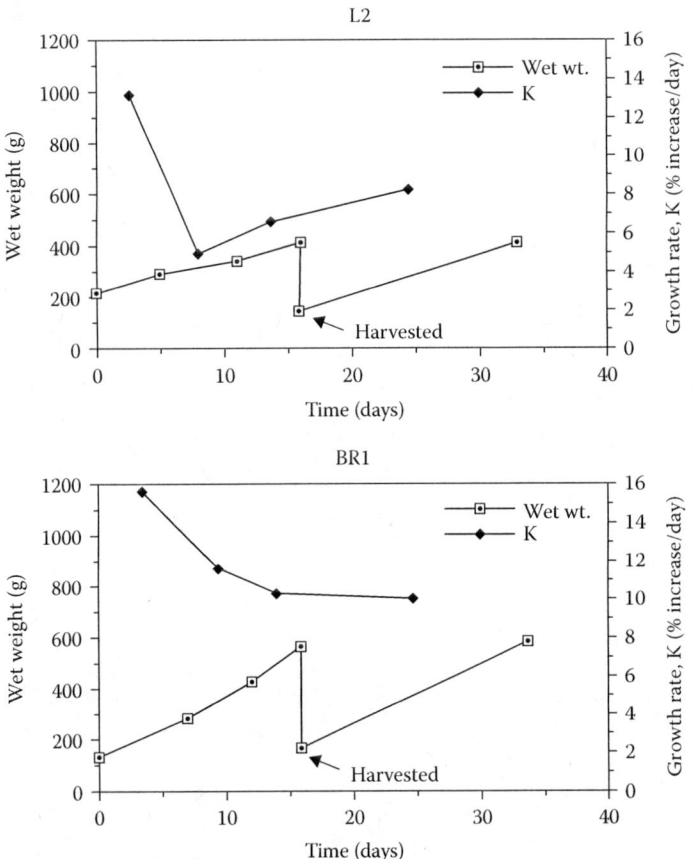

Figure 17.3 Weight increase and growth rates of simultaneous *G. terete* horizontal rope cultures at L2 and BR1 (July 14–August 17, 1987).

indicating the potential of high year-round growth in culture of this species in the back-reef environment.

Highest *G. terete* growth rates (doubling time of 6.4 days) were obtained with vertical rope cultures at L4 (December 16–22, 1987) of unusually densely branching plants from St. Ann's Bay (Table 17.4). However, these cultures, later determined to be female gametophytes, rapidly fragmented after three weeks in culture with the development of cystocarps.

Varying degrees of fouling by *Ceramium* sp. occurred on most *G. terete* cultures at BR1, limiting the long-term viability and vegetative propagation of cultures. The occurrence and abundance of *Ceramium* on *G. terete* cultures appeared to be related to a number of factors, some of which may be amenable to manipulation in order to minimize its occurrence. These factors include the following:

1. *G. terete* strains: More vigorous strains appeared to be more resistant to colonization by *Ceramium*. Viable cultures of vigorous strains of *G. terete* were maintained at BR1 for over one year (April 22, 1988–June 3, 1989), despite long untended periods and with maximum growth rates (relative to original cultures) during tended periods. Low levels of *Ceramium* were always present to varying degrees on some of these cultures.
2. Grazing by amphipods: Amphipods, which commonly live among *Gracilaria* cultures, were observed in two replicate lab experiments to completely remove, by grazing, dense *Ceramium* growth on *G. terete* thalli within 1–2 weeks.

Table 17.4 Characteristic Growth Rates of Unfertilized In Situ Gracilaria Cultures

Species	Origin	Growth Rate (% Increase/Doubling)	Day ± S.D. (N), Time (Days)	Date (No. of Days in Culture)	Site Culture	Method
G. terete	Llandovery		13.99 ± 2.09(2), 7.15	7/21/87 (7)	BR1	Hor. rope
			11.19 ± 2.67 (6), 8.94	7/19/87 (5)	L2	Hor. rope
			11.83 ± 0.65 (5), 8.45	2/22/88 (11)	BR1	Hor. rope
			11.31 ± 0.10 (2), 8.84	5/03/88 (11)	BR1	Hor. Rope
			13.83 ± 0.77 (2), 7.23	6/03/89 (24)	BR1	Hor. Rope
	St. Ann's Bay		15.55 ± 2.38 (2), 6.43	12/22/87 (6)	L4	Vert. Rope
G. domingensis	St. Ann's Bay	20.72	4.83	12/22/87 (6)	L4	Vert. rope
			10.97 ± 3.52 (2), 9.12	12/22/87 (7)	L4	Hor. Rope
			7.83 ± 0.86 (3), 12.77	12/21/87 (7)	BR1	Hor. rope
		12.21	8.19	1/20/88 (11)	BR1	Vert. Rope
	In situ sporeling	14.60	6.85	3/18/88 (11)	BR2	Vexar cage
	EDB[a]	4.97	20.12	12/11/88 (61)	BR1	Vexar cage
		5.96	17.36	12/22/88 (10)	BR1	Vexar cage
			4.60 ± 0.05 (2), 21.74	5/2/89 (13)	BR1	Vexar cage
			3.80 ± 0.38 (2), 26.32	5/2/89 (13)	BR1	Net
			4.46 ± 0.08 (2), 22.42	6/11/89 (16)	BR1	Net
			6.21 ± 1.13 (3), 16.10	4/14/89 (7)	BR1	Vert. rope
			14.27 ± 1.86 (3), 7.01	4/16/89 (7)	L4	Vert. Rope
G. blodgettii	St. Ann's Bay	21.51	4.64	12/22/87 (6)	L4	Vert. Rope
		27.48	3.64	1/1/88 (5)	BR1	Vert. Rope
		20.40	4.90	1/1/88 (5)	BR1	Vexar cage
	Llandovery	11.44	8.74	4/29/88 (10)	L4	Vexar cage

Species	Location			Date (n)		Method
G. cervicornis	DBML lagoon	8.41	11.89	7/22/87 (9)	L2	Vexar cage
		5.40	18.52	7/22/87 (8)	BR1	Vexar cage
		13.87	7.21	5/2/89 (13)	BR1	Net
	Mamee Bay	3.67	27.25	1/20/88 (7)	BR1	Vexar cage
			8.38 ± 1.15 (2), 11.93	1/20/88 (8)	BR1	Hor. Rope
G. crassissima	Ocho Rios	2.7	37.04	7/26/87 (4)	L2	Vexar cage
	Bali Hai[b]	3.48	28.74	4/29/88 (10)	BR2	Vexar cage
			5.15 ± 1.70 (3), 19.42	2/27/89 (13)	BR1	Vert. Rope
			7.21 ± 1.50 (3), 13.87	4/30/89 (6)	L4	Vert. Rope
	EDB		6.51 ± 2.51 (4), 15.36	5/5/89 (6)	BR1	Vert. rope
G. damaecornis	St. Ann's Bay		10.14 ± 3.31 (2), 9.86	12/22/87 (6)	L4	Vexar cage
G. mammillaris	DBML lagoon	4.86	20.58	7/22/87 (9)	L2	Vexar cage
			7.96 ± 0.65 (2), 12.56	2/14/89 (12)	BR1	Vert. rope

[a] East Discovery Bay.
[b] ~1 mile west of Discovery Bay.

3. Depth of culture: Replicates of shallow (within 15 cm of the surface) and deep (30–45 cm) horizontal rope cultures were maintained at BR1 for two months (February 3–April 7, 1989) but were untended during the second month. Although the shallow cultures initially grew twice as fast as the deeper cultures (mean growth rates of 11.6% increase per day for shallow cultures compared to 6.7% increase per day for deep cultures), they became more rapidly fouled with *Ceramium*. At the end of two months, the shallow cultures had a stunted growth with much epiphyte cover and fragmentation, while deeper cultures were much cleaner, with healthier fronds (during the second month, mean growth rate of the shallow cultures was −2.6% increase per day compared to 2.2% increase per day for the deep cultures). Deeper cultures also had a much denser population of amphipods living among the fronds relative to the shallow cultures.

G. domingensis

G. domingensis had flattened central fronds, 2–5 mm wide with dichotomous or pinnate branching. Branching morphologies ranged from elongated moderately branched plants up to 30 cm in length, to densely branched, bushy plants up to 15 cm or more in diameter. Fronds were flexible and moderately resilient and were amenable to rope culture if tied on with string. Densely branched morphologies were adaptable to vexar cage or net culture.

Very high to moderate growth rates (doubling times of 4.8–12.8 days) were obtained with rope cultures of varieties from St. Ann's Bay at L4 and BR1 (Table 17.4). Highest growth rates were obtained with a particularly vigorous plant (labeled 2k) in vertical rope culture at L4 (Figure 17.4) that was also more epiphyte-resistant relative to other cultures. Significantly lower growth rates of this strain after approximately two weeks in culture appeared to be related to factors other than self-shading, because growth rates subsequent to harvesting, as well as growth rates of replicate cultures using harvested material, were also much lower. Other vertical and horizontal rope cultures had initial growth rates comparable to *G. terete* cultures (Table 17.4) but tended to fragment or become necrotic more rapidly with associated dense growths of green and red filamentous epiphytes (including *Enteromorpha* and *Ceramium* sp.). The original culture of strain 2k remained mostly free of such epiphytes during almost five months in culture (December 16, 1987–May 9, 1988).

Spore release by these cultures was often indicated by the appearance and growth of young sporelings attached to the rafts or to loose ends of rope. A loose sporeling (later found to be a female gametophyte) cultured in a vexar cage at BR2 had very high growth rates (doubling time of 6.9 days; Table 17.4) while remaining epiphyte-free for up to three weeks. Subsequent reduction in growth appeared to be associated with carposporophyte development and dense fouling of the cage during a period of unusually calm weather.

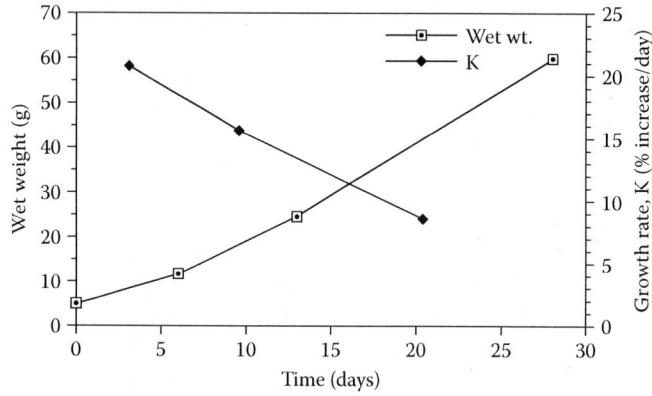

Figure 17.4 Weight increase and growth rate of a vigorous *G. domingensis* vertical rope culture at L4 (December 16, 1987–January 13, 1988).

Table 17.5 Initial Density and Corresponding Growth Rates of G. domingensis Net Cultures at BR1 after 15 Days in Culture (May 26–June 11, 1989)

Initial Density (kg/m^2)	Growth Rate (% Increase/Day)
0.61	4.40
1.21	3.47
1.72	2.48

Populations of another strain of G. domingensis were subsequently located on the east coast of Discovery Bay in a shallow, high-wave-energy rocky shore habitat distinctly different from the more protected habitat at St. Ann's Bay, where the plants were growing at depths of 2–3 m on a fine sand or silty substrate. Discovery Bay plants had a more compact and densely branching morphology relative to those from St. Ann's Bay, with resilient thalli more resistant to fragmentation by wave action.

Moderate growth rates (with doubling times of 16.1–21.7 days) in vexar cage and vertical rope culture at BR1 were obtained with this strain (Table 17.4), which were generally lower than those of the previous G. domingensis strains. This strain proved, however, to be very epiphyte resistant and amenable to long-term in situ culture. A few small plants (26.3 g) of this strain (collected after the passing of hurricane Gilbert, which decimated local macroalgal populations) grew to over 100 times the original weight (2776 g) during six months of largely untended vexar cage culture at BR1 (October 11, 1988–April 17, 1989). During this period, the plants remained mostly free of epiphytes despite dense fouling of the cages by filamentous and frondose epiphytes.

Relatively low growth rates (with doubling times of 22.4–26.3 days) were obtained with larger scale net cultures of this strain (Table 17.4). Although the plants in net culture remained healthy and appeared more vigorous relative to simultaneous cage cultures, low growth rates may have been partly due to self-shading, as the plants tended to be compacted to one side of the net sections by prevailing unidirectional wave action. In one experiment (May 26–June 11, 1989), growth rate was inversely related to biomass of culture per net section (Table 17.5). The wave-driven brushing action of G. domingensis thalli tended to maintain the nets free of diatom accumulation and subsequent colonization by epiphytes.

G. blodgettii

Tentatively identified, plants of this species had densely branching terete fronds (1–2 mm in diameter) that formed bushy clumps up to 15 cm in diameter. Fronds were fleshy, soft, and flexible, and although fragile, they were quite resistant to fragmentation, partly due to their tendency to fuse together at points of contact as they grew out. Thalli were somewhat amenable to rope and cage culture.

Very small and isolated populations of this species were observed in the field at St. Ann's Bay (December 11, 1987) and Llandovery (April 1988) and were not found on subsequent trips to these locations. Thus, little material was available for experimental culture.

Very high growth rates (with doubling times of 3.6–5.4 days) were obtained with vertical rope and vexar cage culture at three locations (L4, BR2, and BR1; Table 17.4) of plants collected from St. Ann's Bay. These rates were the highest among all Gracilaria cultures. An initial plant of 6 g was grown in vertical rope culture at L4 to several hundred grams (Figure 17.5), providing material for replicate cage and rope cultures. During rapid growth periods, densely branching clumps remained mostly free of epiphytes. Cultures were more susceptible to fragmentation and fouling (by microalgae on inner thalli and filamentous green and red species on lower or shaded surfaces) after 4–6 weeks in culture, at which time growth rates were much lower. Significant grazing of many cultures was also apparent.

Vexar cage cultures of another strain of this species (collected from Llandovery) at L4 also had high growth rates (with a doubling time of 8.7 days; Table 17.4), but these were much lower relative

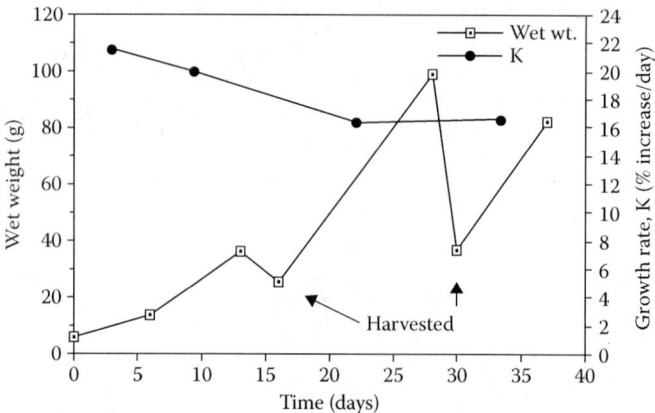

Figure 17.5 Weight increase and growth rates of G. *blodgettii* vertical rope culture at L4 (December 16, 1987–January 22, 1988).

to those of the first strain. These plants were also more susceptible to fouling by microfilamentous epiphytes and rapidly fragmented under rope culture.

G. cervicornis

G. *cervicornis* had densely branched terete fronds, 1–3 mm in diameter. Branching morphologies ranged from dense, compact plants to spreading, more elongate plants growing up to 20 cm or more in diameter or length. Fronds were stiff and brittle and clumps easily fragmented into smaller pieces, but were somewhat amenable to vexar cage and net culture.

Moderate to low growth rates (with doubling times of 11.9–27.3 days) were initially obtained with both vexar cage and rope cultures at three locations (L2, L4, and BR1; Table 17.4). Cultures tended to fragment after two to four weeks, associated with colonization by epiphytes that commonly included *Hypnea*, *Enteromorpha*, *Ulva*, and *Ceramium* species. Untended cage cultures at BR1 remained viable for up to two months (October 11–December 11, 1988).

Horizontal rope cultures at BR1 of plants collected from Mamee Bay had much higher growth rates relative to vexar cage cultures (Figure 17.6 and Table 17.4) despite fragmenting. These cultures appeared particularly vigorous relative to the large number of strains of this species collected from several locations.

Highest growth rates, comparable to maximum G. *terete* growth rates, were obtained with BR1 net culture of plants collected from the DBML lagoon (Table 17.4), but this was maintained only for two weeks, followed by much lower growth and fragmentation of the culture.

G. crassissima

Fronds of this species were cylindrical or mostly flattened (up to 5 mm thick and 2 cm broad) and closely irregularly or alternately branched. Branching was tangled or often recurved. Fronds were cartilaginous, highly resilient, and somewhat amenable to rope culture.

Low growth rates (with doubling times of 19.4–37.0 days) were obtained with vexar cage cultures at three locations (L2, BR1, and BR2; Table 17.4). Although healthy growth of new fronds was observed for some of these cultures, older fronds tended to become necrotic and fragment after two to three weeks in culture. Much higher growth rates (with doubling times of 13.9–15.4 days) and longer-term viability were subsequently obtained with vertical rope cultures at two locations (L4 and BR1; Table 17.4). Vertical rope cultures at BR1 remained viable for up to 2.5 months (February 14–April 29, 1989), increasing up to six times in weight despite an untended period of one month. Subsequently, these cultures tended to become necrotic and fragment.

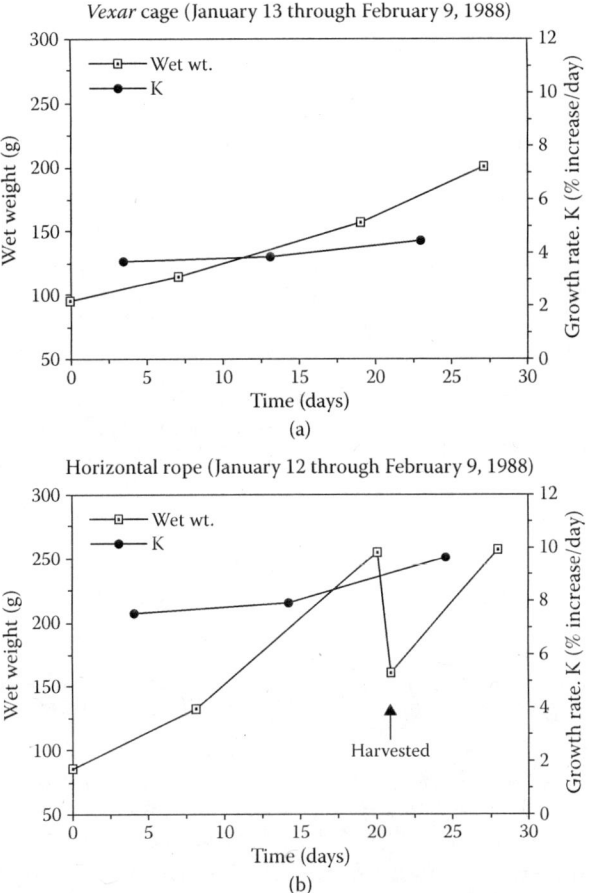

Figure 17.6 Weight increase and growth rates of *G. cervicornis* vexar cage and horizontal rope cultures at BR1 (January 12–February 9, 1988)

Although these cultures became easily fouled, commonly by diatoms and *Ceramium* and *Enteromorpha* species, fouling organisms were easily removed by hand or vigorous rinsing from the large mostly smooth surfaces of some strains (those from east Discovery Bay and Salem). Epiphytes such as *Ceramium* and *Enteromorpha* that became well attached to other *Gracilaria* species appeared to remain only loosely attached to these *G. crassissima* strains.

Horizontal rope cultures with larger amounts of *G. crassissima* were suspended over and in contact with net cultures of *G. domingensis* in order to take advantage of the brushing action of *G. domingensis* fronds during wave action. Relatively low but significant growth rates were obtained (Table 17.4). Although all surfaces in contact with *G. domingensis* cultures remained free of fouling, upper surfaces were densely fouled during untended periods, and this may have contributed to reduced growth.

G. damaecornis

Fronds of this species were terete (2–3 mm diameter) and branched subdichotomously to form compact somewhat horizontally expanded plants up to 10 cm or more in diameter. Fronds were "cartilaginous" and resilient and somewhat amenable to rope culture. Only a small amount of this species (~20 g) was collected on one occasion (December 11, 1989) at St. Ann's Bay.

Relatively high initial growth rates were obtained with vertical rope cultures at L4 (Table 17.4). Subsequent lower growth and fragmenting were associated with fouling by green filamentous epiphytes and cystocarp development. One replicate remained relatively epiphyte-free and viable in culture for up to six weeks, increasing up to nine times in weight.

G. mammillaris

Fronds of this species were flattened (up to 1 cm or more broad) and were repeatedly dichotomously branched above a short compressed or subterete stalk, forming a relatively compact plant up to 15 cm or more in diameter. Fronds were tough and somewhat leathery but tended to fragment easily near the base or central portion of the plant.

Low or negative growth rates were obtained with vexar cage cultures at L2 and BR1 (Table 17.4). The plants tended to rapidly fragment within the first two weeks of culture. Higher growth rates (Table 17.4) and longer culture viability of up to two months (February 2–April 3, 1989) were obtained with vertical rope cultures at BR1. Although these cultures remained relatively epiphyte-free, thalli became necrotic and fragmented at initially low rates that increased significantly toward the end of the culture period.

G. sjoestedtii

Plants of this species had a similar morphology to *G. terete* thalli but with thinner (<2 mm diameter) and more delicate fronds. Samples were collected on only one occasion (July 1987) on the east coast of Jamaica, and only one culture experiment was carried out.

Shallow vertical rope cultures at L2 (20 replicates on a horizontal rope) had very low or negative mean growth rates that appeared to be associated with intense grazing and with fouling by diatoms and filamentous epiphytes at this location. A few replicate cultures had unusually high growth rates (up to 17% increase/day) between July 20 and July 26, 1987, reflected in the high standard deviation of mean growth rate during this period ($-1.21 \pm 12.58\%$ increase/day). Fast-growing plants may initially have escaped grazing, since these along with all other replicates had subsequent negative growth. This may have been an indication of potentially high productivity in this species if cultured at locations with minimal grazing and less intense fouling.

G. folifera

Plants of this species had flattened to terete fronds (1–10 mm wide) that branched irregularly to form elongate, somewhat flattened thalli up to 20 cm or more in length. Fronds were cartilaginous and fragmented easily and were non-amenable to *in situ* culture methods. Vertical rope cultures at BR1 (January 1–20, 1988) were almost completely fragmented and lost from the ropes within the first week of culture.

Agar Contents of Collected *Gracilaria* Species

A wide range of agar contents was found for the *Gracilaria* species collected (Table 17.6). Four species (*G. domingensis*, *G. crassissima*, *G. blodgettii*, and *G. sjoestedtii*) had high agar contents in the range of 30–40%, while two species (*G. terete* and *G. cervicornis*) had moderate agar contents in the range of 20–25%. *G. mammillaris* had a low agar content of <5%. Agar contents of the three other collected species were not determined due to the small quantity of the collected material. Although quantitative measurements were not made, the extracted gels displayed a wide range of characteristics observed as relative gelling temperatures and gel rigidity. (Extracted gels have been saved for subsequent analysis of physical properties.)

Table 17.6 Wet/Dry Weight Ratios and Agar Contents of Collected *Gracilaria* Species

Species	Sample Origin	W/D Ratio	% Agar	Date
G. terete	Llandovery	8.6	23.2 (unbl) 25.5 (bl)	May 9, 1988
		10.2[a]		June 19, 1989
	Salem	6.8		May 7, 1989
G. domingensis	St. Ann's Bay	8.7	30.6	May 9, 1988
	E. Disc. Bay	7.1[a]		June 15, 1989
		6.5		June 24, 1989
G. cervicornis	Mamee Bay	8.2	17.7	May 9, 1988
	DBML lagoon		25.6	May 9, 1988
	Llandovery	7.5		May 7, 1989
G. crassissima	Bali Hai	5.1	38.9	May 9, 1988
		8.4[a]		May 14, 1989
	E. Disc. Bay	8.0		April 15, 1989
	Salem	8.6		May 7, 1989
G. blodgettii	St. Ann's Bay	9.4	32.3	May 9, 1988
	Llandovery	14.5	37.4	May 9, 1988
G. mammillaris	DBML lagoon		2.5 4.0 (+NaOH)	July 15, 1987
G. sjoestedtii	Bowden		12.9 30.5 (+NaOH)	July 15, 1987

[a] W/D ratios of back-reef (BR1) cultures.
unbl, unbleached; bl, bleached.

Between-Habitat Variability

Salinity/Nutrient Relationships of Culture Sites

Nitrate concentrations at culture sites ranged from highs of 20 µM (L2) to lows of 1 µM (BR1). Associated salinity and temperature gradients due to cold freshwater input ranged from 28 ppt and 26°C (L2) to 37 ppt and 29°C (BR1) (Table 17.7, A and B). A highly significant negative correlation between nitrate concentration and salinity was found at five locations on this gradient during January 1988 (Figure 17.7a), which confirmed that nitrate concentration was primarily associated with freshwater input.

Phosphate concentrations at culture locations ranged from ~1.3 to ~0.2 µM (Table 17.7, A and B). A significant negative correlation between phosphate concentration and salinity at five locations during January 1988 (Figure 17.7b) and more frequently higher phosphate concentrations at L2 appeared to indicate an association between phosphate concentration and freshwater input. Atomic N:P ratios (NO_3^-: PO_4^{-3}) ranged from a high of ~80 (L2) to a low of 5 (BR1) (Table 17.7, A and B).

Occasionally, higher-than-usual phosphate and ammonium concentrations at some locations may have been due to nutrient inputs via rainfall that had relatively high phosphate and ammonium concentrations (Table 17.7, D).

Density stratification due to freshwater springs also resulted in distinct depth gradients of salinity, temperature, and nitrate concentration at *in situ* culture locations. The range of the depth gradient of these variables is indicated by their values at surface and bottom layers on December 1, 1987 (Table 17.7, C), and in the Lagoon (L2), was similar to that of the surface gradient from high to low groundwater input locations (more detailed depth profiles of these variables were not obtained due to the unavailability of appropriate sampling equipment). Periodic measurements of temperature

Table 17.7 Temperature and Salinity Range and Concentration of Dissolved Nitrate, Ammonium, and Phosphate at *In Situ* Culture and Sampling Locations

Location	Temp. (°C)	Salinity (ppt)	Dissolved Nutrients (µM)			(n), (atoms)	N:P
			NH_4^+	NO_3^-	PO_4^{3-}		
A. L2	26.0–26.5	28.4	0.48 ± 0.01	17.32 ± 2.84	0.27 ± 0.10	2	58.2–79.0
L3	27.2–28.0	30.3–33.8	0.63 ± 0.02	11.73 ± 2.96	0.20 ± 0.06	2	60.1–64.3
L4	26.6–27.5	31.8–33.0	0.68 ± 0.31	9.10 ± 1.71	0.20 ± 0.02	4	38.4–60.2
BR2	26.8–27.5	33.0–34.7	2.00 ± 1.73	5.27 ± 1.73	0.27 ± 0.13	3	15.2–50.3
BR1	27.2–28.5	34.6–36.5	0.49 ± 0.12	2.29 ± 2.57	0.17 ± 0.03	5	5.1–41.4
B. L2	26.1–26.8	28.9–30.2	0.79 ± 0.27	18.41 ± 0.95	0.43 ± 0.02	40	43.9–45.0
L4	26.4–28.7	31.3–35.0	0.87 ± 0.53	10.29 ± 3.49	0.37 ± 0.07	9	17.1–40.4
BR1	26.4–29.4	34.0–36.3	2.68 ± 5.32	2.99 ± 2.67	0.45 ± 0.28	7	5.0–22.5
C. L1 (0.2)	25.0	29.5		17.64			0.23
(2.0)	28.8	35.4		2.75			0.35
L2 (0.2)	26.5	28.4		15.31			0.20
(1.5)	29.0	35.2		2.71			0.15
BR1 (0.2)	28.5	36.5		6.55			0.17
(2.0)	29.0	36.0		1.10			0.14
D.[a] L4	28.0	32.6	0.57	10.52	0.34		32.6
BR1	28.3	34.2	14.72	8.28	1.02		22.5
DBML rain							
11/22/87			14.92	9.20	1.01		
12/19/88			17.15	8.37	0.50		
E. EDB							
12/12/88	27.6	36.3	5.74	2.74	0.43		
06/24/89	28.4	35.6	0.22	0.96	0.29		

Note: (A) During initial culture trials (November 1987–May 1988). (B) During nutrient pulse experiments (April–June 1989). Values represent means ± 1 SD (n in PO_4^{3-} column applies to all nutrient measurements). (C) Surface and deep values on December 1, 1987 (numbers in parentheses indicate depth in meters). (D) On May 27, 1989, and nutrient concentrations in rainwater at DBML. (E) Values at East Discovery Bay (near natural populations of *G. domingensis*).

[a] Rain showers from 7:30 to 9:30 AM; water samples are taken from 10:30 to 11:30 AM.

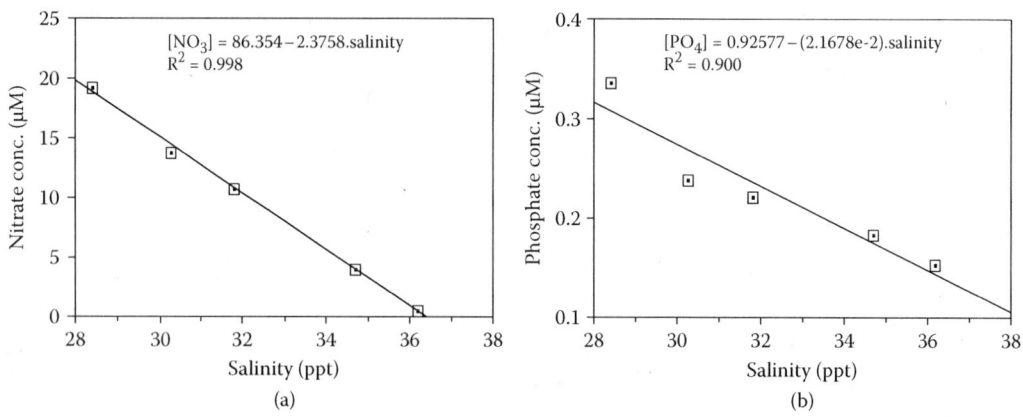

Figure 17.7 (a) Nitrate–salinity and (b) phosphate–salinity relationships at culture locations in Discovery Bay, January 26, 1988.

and salinity indicate that such gradients were relatively stable at locations receiving high to intermediate groundwater input but were reduced during turbulent conditions when mixing of the water column increased.

Growth, Epiphyte, and Grazing Differences between Sites

G. terete

Growth rates of the initial cultures at BR1 (July 14–August 17, 1987) were generally higher and declined less rapidly than those at L2 (Figures 17.3 and 17.8). These differences were significant (F-test, $p < .025$) during the second week of culture.

The type of epiphyte accumulation on the cultures differed between these two locations and appeared to be a significant factor influencing growth. Cultures at L2 were, within a few days, consistently covered with a dense film of microalgae, consisting mostly of diatoms (of which the major genera were *Licmorpha*, *Nitzschia*, *Fragilaria*, *Navicula*, *Thalassionema*, and *Asterionella*). Those at BR1 remained relatively clean of such fouling, but sporadically accumulated red and green macrophytic or filamentous epiphytes or drift algae. These included the following: *Chaetomorpha linum* (a loose drifting filamentous green alga that tended to form massive blooms in the back-reef during the summer months), *Enteromorpha*, *Ulva*, *Hypnea*, *Laurencia*, and *Acanthophora spicifera*.

Grazing by fish also appeared to be a significant factor affecting growth rates, with much more evidence of grazing (as cut tips of cultures) observed on cultures at L2 relative to those at BR1. Amounts of cut tips indicated that cultures at a depth of 30–40 cm at L2 were more heavily grazed relative to shallow cultures within 15 cm of the surface. Deeper cultures at L2 had significantly lower growth rates than shallow cultures (F-test, $p < .01$) during the third week of culture.

During controlled grazing experiments at L4 and BR1 (May 21–June 19, 1989), growth rates of shallow caged cultures at L4 were significantly higher than those at BR1 during the second week of culture (F-test, $p < .05$; Figure 17.9). Visual estimates of epiphyte abundance indicated that cultures at BR1 were more densely fouled by *Ceramium* relative to those at L4 after four weeks in culture.

Growth rate differences between caged and exposed cultures (Table 17.8) were not indicative of grazing intensity due to the apparent influence of cages on growth rates. Caged and exposed cultures at L4 had similar growth rates despite much observed grazing of the exposed cultures, while exposed cultures at BR1 had significantly higher growth rates than caged cultures during the second and third

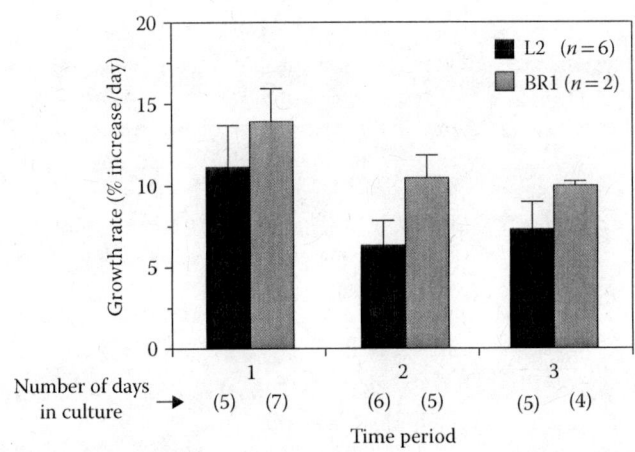

Figure 17.8 Mean growth rates of simultaneous *G. terete* horizontal rope cultures at L2 and BR1 (July 14–30, 1987). Error bars represent +1 SD.

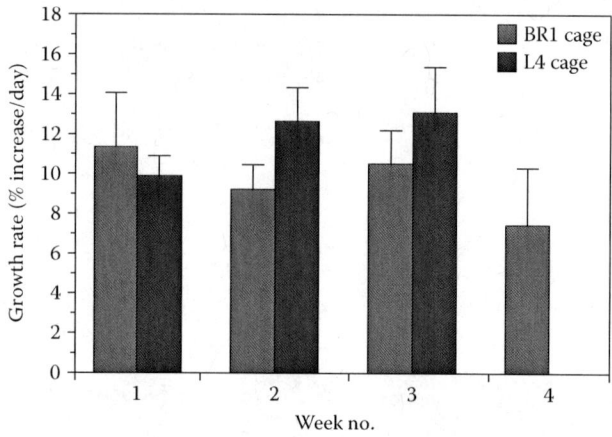

Figure 17.9 Mean growth rates of simultaneous *G. terete* shallow caged cultures at BR1 and L4 (May 21–June 19, 1989). Error bars represent +1 SD ($n = 3$).

Table 17.8 Weekly Mean Growth Rate Differences [k(Caged) – k(Exposed), Where k = Growth Rate as % Increase or Decrease/Day] between Caged and Exposed *G. terete* Cultures at Two Locations (L4 and BR1; May 21–June 19, 1989)

Date	L4 Shallow	BR1 Shallow	BR1 Deep
May 28, 1989		−0.9	0.5
May 30, 1989	−1.5		
June 4, 1989		−5.2**	−1.9
June 6, 1989	−0.5		
June 11, 1989		−3.6*	−3.0
June 14, 1989	0.8		
June 19, 1989		−0.3	−3.5

Note: Negative numbers indicate higher growth of exposed cultures.
F-test significance levels of these differences are indicated as follows: *$p < .025$; **$p < .05$.

weeks of culture despite some observed grazing on the exposed cultures. Frequency and amount of cut tips of exposed cultures indicated that grazing was more intense at L4 than at BR1. Equivalent or higher growth rates of exposed relative to caged cultures may have been related to stronger water movement outside the cage. During the culture period, the cages became increasingly fouled with tenacious epiphytes (mostly *Ceramium* sp. and *Acanthophora spicifera*), thus further restricting water movement. Shallow exposed cultures at BR1 had significantly higher growth rates than deep exposed cultures (F-test, $p < .025$) during the second and third weeks of culture. Frequency of cut tips was somewhat higher on the deeper cultures, but grazing of all cultures was generally low.

G. domingensis

During controlled grazing experiments at three locations (L2, L4, and BR1; April 5–May 8, 1989), initial growth rates of shallow caged cultures at L4 were significantly higher than those of the cultures at both L2 and BR1 (Figure 17.10 and Table 17.9) and were the highest among all cultures at the three locations during the experimental period. During the third week, growth rates of cultures at L4 and BR1 were similar and were significantly higher than those of the cultures at L2. During

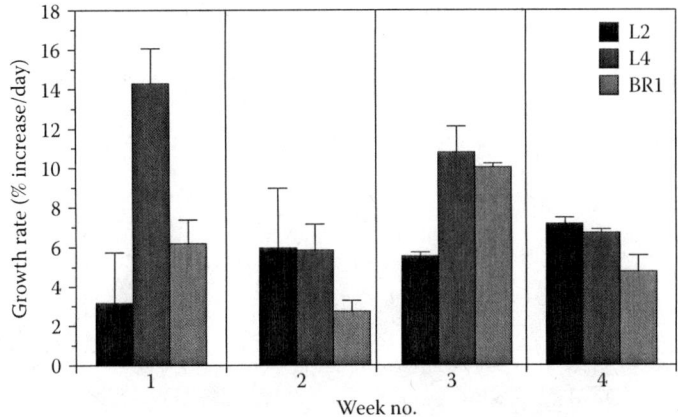

Figure 17.10 Mean growth rates of *G. domingensis* shallow caged cultures at three locations (L2, L4, and BR1) during experiment 1 (April 5–May 7, 1979). Error bars represent +1 SD ($n = 3$).

Table 17.9 Results of Single-Classification ANOVA of Growth Rates of *G. domingensis* Cultures as a Function of Culture Location (L2, L4, and BR1; April 5–May 8, 1989) (Model I: Effects of Culture Conditions [Nutrients, Salinity, Temperature, and Epiphytes] of Different Locations on Growth Rate Means)

Culture Period	Fs
Week 1	14.54***
A priori tests	
L4 vs. (BR1, L2)	47.13***
BR1 vs. L2	3.70 ns
Week 2	1.63 ns
Week 3	34.4***
A priori test	
(L4, BR1) vs. L2	67.42***
Week 4	10.14*
A priori test	
(L2, L4) vs. BR1	19.95**

ns = not significant; *$p < .05$; **$p < .025$; ***$p < .005$.

the fourth week, growth rates of cultures at L2 and L4 were similar and significantly higher than those of the cultures at BR1 (Figure 17.10 and Table 17.9).

The cultures at L2 were most densely fouled by the common epiphytes, *Ceramium* and *Enteromorpha* spp. Cultures at BR1 were least fouled, whereas those at L4 were intermediately fouled by these species. Fouling by diatoms of the cultures at L4 was also relative to that at L2 (most densely fouled) and BR1 and was more variable at L4 and BR1 (Table 17.10).

The amount of cut tips on exposed cultures indicated that those at L2 were most heavily grazed, whereas those at BR1 were the least grazed. Grazing was more intense on the deeper cultures at both L2 and L4, but not on those at BR1. At both L2 and L4, growth rates of shallow exposed cultures were significantly higher than those of deep exposed cultures during some of the culture period (during the third week at L2 [F-test, $p < .005$] and second, fourth, and fifth weeks at L4 [F-test, $p < .05$]). There were no significant differences between growth rates of shallow and deep exposed cultures at BR1.

Growth rate differences between caged and exposed cultures at the three locations were highly variable and did not appear to provide a quantitative measure of grazing intensity (Table 17.11).

Table 17.10 Quantitative Estimates of Diatom Fouling of G. domingensis Cultures at Three Locations during Experiments 1 and 2

Date	% Diatom Dry Weight/G. domingensis Dry Weight [Mean ± S.D. (n)]		
	L2	L4	BR1
4/20/89	75.08 ± 13.97 (2)		
4/22/89			86.15 ± 0.00 (2)
4/24/89		47.71 ± 2.39 (2)	
5/4/89	84.64 ± 4.16 (3)		
5/5/89			35.57 ± 1.55 (3)
5/8/89		93.03 ± 3.18 (3)	
Mean ± S.D.	79 ± 6.76	70.37 ± 32.04	60.86 ± 35.73

Table 17.11 Weekly Mean Growth Rate Differences [k(Caged) − k(Exposed), Where k = Growth Rate as % Increase or Decrease/Day] between Caged And Exposed G. domingensis Cultures at Three Locations (L2, L4, and BFl1; April 5–May 15, 1989)

	L2		L4		BR1	
Date	Shallow	Deep	Shallow	Deep	Shallow	Deep
April 14, 1989					−1.7	−5.5***
April 16, 1989	−5.7*	−7.7**	0.4	−1.3		
April 21, 1989					4.5**	6.7**
April 23, 1989	4.1	6.8***	0.9	1.3		
April 28, 1989					8.6*	1.9
April 30, 1989	−4.2***	−1.8	0.5	1.2		
May 5, 1989					−0.4	−1.2
May 7, 1989	21.7**	22.5***	−1.7	0.7		
May 15, 1989			0.1	6.1		

F-test significance levels of these differences are indicated as follows: *$p < .05$; **$p < .025$; ***$p < .005$.

Significant negative growth rate differences at L2 and BR1 correspond to much lower growth of caged relative to exposed cultures, which may have resulted from reduced water motion inside the cages. Highest positive growth rate differences at L2 during the fourth week of culture (Table 17.11) indicate that grazing was most intense at this location. Positive growth rate differences at BR1 indicate that significant grazing occurred at this location. The absence of significant growth rate differences at L4 may have been due to a balance between opposed effects of grazing (as indicated by cut tips) and higher water motion on exposed cultures.

G. crassissima

During controlled grazing experiments at two locations (L4 and BR1; April 7–May 15, 1989), growth rates of simultaneous shallow caged cultures were similar except during the second week of culture (third week of culture at BR1), when the cultures at L4 had significantly higher growth rates (F-test, $p < .025$), which were the highest among all cultures (including nutrient-enriched cultures at BR1) during the experimental period (Figure 17.11).

Cultures at L4 were more densely fouled with *Ceramium* sp. relative to those at BR1, particularly during the latter two weeks of culture, when low or negative growth rates were obtained. Grazing was not observed on exposed *G. crassissima* cultures, and negative growth rates appeared to be mainly associated with fragmenting of cultures.

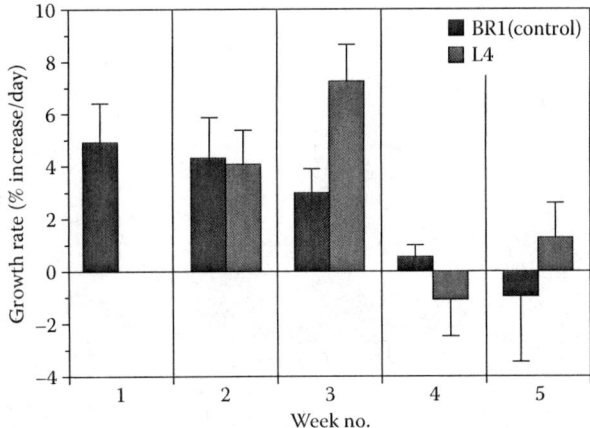

Figure 17.11 Mean growth rates of simultaneous *G. crassissima* caged cultures at two locations (L4 and BR1; April 15–May 15, 1989). Error bars represent +1 SD ($n = 3$).

Nutrient Enrichment Experiments

G. terete

Culture Location: BR1 (One Experiment; May 21–June 19, 1989)

Plants used in this experiment were collected from Llandovery (May 8, 1989) and acclimated to culture conditions in vexar cages at BR1 for two weeks before the start of the experiment. Growth was significantly enhanced by medium- and high-phosphate enrichment subsequent to the first and second pulse treatments but not by low-phosphate or groundwater enrichment (Figure 17.12 and Table 17.12). There was no significant difference between the effects of medium- and high-phosphate enrichment on growth rates. During the last week of culture (subsequent to the third pulse treatment) there were no significant differences between the growth rates of all cultures. This was related to a general decline in all growth rates associated with increasingly dense fouling of all cultures by *Ceramium* sp. No effect of pulse treatments on the relative abundance of *Ceramium* on *G. terete* could be discerned. Significant variation of growth rate with respect to time occurred only for cultures receiving medium- and high-phosphate enrichment and was primarily related to increased growth rates subsequent to the first two pulse treatments (Figure 17.12 and Table 17.12).

G. domingensis

Culture Location: BR1 (Two Replicate Experiments)

Experiment 1 (April 7–May 12, 1989)

Plants used in this experiment were obtained from vexar cage cultures maintained at BR1 for the previous six months. These cultures were originally obtained from east Discovery Bay (October 1988).

Phosphate enrichment at all concentrations significantly increased the growth rates of cultures, whereas groundwater nitrate enrichment with no phosphate additions had no effect on growth rate (Figure 17.13 and Table 17.13). Medium and high levels of phosphate enrichment significantly increased growth rates after two pulse treatments (three weeks in culture), whereas low-phosphate enrichment increased growth rates after three pulse treatments and required four treatments to have a similar effect on growth as the higher concentrations. There was no significant difference between the effects of medium- and high-phosphate treatments on growth rates (Table 17.13).

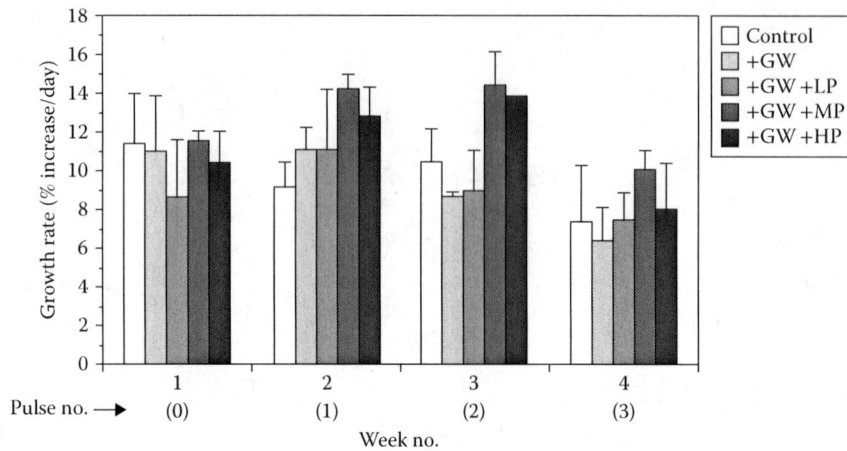

Figure 17.12 Mean growth rates of back-reef (BR1) *G. terete* cultures receiving different nutrient enrichment treatments during the experimental period May 21–June 19, 1989. Error bars represent +1 SD ($n = 3$).

Table 17.12 Results of Single-Classification ANOVA of Growth Rates of *G. terete* Back-Reef (BR1) Cultures as Functions of N and P Enrichment (Model I: Effects of Fixed Nutrient Enrichment Treatments on Group Means) and Environmental Variables (Model II: Effects of Random Environmental Variables over Time [e.g., Water Motion and Natural Nutrient Input] on Means of Each Group Receiving a Fixed Treatment)

Factor	Fs
Nutrient Enrichment	
(Variability among Treatments Subsequent to Each Pulse)	
Pulse #1	3.63*
A priori tests	
(C,+GW,+LP) vs. (+MP,+HP)	1 0.89**
Pulse #2	10.07***
A priori tests	
C vs. +GW vs. +LP	1.27 ns
(c+GW,+LP) vs. (+MP,÷HP)	37.43***
+MP vs. +HP	0.33 ns
Pulse #3	1.47 ns
Environmental Variables	
(Variability within Treatments through Time)	
Control	1.79 ns
+GW	4.18*
+GW,+LP	1.15 ns
+GW,+MP	10.54***
+GWI+HP	7.79**

Significance levels are denoted as follows: ns = not significant; *$p < .05$; **$p < .025$; ***$p < .005$.

After four weeks in culture, the plants receiving phosphate enrichment developed a more densely branching morphology than those receiving no phosphate. This was most pronounced in high-phosphate-treated cultures, which developed highly branched new growth.

Nutrient pulse treatments had no discernible effect on epiphyte colonization of the cultures. Common epiphytes, which remained at low densities, included *Ceramium* and *Enteromorpha* species, with some accumulation of *Chaetomorpha linum*.

MARICULTURE POTENTIAL OF *GRACILARIA* SPECIES [RHODOPHYTA]

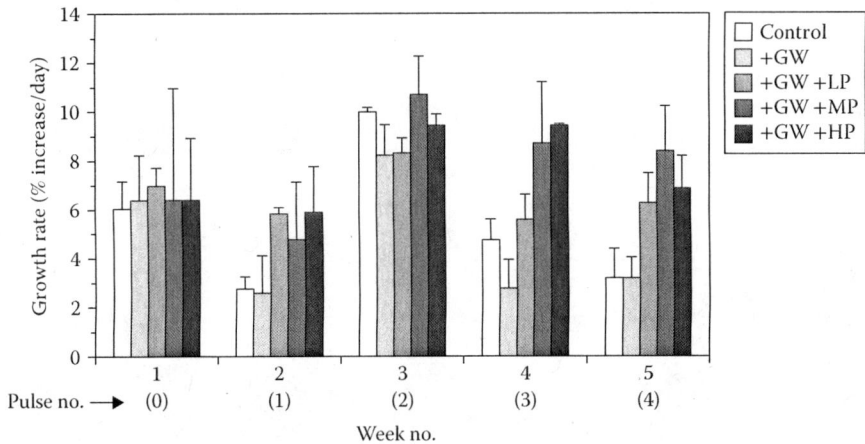

Figure 17.13 Mean growth rates of back-reef (BR1) *G. domingensis* cultures receiving different nutrient enrichment treatments during experiment 1 (April 7–May 12, 1989). Error bars represent +1 SD ($n = 3$).

Table 17.13 Results of Single-Classification ANOVA of Growth Rates of Back-Reef (BR1) *G. domingensis* Cultures as Functions of N and P Enrichment (Model I: Effects of Fixed Nutrient Enrichment Treatments on Group Means; Significant Group Mean Differences Compared by A Priori Tests) and Environmental Variables (Model II: Effects of Random Environmental Variables over Time [e.g., Water Motion and Natural Nutrient Input] on Means of Each Group Receiving a Fixed Treatment)

	Fs	
Factor	Experiment 1	Experiment 2
Nutrient Enrichment		
(Variability among Treatments Subsequent to Each Pulse)		
Pulse #1	2.84 ns	2.90 ns
Pulse #2	3.52*	24.87***
A priori tests		
C vs. +GW		2.26 ns
(C,+GW) vs. (+LP,+MP,+HP)		76.55***
+LP vs. (+MP,+HP)		19.78***
+MP vs. +HP		0.87 ns
Pulse #3	8.94***	11.78***
A priori tests		
C vs. +GW	2.10 ns	0.30 ns
(C,+GW) vs. (+LP,+MP,+HP)	17.20***	35.93***
+LP vs. (+MP,+HP)	16.16***	10.67**
+MP vs. +HP	0.28 ns	0.21 ns
Pulse #4	9.05***	
A priori tests		
C vs. +GW	0 ns	
(C,+GW) vs. (+LP,+MP,+HP)	32.55***	
+LP vs. (+MP,+HP)	1.87 ns	
+MP vs. +HP	1.75 ns	

(Continued)

Table 17.13 Results of Single-Classification ANOVA of Growth Rates of Back-Reef (BR1) *G. domingensis* Cultures as Functions of N and P Enrichment (Model I: Effects of Fixed Nutrient Enrichment Treatments on Group Means; Significant Group Mean Differences Compared by A Priori Tests) and Environmental Variables (Model II: Effects of Random Environmental Variables over Time [e.g., Water Motion and Natural Nutrient Input] on Means of Each Group Receiving a Fixed Treatment) *(Continued)*

Factor	Fs	
	Experiment 1	Experiment 2
Environmental Variables *(Variability within Treatments through Time)*		
Control	33.03***	6.86**
+GW	10.17***	0.51 ns
+GW,+LP	4.48*	
+GW,+MP	1.97ns	2.32ns
+GW,+HP	3.45ns	6.64**

ns = not significant; *$p < .05$; **$p < .025$; ***$p < .005$.

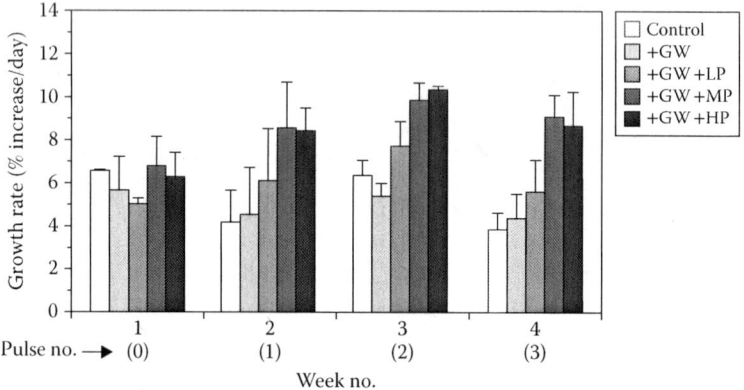

Figure 17.14 Mean growth rates of back-reef (BR1) *G. domingensis* cultures receiving different nutrient enrichment treatments during experiment 2 (May 18–June 15, 1989). Error bars represent +1 SD ($n = 3$).

During the course of the experiment, highly significant variation in growth rates with respect to time occurred for the control and groundwater-treated cultures, whereas little to no significant variation occurred for the phosphate-treated cultures (Figure 17.13 and Table 17.13). Considerable variation in weather and water conditions occurred during the experimental period, which may be related to the variations in growth rate with respect to time.

Experiment 2 (May 18–June 15, 1989)

Plants used in this experiment were collected from east Discovery Bay (April 29, 1989) and acclimated to culture conditions in vexar cages at BR1 for over two weeks before the start of the experiment.

The effects of nutrient pulse treatments on growth rates were essentially the same as those found in Experiment 1 (Figure 17.14 and Table 17.13). However, phosphate enrichment at all concentrations significantly increased growth rates subsequent to two pulse treatments during this experiment (Table 17.13). With increasing phosphate enrichment, these cultures also developed increasingly dense branching. Fouling of cultures by *Ceramium* spp. was much more intense during this experiment, as estimated by visual observations. However, after four weeks in culture, all cultures receiving

MARICULTURE POTENTIAL OF *GRACILARIA* SPECIES [RHODOPHYTA]

phosphate enrichment were much less fouled by *Ceramium* than the control and nitrate-enriched cultures. Medium- and high-phosphate-treated cultures had the least amounts of *Ceramium*.

During the course of this experiment, significant variations in growth rates with respect to time occurred for the control and high-phosphate-enriched cultures but not for the groundwater-enriched and low- and medium-phosphate-enriched cultures (Figure 17.14 and Table 17.13). While this variation of growth of the control cultures appears related to weather and water conditions, variation of growth of the high-phosphate-enriched cultures appears primarily related to increased growth subsequent to pulse treatments.

Culture Location: L2 (One Experiment; April 5–May 4, 1989)

Plants used in this experiment had the same origin as those used for Experiment 1 at BR1.

Although the mean growth rates of all phosphate-treated cultures were higher than the control subsequent to the first and third pulse, these differences were not significant (Figure 17.15 and Table 17.14). These cultures were densely fouled by masses of diatoms (as previously observed for

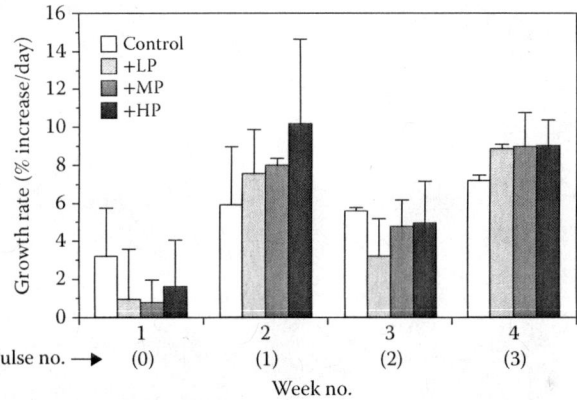

Figure 17.15 Mean growth rates of lagoon (L2) *G. domingensis* cultures receiving different nutrient enrichment treatments during the course of the experiment April 5–May 4, 1989. Error bars represent +1 SD ($n = 3$).

Table 17.14 Results of Single-Classification ANOVA of Growth Rates of Lagoon (L2) *G. domingensis* Cultures as Functions of Phosphate Enrichment (Model I) and Environmental Variables (Model II: Effects of Random Environmental Variables over Time on Means of Each Group Receiving a Fixed Treatment)

Factor	Fs
Phosphate Enrichment	
(Variability among Treatments Subsequent to Each Pulse)	
Pulse #1	0.96 ns
Pulse #2	0.83 ns
Pulse #3	1.08 ns
Environmental Variables	
(Variability within Treatments through Time)	
Control	1.43 ns
+LP	10.46***
+MP	22.83***
+HP	5.19*

ns = not significant; *$p < .05$; **$p < .025$; ***$p < .005$.

other cultures at this location) and much more epiphytized by *Ceramium* and *Enteromorpha* sp. relative to simultaneous cultures at BR1.

Significant variations in growth rates with respect to time occurred for the phosphate-treated cultures but not for the control cultures (Figure 17.15 and Table 17.14). Much of this variation in the phosphate-treated cultures was associated with increases in growth rates subsequent to the first and third pulse treatments. While this seems to indicate stimulation of growth by phosphate enrichment, similar trends in growth occurred for the control (though less pronounced) and caged deep cultures. These trends therefore appear to be linked to external factors affecting the growth of all cultures at this location.

Flow-Through Culture

Growth rates of the groundwater-enriched flow-through cultures were significantly increased by phosphate enrichment at all concentrations (Figure 17.16 and Table 17.15). Enhancement of growth of these cultures by phosphate enrichment was evident for up to two weeks subsequent to the initial pulse treatments.

G. crassissima

Culture Location: BR1 (One Experiment; April 8–May 14, 1989)

Plants used in this experiment were obtained from vertical rope cultures maintained at BR1 for the previous two months and were originally collected from Bali Hai (~one mile west of Discovery Bay).

During the fourth week of culture (following three pulse treatments), cultures receiving phosphate enrichment at all concentrations had higher growth rates than control and groundwater-treated cultures, which had similar growth rates. Cultures receiving medium-phosphate enrichment had the

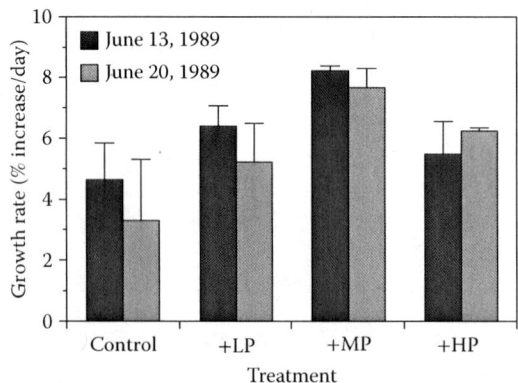

Figure 17.16 Mean growth rates of *G. domingensis* groundwater-enriched flow-through cultures during two weeks subsequent to initial pulsed phosphate enrichment. Error bars represent +1 SD ($n = 2$ or 3).

Table 17.15 Results of Single-Classification ANOVA of Growth Rates of *G. domingensis* Groundwater-Enriched Flow-Through Cultures as a Function of Phosphate Enrichment (Model I)

Factor	Fs
Phosphate Enrichment	
Week 1 (June 13, 1989)	8.54**
A priori test: C vs. (+LP,+MP,+HP)	11.36**
Week 2 (June 20, 1989)	5.94*
A priori test: C vs. (+LP,+MP,+HP)	10.96**

ns = not significant; *$p < .05$; **$p < .025$; ***$p < .005$.

most significant increase in growth rate relative to nonphosphate-treated cultures (Figure 17.17 and Table 17.16). Although the mean growth rates of phosphate-treated cultures were also higher than those of the nonphosphate-treated cultures during the third week of culture, these differences were not significant (Table 17.16).

During the culture period, significant variation in growth rates with respect to time has occurred for the control, groundwater-enriched, and low-phosphate-enriched cultures, but not for the medium- and high-phosphate-enriched cultures (Figure 17.17 and Table 17.16). This variability was primarily related to a consistent reduction in growth of the cultures during the culture period. All cultures generally

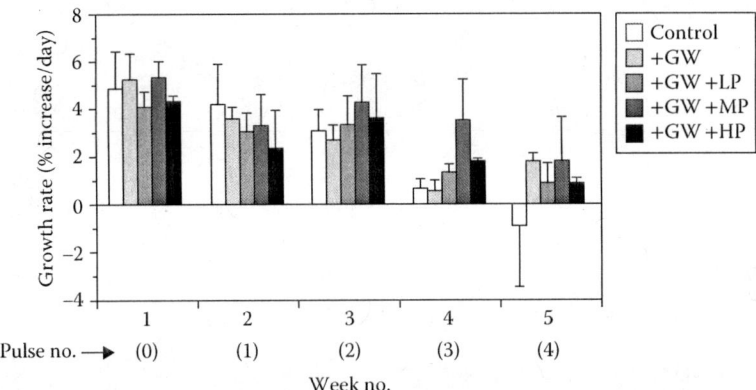

Figure 17.17 Mean growth rates of back reef (BR1) *G. crassissima* cultures receiving different nutrient enrichment treatments during the culture period April 8–May 14, 1989. Error bars represent +1 SD ($n=3$).

Table 17.16 Results of Single-Classification ANOVA of Growth Rates of *G. crassissima* Back-Reef (BR1) Cultures as Functions of N and P Enrichment (Model I: Effects of Fixed Nutrient Enrichment Treatments on Group Means) and Environmental Variables (Model II: Effects of Random Environmental Variables over Time on Means of Each Group Receiving a Fixed Treatment)

Factor	Fs
Nutrient Enrichment	
(Variability among Treatments Subsequent to Each Pulse)	
Pulse #1	0.90 ns
Pulse #2	0.65 ns
Pulse #3	6.24**
A priori tests	
C vs. +GW	0.02 ns
(C,+GW) vs. (+LP,+MP,+HP)	12.77**
+LP vs. (+MP,+HP)	4.91 ns
+MP vs. +HP	7.23*
Environmental Variables	
(Variability within Treatments through Time)	
Control	4.84*
+GW	24.58***
+GW,+LP	6.69**
+GW,+MP	1.40 ns
+GW,+HP	2.42 ns

ns = not significant; *$p < .05$; **$p < .025$; ***$p < .005$.

started to become necrotic and fragment during the last week of culture. Medium- and high-phosphate treatments appeared, therefore, to maintain more constant growth during the culture period.

Although these cultures were fouled with only small amounts of *Ceramium* sp., they tended to be densely fouled by diatoms. Many small branches appeared on all cultures during the experimental period, but growth of these appeared to be stunted by dense accumulation of diatoms.

C, N, and P Contents and Atomic Ratios of Cultured and Natural Populations of *G. domingensis*

In Situ *Cultures: BR1 (Experiment 2; May 18–June 15, 1989)*

With increasing phosphate enrichment, levels of tissue P (in percent of dry weight) increased to more than twice the amount relative to that of P-unenriched cultures (from 0.040% for groundwater-treated cultures to 0.096% for high-phosphate-treated cultures; F-test, $p < .005$); percentages of tissue N showed a slight increase (1.33–1.45%), which was not statistically significant; whereas percentages of tissue C remained relatively constant (Table 17.17). Percentages of tissue N of the groundwater-enriched cultures were slightly higher than those of the control cultures, but this difference was not significant. C:P and N:P ratios significantly decreased (1744 to 713 and 73.9 to 33.2, respectively)

Table 17.17 Levels of Chemical Constituents (in Percent of Dry Weight) and Their Atomic Ratios in *G. domingensis* Natural and Cultured Populations

Population/Treatment	% C	% N	% P	C:N	C:P	N:P
Expt. 2 BR1[a] control	27.7 ± 0.1	1.22 ± 0.05	0.046 ± 0.009	26.4 ± 0.9	1570 ± 296	59 ± 9
+GW	26.9 ± 0.5	1.33 ± 0.08	0.040 ± 0.004	23.6 ± 1.8	1744 ± 141	74 ± 3
+LP	27.2 ± 2.1	1.35 ± 0.03	0.044 ± 0.008	23.6 ± 2.3	1618 ± 166	69 ± 11
+MP	27.2 ± 0.7	1.49 ± 0.32	0.095 ± 0.008	21.7 ± 4.1	744 ± 43	35 ± 5
+HP	26.6 ± 2.6	1.45 ± 0.02	0.096 ± 0.001	21.5 ± 2.3	713 ± 63	33 ± 0
Net culture BR1[b]	26.3 ± 1.7	1.35 ± 0.26	0.033 ± 0.001	23.0 ± 3.0	2041 ± 102	90 ± 16
FT Expt. (end)[c]						
SWFT control	27.3 ± 0.3	0.62 ± 0.10	0.024 ± 0.006	51.8 ± 8.9	3051 ± 770	59 ± 6
+GW	27.6 ± 0.8	0.77 ± 0.00	0.025 ± 0.010	41.5 ± 1.1	3056 ± 1090	74 ± 28
+LP	27.2 ± 0.0	0.59 ± 0.04	0.024 ± 0.012	53.7 ± 3.3	3372 ± 1701	62 ± 30
+MP	27.3 ± 2.5	0.56 ± 0.10	0.045 ± 0.004	57.0 ± 4.9	1509 ± 9	27 ± 2
+HP	26.4 ± 2.1	0.56 ± 0.08	0.059 ± 0.006	55.1 ± 3.2	1160 ± 34	21 ± 1
GWFT control	27.4 ± 1.1	1.71 ± 0.07	0.045 ± 0.008	18.7 ± 0.0	1604 ± 234	86 ± 13
+LP	27.0 ± 0.9	1.56 ± 0.06	0.043 ± 0.009	20.0 ± 1.5	1640 ± 296	81 ± 14
+MP	27.1 ± 0.4	1.61 ± 0.14	0.070 ± 0.000	19.7 ± 1.4	1002 ± 11	51 ± 4
+HP	26.4 ± 0.5	1.96 ± 0.19	0.081 ± 0.004	15.8 ± 1.2	837 ± 24	53 ± 3
FT Expt. (before 2nd Pulse)						
SWFT control	26.8 ± 0.2	0.98 ± 0.02	0.031 ± 0.010	32.0 ± 0.3	2364 ± 772	74 ± 23
+GW	26.7 ± 0.1	0.96 ± 0.00	0.033 ± 0.007	32.5 ± 0.01	2155 ± 476	66 ± 15
+LP			0.042 ± 0.004			
+MP	27.5 ± 0.8	0.93 ± 0.03	0.060 ± 0.002	34.4 ± 2.1	1181 ± 2	34 ± 0
+HP	27.1 ± 0.6	0.92 ± 0.04	0.091 ± 0.001	34.5 ± 2.1	770 ± 5	22 ± 1
GWFT control			0.045 ± 0			
+LP			0.075 ± 0.013			
+MP			0.061 ± 0.006			
+HP	27.5 ± 0.2	1.78 ± 0.01	0.104 ± 0.013	18.0 ± 0.1	686 ± 80	38 ± 5

Table 17.17 Levels of Chemical Constituents (in Percent of Dry Weight) and Their Atomic Ratios in *G. domingensis* Natural and Cultured Populations *(Continued)*

Population/Treatment	% C	% N	% P	C:N	C:P	N:P
Nat. Pop. E. Discovery Bay						
December 17, 1988	19.8 ± 0.1	1.45 ± 0.08	0.68 ± 0.010	16.0 ± 1.0	757 ± 111	47 ± 5
April 29, 1989	21.7 ± 1.1	0.58 ± 0.03	0.043 ± 0.007	43.7 ± 4.1	1305 ± 135	30 ± 3
June 24, 1989	23.5 ± 0.7	0.54 ± 0.20	0.030 ± 0.002	54.2 ± 18.6	2029 ± 105	39 ± 12

Note: Values represent means ± S.D. ($n = 2$).
[a] *In situ* nutrient enrichment experiment (May 18–June 15, 1989).
[b] Original culture, flow-through experiment.
[c] Flow-through nutrient uptake experiment.
SWFT, unenriched seawater flow-through cultures; GWFT, groundwater-enriched flow-through cultures.

with increasing phosphate enrichment, whereas C:N ratios were not significantly affected. C:N ratios decreased, whereas N:P and C:P ratios increased with groundwater enrichment (Table 17.17).

Flowing Aquaria Cultures

Unenriched Seawater Culture

With increasing phosphate enrichment, levels of tissue P were significantly higher at the end of both the first week (F-test, $p < .005$) and second week (F-test, $p < .05$) subsequent to enrichment (from 0.025% for groundwater enriched to 0.059% for high-phosphate-enriched cultures after two weeks; Table 17.17). Corresponding C:P and N:P ratios were significantly lower with increasing phosphate enrichment (3056 to 1160 and 58.5 to 21.1, respectively). Levels of tissue N were highest for groundwater-enriched cultures. These were not significantly different from those of the control cultures, but were significantly higher than those of the phosphate-enriched cultures. This trend was also somewhat evident for the groundwater-enriched flow-through cultures where levels of tissue N of the control cultures were significantly higher than those of the low-phosphate-enriched cultures. Levels of tissue N of all cultures were much lower than that of the original culture at BR1, and this difference was more pronounced after two weeks of flow-through culture.

Groundwater-Enriched Seawater Culture

Levels of tissue P were also significantly increased with increasing phosphate enrichment at the end of both the first week (F-test, $p < .025$) and second week (F-test, $p < .01$) subsequent to enrichment (from 0.045% for control to 0.081% for phosphate-enriched cultures after two weeks; Table 17.17). Corresponding C:P and N:P ratios also decreased significantly with increasing phosphate enrichment (1604 to 837 and 85.6 to 53.1, respectively). Levels of tissue P of these cultures were significantly higher than those of the unenriched seawater flow-through cultures for all treatments, and this difference was more pronounced after two weeks of culture.

Levels of tissue N of these cultures were also much higher than those of the unenriched seawater flow-through cultures and somewhat higher than the original culture at BR1. Levels of tissue C of all flow-through cultures were similar and relatively constant.

Natural Populations: East Discovery Bay

In samples of this population taken over a six-month period (December 1988–June 1989), percentages of tissue N and P were highest during December and lowest during June, whereas percentages of tissue C increased slightly during this period (Table 17.17). Levels of C increased from 19.75% to 23.47% of dry weight, whereas levels of N decreased from 1.45% to 0.54% of dry weight, and levels of P decreased from 0.068% to 0.030% of dry weight. Corresponding C:N and C:P ratios increased to much higher levels during this period (16–54 and 757–2029, respectively), whereas N:P ratios decreased or remained fairly constant.

Nitrate and Phosphate Uptake Kinetics

In Situ *Cultures*

During the course of the nutrient enrichment experiments, nutrients were generally taken up more rapidly and reduced to lower concentrations with increasing biomass of the cultures. Estimates of nutrient uptake rates presented here were calculated from pulse treatments in which final nutrient concentrations were not reduced below several micromolars (1.5 µM for *G. terete* and >2.0 µM for all other cultures).

Nitrate Uptake

G. domingensis had the highest nitrate uptake rates during groundwater enrichment (4.00–5.08 µM/g dry wt/h) compared with *G. terete* (3.87–4.24 µM/g dry wt/h) and *G. crassissima* (2.36–3.20 µM/g dry wt/h) (Table 17.18). Phosphate enrichment did not have any clearly discernible effect on nitrate uptake of cultures at BR1.

G. domingensis cultures at L2 had initial nitrate-uptake rates similar to the cultures at BR1. However, subsequent nitrate uptake during pulse 3 of the L2 cultures was much lower, and phosphate enrichment at all concentrations clearly enhanced nitrate uptake of these cultures (Table 17.18).

Table 17.18 Nitrate Uptake Rates of Three *Gracilaria* Species Estimated from Nutrient Depletion Measurements during Pulse Treatments of *In Situ* Cultures

Species	NO_3 Uptake Rates (µM/g Dry wt/h) (±S.D., $n = 2$)	Initial NO_3 Conc. (µM)	Conditions (Culture Location, Pulse #, Treatment)
G. domingensis	5.08 ± 0.64	19.3	BR1, pulse 1, +GW
	4.00 ± 0.29	16.2	+LP
	4.58 ± 0.29	17.4	+MP
	4.36 ± 0.57	17.4	+HP
	5.90 ± 0.83	23.4	L2, pulse 1, +GW
	4.66 ± 0.83	19.6	+LP
	6.07 ± 0.91	19.7	+MP
	5.49 ± 0.93	18.7	+HP
	0.50 ± 0.75	15.7	Pulse 3, +GW
	2.58 ± 0.42	18.7	+LP
	2.91 ± 0.50	18.8	+MP
	1.66 ± 0.83	17.8	+HP
G. terete	3.96 ± 0.18	17.6	BR1, pulse 1, +GW
	4.24 ± 0.18	17.8	+LP
	3.87 ± 0.18	16.5	+MP
	4.05 ± 0.28	17.4	+HP
G. crassissima	2.61 ± 0.17	18.0	BR1, pulse 3, +GW
	2.36 ± 0.08	20.3	+LP
	3.20 ± 0.08	17.7	+MP
	2.52 ± 0.08	16.7	+HP

Table 17.19 Phosphate Uptake Rates of Three *Gracilaria* Species Estimated from Nutrient Depletion Measurements during Pulse Treatments of *In Situ* Cultures

Species	PO_4^{3-} Uptake Rates (µM/g Dry wt/h) (±S.D., $n = 2$)	Initial PO_4^{3-} Conc. (µM/L)	Conditions (Culture Location, Pulse #, Treatment)
G. domingensis	1.72 ± 0.50	20.8	BR1, pulse 1, +MP
	1.50 ± 0.21	17.9	Pulse 2, +MP
	2.14 ± 0.21	23.5	Pulse 3, +MP
	8.01 ± 2.86	200.8	Pulse 2, +HP
	4.65 ± 2.36	188.0	Pulse 3, +HP
	1.66 ± 0.50	17.9	L2, pulse 3, +MP
	8.40 ± 2.41	154.3	Pulse 3, +HP
G. terete	2.56 ± 0.20	17.6	BR1, pulse 1, +MP
	2.07 ± 0.10	17.4	Pulse 2, +MP
	2.86 ± 0.99	154.9	Pulse 1, +HP
	3.85 ± 1.28	183.3	Pulse 3, +HP
G. crassissima	0.93 ± 0.08	18.4	BR1, pulse 2, +MP
	2.19	19.2	Pulse 4, +MP
	3.45 ± 2.44	150.6	Pulse 2, +HP

Phosphate Uptake

G. domingensis and *G. terete* had similar phosphate uptake rates, whereas *G. crassissima* had somewhat lower phosphate uptake rates during medium-phosphate enrichment (Table 17.19). Uptake rates have ranged from 1.50 to 2.14 µM/g dry wt/h for *G. domingensis*, from 2.07 to 2.56 µM/g dry wt/h for *G. terete*, and from 0.93 to 2.19 µM/g dry wt/h for *G. crassissima* during medium-phosphate enrichment.

During high-phosphate enrichment, *G. domingensis* had much higher uptake rates than those at medium-phosphate enrichment. Although *G. terete* and *G. crassissima* phosphate uptake rates were somewhat higher during high-phosphate enrichment relative to those during medium-phosphate enrichment, these differences were not significant.

Flow-Through Cultures: *G. domingensis*

Nitrate Uptake (Pulse 1)

Nitrate depletion curves for all groundwater-enriched treatments that include varying phosphate enrichment have generally linear trends down to concentrations of 2–4 µM. (Figure 17.18). This pattern is suggestive of saturation uptake kinetics above nitrate concentrations of 2 µM.

Nitrate uptake rates as determined by linear regression analysis of the straightest portion of the depletion curves (Table 17.20) were similar to or somewhat higher than those of the *in situ* cultures. Although the regression uptake rates of most cultures receiving phosphate enrichment were higher than the nonphosphate-enriched cultures, an F-test comparison of the means of replicate nitrate uptake rates was not significant.

Phosphate Uptake

Depletion curves for cultures receiving low-phosphate treatments (Figure 17.19a) are suggestive of saturation type kinetics at low phosphate concentrations. Additional data are required to demonstrate this clearly. Depletion curves and uptake rates calculated at two-hour intervals for the medium

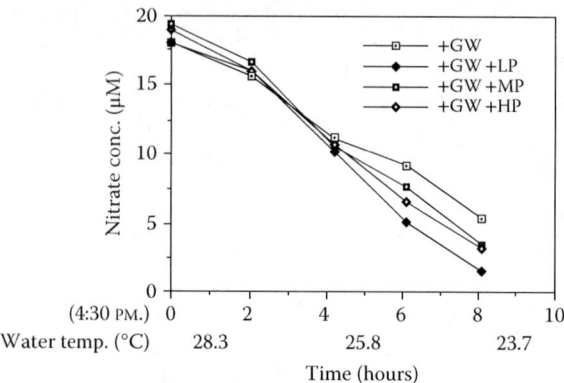

Figure 17.18 Nitrate depletion during groundwater enrichment with varying phosphate enrichment (pulse 1) by *G. domingensis* flow-through cultures. (Depletion curves shown are for the replicates, which were subsequently cultured in unenriched flowing seawater.)

Table 17.20 Nitrate Uptake Rates of *G. domingensis* Flow-Through Cultures during Phosphate and Groundwater Enrichment (Purse #1) Determined by Linear Regression Analysis of the Straightest Portion of Depletion Curves

Replicate (Subsequent Treatment)	NO_3 Uptake Rate (µM/g Dry wt/h)			
	+GW	+GW +LP	+GW +MP	+GW +HP
Unenriched flow-through culture	3.91	5.24	4.29	4.65
Groundwater-enriched flow-through culture	4.16	4.80	5.49	4.04
Mean ± S.D.	4.04 ± 0.18	5.02 ± 0.31	4.89 ± 0.85	4.35 ± 0.43

Fs = 1.69 ns

and high-phosphate-treated cultures (Table 17.21 and Figure 17.19b, c) indicate approximately linear or increasing uptake rates during the pulse.

Phosphate uptake rates at medium and high concentrations as determined by linear regression analysis of the straightest portion of the depletion curves (Table 17.21) were similar to or somewhat higher than those of *in situ* cultures. Nonsaturating uptake kinetics with increasing phosphate concentration are indicated by the much higher uptake rates at high phosphate concentrations relative to uptake rates at the lower concentrations. Increasing uptake rates during the pulse by some cultures receiving medium and high-phosphate enrichment (excepting apparent negative uptake during the third two-hour period) suggests acclimation of the cultures to phosphate uptake at high concentrations. This trend was most pronounced for the seawater flow-through high-phosphate-treated cultures, which suggests that acclimated cultures may have uptake rates at high phosphate concentrations as much as 10 times that at medium phosphate concentrations (Table 17.21).

Nitrate and Ammonium Uptake (Pulse 2)

During pulsed groundwater enrichment, nitrate uptake rates of the unenriched seawater flow-through cultures were much higher than those of the groundwater-enriched flow-through cultures. Uptake rates of the former cultures were similar to those obtained during the first pulse treatment (Table 17.22 and Figure 17.20).

MARICULTURE POTENTIAL OF *GRACILARIA* SPECIES [RHODOPHYTA]

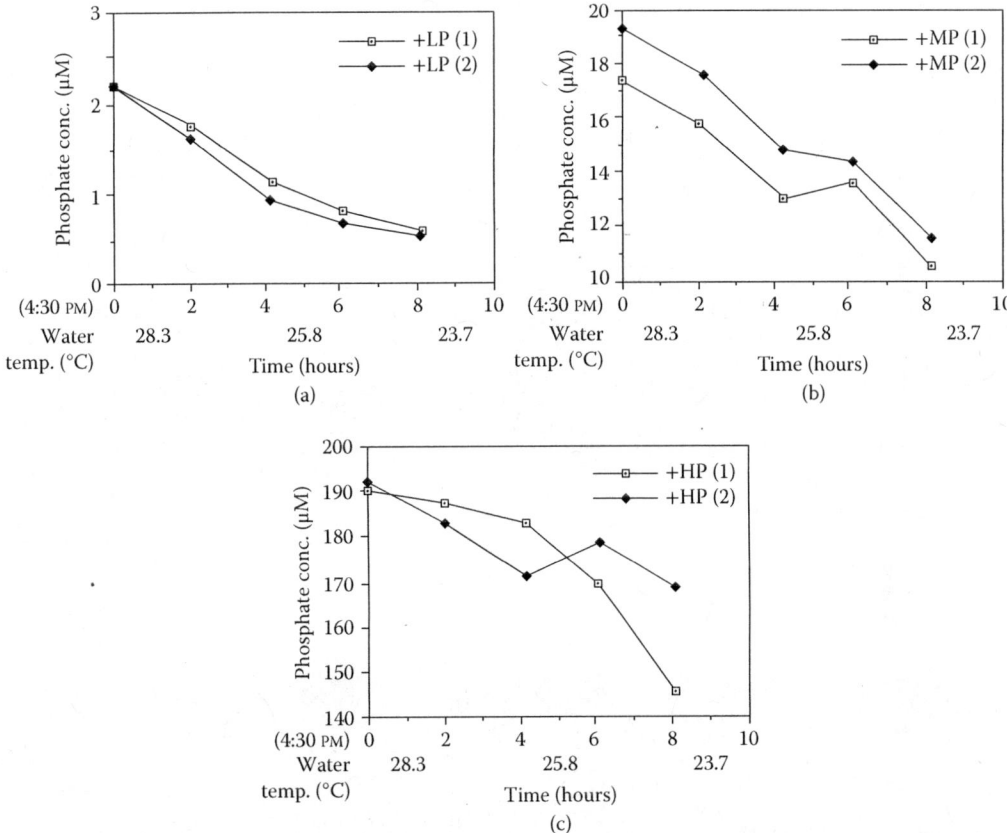

Figure 17.19 Phosphate depletion during groundwater enrichment with varying phosphate enrichment (pulse 1) by *G. domingensis* flow-through cultures: (a) low-phosphate, (b) medium-phosphate, and (c) high-phosphate enrichment. Replicates (1) and (2) correspond to cultures that were subsequently grown in unenriched and groundwater-enriched flowing seawater, respectively.

Table 17.21 Phosphate Uptake Rates of *G. domingensis* Flow-Through Cultures, during Groundwater and Phosphate Enrichment (Pulse #1), Determined at Two-Hour Intervals and by Linear Regression Analysis of the Straightest Portion of Depletion Curves

Replicate (Subsequent Treatment)	PO_4^{3-} Uptake Rate (µM/g Dry wt/h)		
	+LP	+MP	+HP
Unenriched Flow-Through Culture			
4:30–6:30 PM	0.50	1.70	3.09
6:30–8:40 PM	0.63	2.59	4.80
8:40–10:35 PM	0.38	−0.44	15.08
10:35–12:35 AM	0.25	2.97	25.37
Regression analysis	0.57	2.15	7.19
Groundwater-Enriched Flow-Through Culture			
4:30–6:30 PM	0.63	2.02	7.70
6:30–8:40 PM	0.63	3.09	8.90
8:40–10:35 PM	0.25	0.63	−6.37
10:35–12:35 AM	0.19	3.47	8.14
Regression analysis	0.63	2.59	8.33
Mean ± S.D. (regression rates)	0.60 ± 0.04	2.37 ± 0.31	7.76 ± 0.81

Table 17.22 Nitrate Uptake Rates of G. domingensis Flow-Through Cultures, during Groundwater and Ammonium Enrichment (Pulse 2), Determined by Linear Regression Analysis of the Straightest Portion of Depletion Curves

Culture treatment	NO$_3^-$ Uptake Rate (µM/g Dry wt/h)			
	+GW	+LNH$_4^+$	+MNH$_4^+$	+HNH$_4^+$
Unenriched flow-through culture	4.67	3.47	3.79	3.85
Groundwater-enriched flow-through culture	0.63	1.07	1.39	0.88

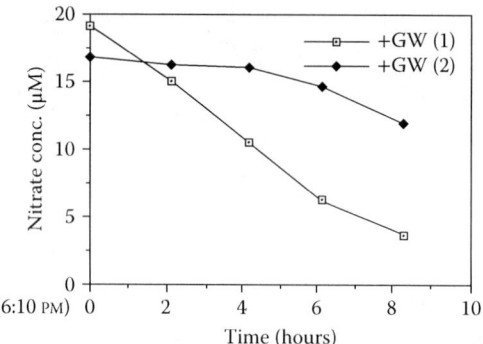

Figure 17.20 Nitrate depletion during groundwater enrichment (pulse 2) by *G. domingensis* flow-through cultures. (1) Replicates cultured in unenriched flowing seawater and (2) replicates cultured in groundwater-enriched flowing seawater.

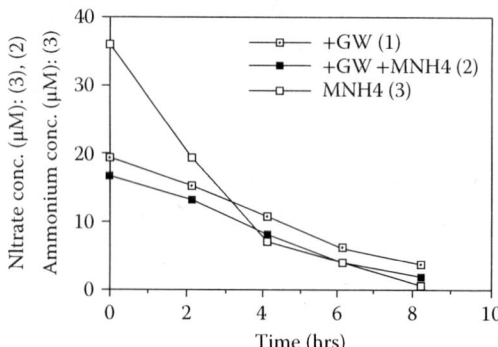

Figure 17.21 Nitrate and ammonium depletion during groundwater and ammonium enrichment (pulse 2) by *G. domingensis* replicates cultured in unenriched flowing seawater. (1) Nitrate depletion by cultures receiving groundwater-only enrichment. (2) Nitrate depletion by cultures receiving groundwater plus medium ammonium (36 µM) enrichment. (3) Simultaneous ammonium depletion by the same cultures as in (2).

Nitrate uptake rates of the seawater flow-through cultures receiving ammonium enrichment, determined by linear regression of the depletion curves, were somewhat less than those receiving only groundwater enrichment. However, these differences did not show inhibition of nitrate uptake with increasing ammonium enrichment (Table 17.22 and Figure 17.21). Cultures receiving high ammonium enrichment had nitrate uptake rates similar to or higher than those receiving low ammonium enrichment. A similar nitrate uptake pattern was shown for the plants cultured in flowing groundwater-enriched seawater, with those receiving ammonium-enriched treatments having similar or higher nitrate uptake relative to those receiving no ammonium enrichment (Table 17.22).

DISCUSSION

Growth and Productivity of *Gracilaria* in Nitrate-Enriched Back-Reef Habitats

The growth rates of the faster growing *Gracilaria* species (*G. terete, G. blodgettii, G. domingensis*, and *G. cervicornis*) in nitrate-enriched back-reef habitats are among the highest known growth rates reported for *in situ Gracilaria* cultures using similar culture methods. Growth rates of *G. crassissima* and *G. terete* rope cultures in Discovery Bay were close to twice or more those obtained, using similar culture methods, for these species in St. Lucia (5.2–7.2%/day compared to 1.5%/day for *G. crassissima* and 11.2–14.0%/day compared to 6.7–9.1%/day for *G. terete*) (Smith et al. 1986). Reported growth rates for other *Gracilaria* species using similar culture methods are as follows: *G. edulis*, 6.9–10.9%/day (Nelson et al. 1980, Micronesia); *G. verrucosa*, 9.9–13.2%/day (Ren-Zhi et al. 1984, China), 6.7%/day (Friedlander and Lipkin 1982, E. Mediterannean), and 8.7–11.5%/day (Rueness et al. 1987, Norway). These growth rates were obtained in regions with temperate or subtropical climates and were the highest obtained during the warmer periods of the year, when conditions were optimum for growth. It is of interest that the reported morphologies of *G. edulis* (Raju and Thomas 1971) and *G. verrucosa* and their ability to adjust to rope culture is very similar to those of *G. terete*.

Growth rates of unfertilized *G. terete* cultures in Discovery Bay were comparable to, or higher than, those obtained with *in situ* cage cultures of a fast-growing strain of *G. tikvahiae* in the Florida Keys, which received growth-optimizing nutrient pulse treatments (~10.3%/day; Lapointe 1985). At optimum densities of 2–5 kg wet wt/m^2, these *G. tikvahiae* cultures provided short-term yields of 35 g dry wt/m^2/day, which was reported as the highest yield achieved with *Gracilaria* in nonintensive culture systems (Lapointe and Hanisak 1984). A conservative estimate of the highest yield obtained with *G. terete* rope cultures was 16 g dry wt/m^2/day at a lower density of ~1.1 kg wet wt/m^2. Growth data for *G. terete* at higher culture densities were not obtained, usually due to fouling by *Ceramium* sp. Stable growth rates of *G. terete* as culture densities increased from 0.5 to 1.2 kg wet wt/m^2 (Figure 17.22) indicate the potential for much higher yields at higher culture densities of this species.

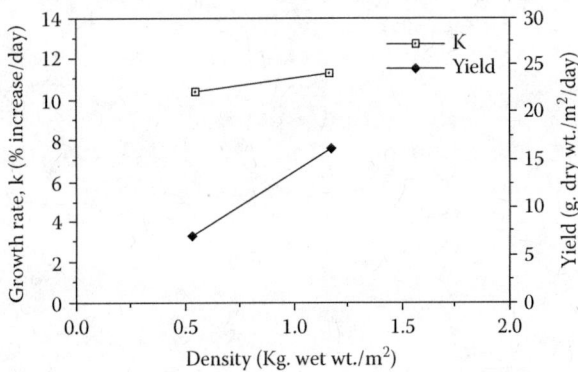

Figure 17.22 Growth rate, culture density, and yield relationships of *G. terete* horizontal rope cultures ($n = 2$) at BRO, April 24–May 15, 1988.

Factors Influencing Growth of *Gracilaria* Cultures in Back-Reef Habitats

Salinity and Temperature

Gracilaria species are known to be generally euryhaline (Shang 1976; Lapointe and Hanisak 1984; Lapointe et al. 1984b; Bird and McLachlan 1986), and the effects of salinity on growth of cultures at the various locations was assumed to be minimal due to the low range of salinities (28–36 ppt) at the culture sites. Some *Gracilaria* species are known to have maximum growth at salinities intermediate between fresh and full seawater (Shang 1976; Bird and Mclachlan 1986). Batch cultures of *G. terete* and *G. domingensis* from St. Lucia were found to have maximum growth in the narrow salinity range of 29–32 ppt (Bird and Mclachlan 1986), which falls within the range of high to intermediate groundwater-enriched locations (L2–L4). Growth of these cultures was 50–89% of maximum in the wider exclusive salinity ranges (i.e., excluding the range of maximum growth) of 18–38 ppt for *G. terete* and 8–38 ppt for *G. domingensis*. Although these results indicate that salinity may be a significant factor related to between-habitat growth rate differences of these species, they can only be cautiously extrapolated to conditions at Discovery Bay due to the wide difference between other experimental parameters such as temperature, light, and culture method, which may influence the range of optimum salinity for growth.

Gracilaria species from warm-water areas are generally known to have maximum growth in the temperature range of 25–30°C (Lapointe and Hanisak 1984; McLachlan and Bird 1984), which includes the temperature range of the back-reef culture locations at Discovery Bay (26–30°C). Thus, temperature per se appears unlikely to contribute to differences in growth rates of cultures at different locations and should permit maximum growth of most *Gracilaria* species at all culture locations. However, batch cultures of *G. terete* from St. Lucia were found to have maximum growth in the temperature range of 28–32°C and at least 50% of maximum growth in the range of 20–28°C (McLachlan and Bird 1984). Thus, *G. terete* may have an increased growth response at higher back-reef temperatures.

Nitrate Concentration

Macroalgae are generally known to have a hyperbolic relationship between growth and external nitrogen concentration in which maximum growth can be maintained by relatively low concentrations if adequate water motion maintains a low diffusion boundary layer (Hanisak 1983). Using a flow-through culture system, the growth rate of *Gracilaria foliifera* (subsequently *G. tikvahiae*) was shown to have a hyperbolic relation to the nitrate concentration of the culture medium, in which growth rapidly increased to close to maximum rates at a nitrate concentration of ~1.5 μM, with maximum growth at nitrate concentrations above ~12 μM (DeBoer 1978). Mean nitrate concentrations in areas of low groundwater input such as BR1 (2–3 μM NO_3) thus appear to be high enough to support maximum or close-to-maximum growth of *Gracilaria* cultures. The lack of enhancement of growth of *Gracilaria* cultures at this location by pulsed groundwater enrichment is also an indication of non-N-limiting conditions. However, fluctuations of nitrate concentration due to the variability of currents, tide, and rainfall, which are associated with groundwater input, often resulted in periods of very low nitrate concentration (<1 μM), which was likely to be below optimum for the growth of *Gracilaria* (particularly of faster-growing species such as *G. blodgettii* or *G. terete*). Nitrate concentrations at culture locations with intermediate groundwater input (e.g., 10.3 ± 3.5 μM at L4 during the 1989 experimental period) or higher were most likely continuously high enough to support maximum growth of *Gracilaria* cultures.

Phosphate Concentration

A negative correlation between phosphate concentration and salinity in the Discovery Bay back-reef, found during January 1988, was also found during July 1987 (Lapointe et al. in preparation) but was not found by D'Elia et al. (1981), and may be an indication of recent groundwater phosphate input to the back-reef. In addition, year-round phosphate concentrations in the back-reef obtained during 1988–1989 (0.3–0.5 µM in nitrate-enriched habitats) were often higher than those reported by D'Elia et al. (1981), which were mostly below 0.25 µM. High atomic N:P ratios in nitrate-enriched habitats, which suggest that growth of back-reef macroalgae is P limited, are mainly due to high nitrate concentrations. Groundwater phosphate input to the back-reef may be a factor contributing to recently increased reef macroalgal biomass and dense epiphyte growth on *Gracilaria* cultures.

Epiphytes

Fouling of *Gracilaria* cultures in mariculture systems by epiphytes is a common and major limitation to growth and viability and requires labor-intensive or innovative solutions (Shang 1976; Nelson et al. 1980; Friedlander and Lipkin 1982; Guo-Zhong et al. 1984; Ren-Zhi et al. 1984). Groundwater in Discovery Bay has a higher silica content than seawater does (Goreau et al. 1988), which results in the enhancement of diatom growth in areas of groundwater enrichment and the association of diatom fouling intensity with degree of groundwater enrichment. Intense diatom fouling appears to be the primary limiting factor to the growth of *Gracilaria* cultures in areas of high groundwater input, masking the growth-enhancing effect of nitrate enrichment and possibly of lowered salinity. Less intense fouling by diatoms at culture sites with intermediate groundwater input (L4 and L3) appears to alter the balance between growth inhibition due to fouling and growth enhancement by high nitrate availability to permit fastest growth of *Gracilaria* at these locations, at least during initial culture periods. In all simultaneous culture experiments, the growth rates of cultures at L4 were equal to or higher than those at BR1 and L2.

The higher incidence of attached filamentous epiphytes on *Gracilaria* cultures in areas of high and intermediate groundwater input may be partly due to stimulation of epiphyte growth by nitrate enrichment. The higher occurrence of *Ceramium* sp. on back-reef *G. terete* cultures was an exception to this trend. *Ceramium*, which appears to be specialized to colonize *G. terete* plants, was observed to be present at low levels on natural populations of *G. terete*. Biomass of *Ceramium* may possibly be controlled under natural conditions by factors such as grazing by amphipods, light intensity, and nutrient availability. A shift in these factors under culture conditions may then serve to enhance *Ceramium* growth.

The occurrence on *Gracilaria* of loose or attached macrophytic algae, which were more amenable to control by direct removal, was more predominant in areas of low groundwater input and appeared to be much less inhibitive to growth and viability of *Gracilaria* cultures than diatoms and filamentous epiphytes. Filamentous epiphytes and diatoms shaded *Gracilaria* cultures and likely inhibited water motion across fronds to a greater extent than macrophytic species, which competed mainly for space with *Gracilaria*.

Herbivory

Consumption of algae by herbivores is known to be intense on undisturbed Caribbean reefs and seagrass beds (Ogden 1976). Herbivorous fish in the families Acanthuridae (surgeonfish) and Scaridae (parrot fish) constitute a major component of the Caribbean herbivore guild (Ogden 1976; Hay 1984; Lewis and Wainwright 1985) and are the major herbivores consuming *Gracilaria* cultures in St. Lucia (Smith et al. 1986) and Discovery Bay.

The intensity of grazing on *Gracilaria* cultures at the various locations appeared to be primarily related to the position of the rafts with respect to refuges of herbivorous fish populations. Surgeonfish and parrot fish take refuge from predators in reef formations, venturing only up to 10 m or so into adjacent seagrass beds during active feeding (Ogden 1976). The lagoon culture location (L2), at which grazing intensity was highest, was within 5 m of a surrounding dense network of limestone formations, which are refuges for dense populations of herbivorous fish supported by the high algal productivity resulting from groundwater enrichment. The intermediate groundwater-enriched location (L4), at which grazing intensity was moderate, was within 5 m of a small patch reef that appeared to be a refuge for some fish populations. The main back-reef location (BR1), at which grazing intensity was low or sporadic, was located at ~75 m from the reef crest (the closest likely refuge), which was beyond the normal venturing range of herbivorous fish.

The mariculture rafts themselves became areas of refuge or habitats for many small fish and invertebrates, including crabs, shrimp, squid, and amphipods. Only small or juvenile fish were observed to remain in the vicinity of the rafts at BR1, and grazing of *Gracilaria* by these appeared to be minimal. Heavy grazing of epiphytes, particularly green filamentous species growing on rafts and cages, was occasionally observed at L2 and L4 when previously dense growths of these species were picked clean within several days.

Small invertebrates living on the rafts or among *Gracilaria* cultures appear to feed primarily on epiphytes, but this was only directly observed in the case of amphipods (the most abundant group) feeding on *Ceramium* sp. Amphipods, which occur in high densities in tropical environments and are subject to heavy fish predation (Young et al. 1976), are generally known to feed on small epiphytic algae, including filamentous species and diatoms (Zimmerman et al. 1979; Brawley and Adey 1981). In a closed microcosm study (Brawley and Adey 1981), grazing by amphipods at high density shifted the algal community structure from a high biomass of mostly filamentous species to a *Hypnea* macroalgal community. Subsequent removal of amphipods by fish predation resulted in rapid and heavy fouling and decline in growth of *Hypnea*. Although some grazing of macroalgae by amphipods has been observed (Nicotri 1977; Brawley and Adey 1981; this study), such losses are generally very small in relation to macroalgal productivity.

These observations suggest a variety of methods for the control of epiphytes in *Gracilaria* mariculture systems in tropical back-reef habitats, such as those at Discovery Bay, through the management of selective grazing by appropriate fish and invertebrate species.

Depth of Culture

Depth gradients of nitrate concentrations at culture sites may have influenced the growth variability of *Gracilaria* cultures at different depths, particularly at sites with relatively low nitrate input to surface layers such as BR1. Growth of deeper *G. terete* cultures at BR1 were often significantly lower than shallow cultures, although grazing was minimal. Lower growth of deeper relative to shallow cultures at L2 and L4 were more clearly related to more intense grazing at the lower depth.

Unshaded light intensity at the lower culture depth of ~40 cm was only slightly lower than near the surface (Table 17.4) and was unlikely to be below saturating levels for photosynthesis and growth of *Gracilaria*. Light-saturating levels for photosynthesis and growth of *G. tikvahiae* are known to be relatively low (<600 $\mu E/m^2/s$, Lapointe and Hanisak 1984; Penniman and Mathieson 1985). Shaded light intensities at the lower depth, due to surface cultures and cages, may have been below saturating levels.

During an extended two-depth culture experiment with *G. terete* at BR1 (February 3–April 7, 1989; described on p. 26), the deeper cultures were partly shaded by both shallow cultures and an attached vexar cage. Higher initial growth rates of the shallow cultures may have resulted from higher surface nitrate concentrations and/or irradiance levels. More rapid fouling of the shallow cultures by *Ceramium* and other epiphytes may have also been related to both higher nutrient and irradiance levels near the surface. Guo-zhong et al. (1984) found that long-term growth of *G. verrucosa* rope cultures on fixed rafts, which were submerged at a lower depth during high tide, was superior to that of cultures on floating rafts, which always remained at a shallow depth, although growth of the shallow cultures was initially higher. This difference was related to the much lower growth of epiphytes (including colonial diatoms, *Ectocarpus*, *Ulva*, and *Polysiphonia*), which appear to require higher light intensities for maximum growth relative to *Gracilaria*, on the periodically deeper cultures. Growth conditions of the fixed raft more closely approximated the natural habitat of *G. verrucosa*.

Culture of *Gracilaria* at deeper levels in intermediate to low groundwater-enriched locations may, therefore, be a useful strategy for reducing epiphytes and maximizing growth, provided that grazing can be minimized by appropriate site selection. See photo below. Photograph is of the nitrate-rich groundwater springs area where the vertical depth culture experiments were made. The clear surface layer is cold brackish water highly enriched with nitrate. The layer below it is warm seawater, with suspended limestone sediment that reflects light. The algae were grown in both layers. (Photograph by Dr. Peter D. Goreau.)

Weather Conditions

The high variability of *Gracilaria* growth rate with respect to time may be related to variations in weather and water movement. Growth of *Gracilaria* cultures is likely to be enhanced by a certain degree of wave action and turbulence due to reduced diatom fouling and enhanced nutrient uptake resulting from reduced boundary diffusion layers (Guerin and Bird 1987). This is supported by the general observation that back-reef cultures were more heavily fouled and had very low growth rates during extended periods of calm weather. Growth variability may also be associated with variation in nutrient input to the back-reef due to tide, current, and rainfall variation.

During the first nutrient enrichment experiment with *G. domingensis* at BR1 (April 7–May 12, 1989), the highest growth rates of the control and nitrate-enriched cultures occurred during weeks with one to several days of strong wind and turbulent conditions. Phosphate enrichment maintained higher and more stable growth during the culture period as indicated by the low growth variability of the P-enriched cultures. Santelices (1978) found that nutrient enrichment compensated for water movement in maintaining maximum growth of three species of Gelidiales (Rhodophyta). Higher growth rates of exposed versus caged cultures of both *G. domingensis* (controlled grazing experiment 2) and *G. terete* at BR1 are also indicative of growth enhancement of *Gracilaria* cultures due to increased water motion.

Assessment of Nutrient Limitation of *In Situ Gracilaria* Cultures and *G. domingensis* Natural Populations

Growth Rates

Growth rate increases of over 100% for *G. domingensis* and over 60% for *G. terete* and *G. crassissima* cultures at BR1 due to phosphate enrichment are strongly indicative of phosphate limitation of growth of *Gracilaria* in the Discovery Bay back-reef. In addition to enhanced growth rates at all levels of phosphate enrichment, *G. domingensis* cultures increased in vigor (as profusely dense branching and increased resistance to attached epiphytes) with increasing phosphate enrichment. Increased branching started to develop only during the latter part (fourth or fifth week) of the experimental periods; thus, the full effects of phosphate enrichment on growth and morphology of *G. domingensis* remain to be documented.

The absence of significant enhancement of growth of *G. domingensis* cultures at the high groundwater-enriched location (L2) by phosphate enrichment appeared to be primarily due to limitation of growth by dense fouling of the cultures by diatoms. The strong increase of nitrate uptake associated with phosphate enrichment during the third pulse treatment of these cultures may be an indication of physiological limitation of nitrate utilization by these cultures associated with phosphate limitation. N-limitation of *G. tikvahiae* in the Florida Keys also appeared to be partly due to physiological stress induced by intense P-limitation (Lapointe 1987).

According to the cell quota model of nutrient limitation in algae (Droop 1973; Smith 1983), photosynthesis and growth are hyperbolic functions of the cellular level of the limiting nutrient. Beyond a critical level, photosynthesis and growth are maximum and show no further increase. The pattern of growth enhancement due to low-, medium-, and high-phosphate enrichment can be explained on this basis if critical or greater cellular P contents of the *Gracilaria* cultures were maintained by medium-phosphate enrichment but not by low-phosphate enrichment.

Although growth of *G. domingensis* cultures was equally enhanced by both medium- and high-phosphate enrichment, increased uptake rates at the high concentrations enabled the accumulation of higher tissue phosphate levels. This is more evident for the flow-through cultures that received only one set of phosphate-enriched treatments. Growth enhancement, due to phosphate enrichment, of cultures maintained in flowing nitrate-enriched seawater was evident for up to two weeks after enrichment. At this time, tissue P levels of the high-phosphate-treated cultures were close to twice those of the nonphosphate-treated cultures. Thus, high-phosphate treatments may have the advantage of permitting uptake of enough phosphate to maintain non-P-limited growth of *G. domingensis* in nitrate-enriched habitats for longer periods between pulse treatments.

The absence of significantly higher phosphate uptake at high relative to medium phosphate concentration by *G. terete* and *G. crassissima* indicates that different phosphate enrichment strategies may be appropriate for these species. Additional work on nutrient uptake under controlled conditions and tissue nutrient levels of these species are required to define appropriate strategies.

In addition to its enhancement of growth at BR1, pulsed phosphate enrichment may be effective in further optimizing growth and long-term viability of *Gracilaria* cultures at intermediate nitrate-enriched locations, since increased phosphate availability appears to permit more efficient utilization of higher ambient nitrate levels while increasing vigor and resistance to filamentous epiphytes (at least for *G. domingensis*).

Nutrient Status: G. domingensis

Very high C:N and C:P ratios found in *G. domingensis* are comparable to those of *G. tikvahiae* (Lapointe 1987) and are characteristic of several macroalgal species that have a high structural carbon content (Atkinson and Smith 1983).

The strong effect of phosphate enrichment on tissue P levels of both *in situ* and flow-through cultures and elevated C:P ratios of nonphosphate-enriched cultures are also highly indicative of P limitation in local groundwater-enriched habitats. A dominant effect of phosphate enrichment on photosynthesis, growth, and tissue levels of C, N, and P of *in situ G. tikvahiae* cultures in the Florida Keys and very high C:P ratios of unfertilized cultures demonstrated severe P-limitation of productivity (Lapointe 1987). C:P ratios of nonphosphate-enriched *G. domingensis* cultures (1744) were comparable to those of P-limited *G. tikvahiae* cultures (>1800), which were threefold greater than the median value reported in a survey of C:N:P ratios of benthic marine plants (Atkinson and

Smith 1983). N:P ratios of nonphosphate-enriched *G. domingensis* cultures (>70) were also much higher than the median value of 30 reported by Atkinson and Smith (1983).

Tissue phosphate levels of medium- and high-phosphate-enriched *G. domingensis* cultures were higher than those of *G. tikvahiae* cultures receiving somewhat similar treatments (Lapointe 1985, 1987) and may be an indication of a higher capacity for phosphate uptake and storage by *G. domingensis*.

Tissue N levels of all *G. domingensis* cultures, including those cultured in flowing nitrate-enriched seawater, were relatively low compared to those of *G. tikvahiae* and critical C:N ratios, above which growth is N-limited, appear to be higher for *G. domingensis* relative to *G. tikvahiae*. Critical C:N ratios of *G. tikvahiae* are known to be in the range of 10–15 (D'Elia and DeBoer 1978; Lapointe and Duke 1984), whereas C:N data for nitrate-enriched *G. domingensis* suggest that critical ratios for this species are above 20.

Levels of tissue N and P and C:N and C:P ratios of natural *G. domingensis* populations at East Discovery Bay during December 1988 were somewhat comparable to those of medium- and high-phosphate-enriched back-reef cultures and thus indicate the absence of either N or P limitation to the growth of natural populations at this time of year. However, the very low levels of tissue N and P and very high C:N and C:P ratios of this population during June 1989 are indicative of both N and P limitation at this time of year. Levels of ammonia, nitrate, and phosphate in seawater samples from the vicinity of this population at both times of year (Table 17.7E) confirm the presence of higher levels of these nutrients during December 1988. No sources of groundwater input are known to be in this vicinity, and nutrient replete conditions during the winter months may be associated with the more frequent occurrence of storms and strong northerly winds, which are likely to increase nutrient inputs via rainfall and oceanic upwelling. The passage of hurricane Gilbert in September 1988 resulted in very strong mixing of deep and surface oceanic layers along the coasts of Jamaica, and associated nutrient inputs may have also contributed to subsequent nutrient-replete conditions of macroalgal populations.

Seasonal variations in chemical composition are known to occur in populations of several *Gracilaria* species at Holland Bay on the east coast of Jamaica (Devi Prasad 1986), a pattern somewhat similar to the trend found in *G. domingensis* in Discovery Bay. Dry weight yields (dry weight as a percentage of fresh weight) of *G. domingensis* were lower between November and January than the rest of the year and may be an indication of higher growth during the former period (Devi Prasad 1986). Protein levels of *G. cervicornis* and *G. domingensis* were highest during August and September, with gradual decrease to a minimum during March. Seasonal trends in nutrient status of these *Gracilaria* populations were also likely to be influenced by inputs of freshwater and agricultural runoff via the river that opens into Holland Bay.

Nitrate and Phosphate Uptake

Nitrate uptake by macroalgae has been generally found to exhibit saturation kinetics that can be modeled by Michaelis–Menten equations (D'Elia and DeBoer 1978; Haines and Wheeler 1978; Hanisak and Harlin 1978). Nitrate depletion curves for *G. domingensis* are suggestive of such rate-limiting kinetics with saturated nitrate uptake above concentrations of 2–6 µM. Estimates of maximum uptake rate (V max) by linear regression were relatively low compared to those of *G. tikvahiae* [4.0–5.5 µM/g dry wt/h for *G. domingensis* compared to 9.7 (D'Elia and DeBoer 1978) and 16–35 (Lapointe et al. 1984a) µM/g dry wt/h for *G. tikvahiae*].

Several factors have been shown to influence nitrate uptake rates of *Gracilaria* and other macroalgal species. These include light, temperature (Lapointe et al. 1984a), acclimation to ambient concentrations (Hwang et al. 1987), and the concentration of other ions, such as ammonia, in the medium (D'Elia and DeBoer 1978).

Nitrate and ammonium uptake rates are commonly higher in the light than in the dark (DeBoer 1981), which is likely due to the energy requirements of active uptake of these ions being supplied via photophosphorylation (Falkowski and Stone 1975). Nitrate uptake rates of *G. tikvahiae* at high light intensity were close to three times the rates at low light intensity at optimum temperatures (25–30°C) for uptake (Lapointe et al. 1984a). The nitrate uptake rates found for the three *Gracilaria* species in this study were determined primarily during dark periods and may therefore be much lower than their uptake rates in daylight.

A high affinity for nitrate uptake and utilization by *G. domingensis* is indicated by the lack of significant inhibition of nitrate uptake by ammonia, even at concentrations as high as 300–400 µM NH_4^+. Nitrate uptake by many macroalgae is substantially suppressed by the presence of NH_4^+ in the medium (D'Elia and DeBoer 1978; Haines and Wheeler 1978; Hanisak and Harlin 1978). Nitrate uptake by *G. tikvahiae* was substantially suppressed by NH_4^+ at concentrations above 5 µM (D'Elia and Deboer 1978). However, nitrate uptake by low intertidal *G. pacifica* plants was unaffected by the presence of ammonia at concentrations up to 15 µM (Thomas et al. 1987). These plants also exhibited a high affinity for nitrate uptake and utilization as higher uptake and assimilation rates and larger nitrate reserves relative to other populations of the same species. They also appeared to exhibit an unsaturated diffusion component of nitrate uptake at high nitrate concentrations not reported for other *Gracilaria* species. This nitrogen physiology was thought to be an adaptation to the influx of nitrate-rich deep water, likely to be a major source of nitrogen for this population of *G. pacifica*. *Gracilaria* species cultured in or endemic to nitrate-enriched back-reef habitats (such as *G. cervicornis*) may therefore exhibit similar adaptations, as indicated by the results for *G. domingensis*.

Nitrate uptake rates of groundwater enriched relative to unenriched flow-through *G. domingensis* cultures were inversely related to levels of tissue N of these cultures. Ammonia and nitrate uptake rates of macroalgae are known to decrease sharply when internal N reserves increase beyond critical levels, at which plants are no longer N-limited. (D'Elia and DeBoer 1978). N reserves of groundwater-enriched *G. domingensis* flow-through cultures therefore appear to be above this critical level. Corresponding C:N ratios of these cultures (18) are also indicative that critical C:N ratios are relatively high for this species.

Some evidence indicates that N and P uptake and metabolism are interactive in microalgae (Rhee 1974; Davies 1988) and macroalgae (Lapointe 1987). In microalgal species, increased nitrate availability is known to enhance phosphate uptake (Rhee 1974), while NH_4^+ incorporation is regulated by the phosphorus cell quota (Davies 1988). Enhancement of nitrate uptake and metabolism by *G. tikvahiae* under P-limiting conditions by simultaneous phosphate enrichment was implied by resulting tissue levels of N and P (Lapointe 1987), but such an effect was not directly measured. Such interactive effects appear to be related to the energy requirements of active uptake and transport of inorganic ions (NO_3^-, NH_4^+, and PO_4^{3-}), which is primarily supplied by cyclic photophosphorylation (Kuhl 1974; Falkowski and Stone 1975). Nitrogen (in chlorophyll biosynthesis) and phosphorus (in ATP production) are both implicated in cyclic photophosphorylation.

Although enhancement of nitrate uptake by phosphate enrichment of *Gracilaria* cultures is indicated by some of the nitrate-uptake data, results from the flow-through experiments were inconclusive, and additional experimental work is required to more clearly demonstrate and quantify such an effect. Enhancement of nitrate uptake by phosphate enrichment may be more evident if phosphate enrichment is carried out before nitrate enrichment, permitting prior phosphate uptake and metabolism. Such enhancement may also be more pronounced under light due to more immediate enhancement of ATP production via photophosphorylation.

G. domingensis appears to exhibit multiple-phase phosphate uptake kinetics, with a saturation phase at low concentrations and a linear phase at high concentrations. This is similar to that found for *G. tikvahiae* by Friedlander and Dawes (1985). Further uptake experiments with more frequent sampling under controlled conditions are required to define adequately these phases.

Phosphate uptake rates of *G. domingensis* at medium and high concentrations were much higher than those of *G. tikvahiae* at similar or higher concentrations (Friedlander and Dawes 1985; Lapointe 1985). Maximum phosphate uptake rates of *G. tikvahiae* have been 0.28 µM/g fresh wt/h (Friedlander and Dawes 1985) compared to 1.32 µM/g fresh wt/h (8.33 µM/g dry wt/h) for *G. domingensis*.

CONCLUSIONS

1. Three *Gracilaria* species (*G. terete*, *G. domingensis*, and *G. blodgettii*) are highly suitable for mariculture due to their high growth rates and agar contents, adaptability to culture, and resistance to epiphytes. In addition, the nutrient physiology of *G. domingensis*, characterized by high phosphate uptake rates at high substrate concentrations, a relatively high phosphate storage capacity, and a high affinity for nitrate uptake, is well adapted to culture in nitrate-enriched back-reef habitats where productivity can be maximized by periodic phosphate enrichment. Four other *Gracilaria* species (*G. crassissima*, *G. cervicornis*, *G. damaecornis*, and *G. sjoestedtii*) are less suitable for mariculture or require further experimental work to fully assess their mariculture potential and elaborate appropriate culture methods.
2. Growth of *Gracilaria* cultures in nitrate-enriched habitats at Discovery Bay is phosphate limited. Pulsed phosphate enrichment significantly increased growth rates and tissue phosphate levels of *Gracilaria* cultures and may permit more efficient utilization of high nitrate levels while reducing epiphyte loads. P-replete *G. domingensis* cultures exhibited increased vigor and epiphyte resistance.
3. In high-nitrate-enriched locations, dense fouling by diatoms is a major limitation to growth and viability of *Gracilaria* cultures. Less intense fouling by diatoms in areas of intermediate nitrate enrichment appears to permit maximum growth of *Gracilaria* cultures during initial culture periods due to nitrate enrichment.
4. Mariculture site selection is an important consideration with respect to the management of epiphytes. The type and degree of fouling of *Gracilaria* cultures varies with the location of the cultures on a gradient of nitrate enrichment. The occurrence of loose or attached macrophytic epiphytes, which are the easiest to remove and least inhibitive to growth of *Gracilaria*, was generally highest in back-reef locations with low groundwater input, while the occurrence of diatoms and attached filamentous epiphytes generally increased with increasing nitrate enrichment.
5. The proximity of mariculture sites to herbivorous fish refuges is an important consideration in limiting herbivory on *Gracilaria* cultures. Herbivory can be minimized by choosing sites beyond the normal foraging range of herbivorous fish species that take refuge in reef habitats. However, selective grazing of epiphytes by some herbivorous fish and invertebrate species suggests a variety of management methods for the biological control of epiphytes on *Gracilaria* cultures.
6. Culture of *Gracilaria* at deeper levels in high- to intermediate-groundwater-enriched locations may be a useful strategy for optimizing growth and reducing epiphytes, provided that grazing can be minimized by appropriate site selection.
7. Variability of weather conditions is an important factor affecting the growth and viability of back-reef *Gracilaria* cultures. Conditions for *Gracilaria* mariculture may be better during periods of the year with weather-related increased water movement and nutrient input.
8. Additional research on the mariculture potential of the *Gracilaria* species studied should include the following:
    a. Further study of the factors related to the occurrence and abundance of *Ceramium* sp. on *G. terete* cultures and the testing of culture and management methods that may minimize its occurrence and growth. This includes possible selection of more *Ceramium*-resistant *G. terete* strains.
    b. Further experimental net culture of appropriate *G. domingensis* and *G. cervicornis* strains and the development of net designs that maintain a more even distribution of seeded material.
    c. Experimental net culture of *G. blodgettii* and *G. damaecornis* that will be dependent on the location and management of natural populations of these species and adequate facilities for maintaining stock cultures. Initial growth rates and branching morphology of these species suggest a high adaptability to net culture.
    d. Experimental culture of *G. crassissima* at sites with increased wave action and/or lower nitrate input as well as additional growth experiments in conjunction with *G. domingensis* net cultures in which the reduction of epiphytes on *G. crassissima* by the brushing action of *G. domingensis* fronds may be maximized by improved net design.

ACKNOWLEDGMENTS

I wish to thank Dr. Larry Brand for his invaluable assistance, guidance, and support; Dr. Thomas Goreau for his encouragement and support, and his advice and assistance during my research at Discovery Bay; and Dr. Brian Lapointe for many helpful suggestions on experimental design and for providing me with valuable research experience on macroalgal culture and physiology. I would also like to thank Dr. Alina Szmant for providing me with extended use of the autoanalyzer and carbon, hydrogen and nitrogen content analyzer, and I am very grateful for the generous assistance of Luigi Ferrer, Lisa Fitzgerald, and Nancy Gassman in the use of this equipment. I am very grateful to Allan Smith of the Eastern Caribbean Natural Area Management Programme for sharing his expertise on *Gracilaria* taxonomy and field culture techniques, and to Dr. Edward Maly, at Concordia University, Montreal, for providing me with lab space and equipment for determining the phosphorus content of *Gracilaria* samples. I wish to thank the staff of the DBML for their assistance and support during my research in Jamaica. I am particularly grateful to Lester Anderson for his valuable assistance in assembling an experimental seawater flow-through system. I also wish to thank Dr. Peter Goreau, Gillian Elliot, and Juli Anne Royes for collecting data and maintaining algal cultures at Discovery Bay during my absence. This research at the Rosenstiel School of Marine and Atmospheric Science (RSMAS) has been generously supported by an RSMAS fellowship. My research in Jamaica has been supported by a fellowship from the Caribbean Resources Development Foundation made possible by the generous donation of Mr. Alberto Vadia, President of the Pan American Land Development Corporation.

REFERENCES

Abbot, L. A., and J. N. Norris (Eds.). 1984. *Taxonomy of Economic Seaweeds with Reference to Some Pacific and Caribbean Species*. California Sea Grant College Program Report # T-CSGCP-O11. University of California, La Jolla, CA.

Atkinson, M. J., and S. V. Smith. 1983. C:N:P ratios of benthic marine plants. *Limnol. Oceanogr.* 28:568–574.

Bird, K. T., C. Habig, and T. Debusk. 1982. Nitrogen allocation and storage patterns in *Gracilaria tikvahiae* (Rhodophyta). *J. Phycol.* 18:344–348.

Bird, C. J., and J. McLachlan. 1986. The effect of salinity on distribution of species of *Gracilaria* (Rhodophyta, Gigartinales): An experimental assessment. *Bot. Mar.* 29:231–238.

Brawley, S. H., and W. H. Adey. 1981. The effect of micrograzers on algal community structure in a coral reef microcosm. *Mar. Biol.* 61:167–177.

Brinkhuis, B. H. 1984. Growth patterns and rates. In M. M. Littler and D. S. Littler (Eds.). *Handbook of Phycological Methods—Ecological Field Methods: Macroalgae*. Cambridge: Cambridge University Press.

Caughley, G. 1976. Plant-herbivore systems. In R. M. May (Ed.). *Theoretical Ecology, Principles and Applications*, 94–113. Oxford, UK: Blackwell Scientific Publications.

Chapman, V. J., 1963. The Marine Algae of Jamaica. Part 2. Phaeophyceae and Rhodophyceae. *Bulletin of the Institute of Jamaica*, Kingston, Jamaica.

Chapman, V. J., and D. J. Chapman. 1980. *Seaweeds and Their Uses*. London and New York: Chapman & Hall.

Chiang, Y.-M. 1981. Cultivation of *Gracilaria* in Taiwan. In T. Levring (Ed.). *Proceeding of the International Seaweed Symposium* 10:569–574, Goteborg (Sweden).

Craigie, J. S., Z. C. Wen and J. P. Van Der Meer, 1984. Interspecific, Intraspecific and Nutritionally-Determined Variations in the Composition of Agars from Gracilaria spp. *Botanica Marina.* 27: 55–61

Davies, A. G. 1988. Nutrient interactions in the marine environment. In L. J. Rogers and J. R. Gallon (Eds.). *Biochemistry of the Algae and Cyanobacteria*, 241–256. Oxford, UK: Clarendon Press.

DeBoer, J. A. 1978. Effects of nitrogen enhancement on growth rate and phycocolloid content in *Gracilaria foliifera* & *Neoagardhiella baileyi*. In A. Jensen and J. R. Stein (Eds.) *Proceeding of the International Seaweed Symposium* 9:263–271, Santa Barbara, California.

DeBoer, J. A. 1981. Nutrients. In C. S. Lobban and M. J. Wynne (Eds.). *Biology of Seaweeds*, 356–391. Oxford, UK: Blackwell Scientific Publications.

DeBoer, J. A., H. J. Guigli, T. L. Israel, and C. F. D'Elia. 1978. Nutritional studies of two red algae. I. Growth rate as a function of nitrogen source and concentration. *J. Phycol.* 14:261–266.

D'Elia, C. F., and J. A. DeBoer. 1978. Nutritional studies of two red algae. II. Kinetics of ammonium and nitrate uptake. *J. Phycol.* 14:266–272.

D'Elia, C. F., K. L. Webb, and J. W. Porter. 1981. Nitrate rich groundwater inputs to Discovery Bay, Jamaica: A significant source of N to local coral reefs? *Bull. Mar. Sci.* 31:903–910.

DeOliveira, E. C., C. J. Bird, and J. McLachlan. 1983. The genus *Gracilaria* (Rhodophyta, Gigartinales) in the Western Atlantic. *Gracilaria domingensis, G. cervicornis* and *G. ferox. Can. J. Bot.* 61:2999–3008.

Devi Prasad, P. V. 1986. A seasonal study of the red seaweeds *Solieria tenera* and three species of *Gracilaria* from Jamaica. *Hydrobiologia* 140:167–171.

Doty, M. S. 1978. Status of marine agronomy with special reference to the tropics. *Proc. Int. Seaweed Symp.* 9:35–38.

Doty, M. S. 1980. Outplanting *Eucheuma* species and *Gracilaria* species in the tropics. In I. A. Abbott, M. S. Foster, and L. F. Eklund (Eds.). *Pacific Seaweed Aquaculture*, 19–22. Pub. by the California Sea Grant College Program. Institute of Marine Resources. University of California, La Jolla, CA.

Doty, M. S., and G. A. Santos. 1983. Agar from *Gracilaria cylindrica. Aquat. Bot.* 15:299–306.

Droop, M. R. 1973. Some thoughts on nutrient limitation in algae. *J. Phycol.* 9:264–272.

Duckworth, M., C. K. Hong, and W. Yaphe. 1971. The agar polysaccharides of *Gracilaria* species. *Carbohyd. Res.* 18:1–9.

Edelstein, T., C. J. Bird, and J. McLachlan. 1976. Studies on *Gracilaria*. 2. Growth under greenhouse conditions. *Can. J. Bot.* 54:2275–2290.

Falkowski, P. G., and D. P. Stone. 1975. Nitrate uptake in marine phytoplankton: Energy sources and the interaction with carbon fixation. *Mar. Biol.* 32:77–84.

Friedlander, M., and C. J. Dawes. 1985. *In situ* uptake kinetics of ammonium and phosphate and chemical composition of the red seaweed *Gracilaria tikvahiae. J. Phycol.* 21:448–453.

Friedlander, M., and Y. Lipkin. 1982. Rearing of agarophytes and carrageenophytes under field conditions in the Eastern Mediterranean. *Bot. Mar.* 25:101–105.

Goreau, T. J., P. D. Goreau, S. H. Goreau, A. H. Macfarlane, P. V. Devi Prasad, B. E. Lapointe, and J. H. Ryther. 1986. Jamaican back-reef springs: Some aspects of their ecology and mariculture potential. *Proc. Assoc. Isl. Mar. Lab. Carib.* 19: 22

Goreau, T. J., B. Lapointe, J. O'Connel, P. D. Goreau, and A. H. Macfarlane. 1988. Groundwater nutrient inputs to Jamaican reefs. *Proc. Assoc. Isl. Mar. Lab. Carib.* 21:40.

Guerin, J. M., and K. T. Bird. 1987. Effects of aeration period on the productivity and agar quality of *Gracilaria* sp. *Aquaculture* 64:105–110.

Guo-Zhong, R., J.-C. Wang, and M. Q. Chen. 1984. Cultivation of *Gracilaria* by means of low rafts. *Hydrobiologia* 116/117:72–76.

Haines, K. C., and P. A. Wheeler. 1978. Ammonium and nitrate uptake by the marine macrophytes *Hypnea musciformis* (Rhodophyta) and *Macrocystis pyrifera* (Phaeophyta). *J. Phycol.* 14:319–324.

Hanisak, M. D. 1983. The nitrogen relationships of marine macroalgae. In E. J. Carpenter and D. G. Capone (Eds.). *Nitrogen in the Marine Environment*, 699–728. New York: Academic Press.

Hanisak, M. D., and M. M. Harlin. 1978. Uptake of inorganic nitrogen by *Codium fragile* sub sp. tomentosoides (Chlorophyta). *J. Phycol.* 14:450–454.

Hanisak, M. D., and J. H. Ryther. 1984. Cultivation biology of *Gracilaria tikvahiae* in the United States. *Hydrobiologia* 116/117:295–298.

Hansen, J. E. 1984. Strain selection and physiology in the development of *Gracilaria* mariculture. *Hydrobiologia* 116/117:89–94.

Hansen, J. E., J. E. Packard, and W. T. Doyle. 1981. Mariculture of red seaweeds, Report # T-CSGCP-002. California Sea Grant College Program. University of California, La Jolla, CA.

Hay, M. E. 1984. Patterns of fish and urchin grazing on Caribbean coral reefs: Are previous results typical? *Ecology* 65:446–454.

Hwang, S.-P. L., S. L. Williams, and B. H. Brinkhuis. 1987. Changes in internal dissolved nitrogen pools as related to nitrate uptake and assimilation in *Gracilaria tikvahiae* McLachlan (Rhodophyta). *Bot. Mar.* 30:11–19.

Kuhl, A. 1974. Phosphorus. In W. D. P. Stewart (Ed.). *Algal Physiology and Biochemistry*, 636–654. Oxford, UK: Blackwell Scientific Publications.

Lahaye, M. J., F. Revol, C. Rochas, J. Mclachlan, and W. Yaphe. 1988. The chemical structure of *Gracilaria crassissima* and *Gracilaria tikvahiae* cell wall polysaccharides. *Bot. Mar.* 31:491–501.

Lapointe, B. E. 1985. Strategies for pulsed nutrient supply to *Gracilaria* cultures in the Florida keys: Interactions between concentration and frequency of nutrient pulses. *J. Exp. Mar. Biol. Ecol.* 93:211–222.

Lapointe, B. E. 1987. Phosphorus and nitrogen limited photosynthesis and growth of *Gracilaria tikvahiae* (Rhodophyceae) in the Florida Keys: An experimental field study. *Mar. Biol.* 93:561–568.

Lapointe, B. E., C. J. Dawes, and K. R. Tenore. 1984a. Interactions between light and temperature on the physiological ecology of *Gracilaria tikvahiae* (Rhodophyta). II. Nitrate uptake and levels of pigments and chemical constituents. *Mar. Biol.* 80:171–178.

Lapointe, B. E., and C. S. Duke. 1984. Biochemical strategies for growth of *Gracilaria tikvahiae* in relation to light intensity and nitrogen availability. *J. Phycol.* 20:488–495.

Lapointe, B. E., and D. Hanisak. 1985. Productivity and nutrition of marine biomass systems in Florida. in D. L. Klass (ed.) *Symposium Papers Energy from Biomass and Wastes* 1X: Lake Buena Vista, Florida, Jan. 28-Feb. 1, 1985 (pp. 111–126) Chicago, IL: Institute of Gas Technology.

Lapointe, B. E., D. L. Rice, and J. M. Lawrence. 1984b. Responses of photosynthesis, respiration, growth and cellular constituents to hypo-osmotic shock in the red alga *Gracilaria tikvahiae*. *Comp. Biochem. Physiol.* 77A:127–132.

Lapointe, B. E., and J. H. Ryther. 1978. Some aspects of the growth and yield of *Gracilaria tikvahiae* in culture. *Aquaculture* 15:185–193.

Lapointe, B. E., and J. H. Ryther. 1979. The effects of nitrogen and seawater flow rate on the growth and biochemical composition of *Gracilaria foliifera* in mass outdoor cultures. *Bot. Mar.* 22:529–537.

Lewis, S., and P. C. Wainwright. 1985. Herbivore abundance and grazing intensity on a Caribbean coral reef. *J. Exp. Mar. Biol. Ecol.* 87:215–228.

Macfarlane, A. H., T. J. Goreau, A. Smith, P. D. Goreau, S. Goreau, and B. Lapointe. 1988. Algal mariculture in Jamaican back-reef springs. *Proc. Assoc. Isl. Mar. Lab. Carib.* 21:37.

Mathieson, A. C., and W. J. North. 1982. Algal aquaculture: Introduction and bibliography. In J. R. Rosowski and B. C. Parker (Eds.). *Selected Papers in Phycology II*, 773–787. Lawrence, KS: Phycological Society of America.

McHugh, D. J. 1984. Marine phycoculture and its impact on the seaweed colloid industry. *Hydrobiologia* 116/117:351–354.

McLachlan, J. 1979. *Gracilaria tikvahiae* sp. nov (Rhodophyta, Gigartinales, Gracilariaceae) from the northwestern Atlantic. *Phycologia* 18:19–23.

McLachlan, J., and C. J. Bird. 1983. *Gracilaria* in the Seaweed market: A prospectus. *International Symposium of Aquaculture, Coquimbo, Chile*. 133–157.

McLachlan, J., and C. J. Bird. 1984. Geographical and experimental assessment of the distribution of *Gracilaria* species (Rhodophyta, Gigartinales) in relation to temperature. *Helgol. Meeresunters.* 38:319–334.

McLachlan, J., C. J. Bird, and M. Greenwell. 1986. Seaweed resources for the extractive industry: What are the options? *Monogr. Biol.* 4:1–12.

Menzel, D. W., and N. Corwin. 1965. The measurement of total phosphorus in seawater based on the liberation of organically bound fractions by persulfate oxidation. *Limnol. Oceanogr.* 19:280–282.

Morris, J. W., and J. P. Riley. 1963. The determination of nitrate in seawater. *Anal. Chim. Acta.* 29:272–279.

Murphy, J., and J. P. Riley. 1962. A modified single solution method for the determination of phosphate in natural waters. *Anal. Chim. Acta* 27:31–36.

Nelson, S. G., R. N. Tsutsui, and B. R. Best. 1980. Preliminary evaluation of the mariculture potential of *Gracilaria* in Micronesia: Growth and ammonium uptake. In, I. A. Abbot and M. S. Foster (eds.), *Seaweeds of the Warm Pacific*. California Sea Grant Program, University of California, La Jolla, CA. 72–79.

Nicotri, M. E. 1977. The impact of crustacean herbivores on cultured seaweed populations. *Aquaculture* 12:127–136.

Ogden, J. C. 1976. Some aspects of herbivore-plant relationships on Caribbean reefs and seagrass beds. *Aquat. Bot.* 2:103–116.

Penniman, C. A., and A. C. Mathieson. 1985. Photosynthesis of *Gracilaria tikvahiae* McLachlan (Rhodophyta) from the Great Bay estuary, New Hampshire. *Bot. Mar.* 28:427–435.

Raju, P. V., and P. C. Thomas. 1971. Experimental field cultivation of *Gracilaria edulis*. *Bot. Mar.* 25:101–105.

Ren-Zhi, L., C. Ren-Yi, and M. Zhao-Cai. 1984. A preliminary study of raft cultivation of *Gracilaria verrucosa* and *Gracilaria sjoestedtii*. *Hydrobiologia* 116/117:252–254.

Rhee, G.-Y. 1974. Phosphate uptake under nitrate limitation by *Scenedesmus* sp. and its ecological implications. *J. Phycol.* 10:470–475.

Rueness, J., H. A. Mathiesen, and T. Tananger. 1987. Culture and field observations on *Gracilaria verrucosa* from Norway. *Bot. Mar.* 30:267–276.

Ryther, J. H., N. Corwin, T. A. DeBusk, and L. D. Williams. 1982. Nitrogen uptake and storage by the red alga *Gracilaria tikvahiae*. *Aquaculture* 26:107–115.

Ryther, J. H., J. A. DeBoer, and B. E. Lapointe. 1979. Cultivation of seaweeds for hydrocolloids, waste treatment and biomass for energy conversion. *Proc. Int. Seaweed Symp.* 9:1–16.

Santelices, B. 1978. Multiple interaction of factors in the distribution of some Hawaiian Gelidiales (Rhodophyta). *Pac. Sci.* 32:119–147.

Shang, Y. C. 1976. Economic aspects of *Gracilaria* culture in Taiwan. *Aquaculture* 8:1–7.

Smith, V. H. 1983. Light and nutrient dependence by algae. *J. Phycol.* 19:306–313.

Smith, A. H., A. Jean, and K. Nichols. 1986. An investigation of the potential for commercial mariculture of seamoss (*Gracilaria* spp., Rhodophyta) in St. Lucia. *Proc. Gulf Caribb. Fish. Inst.* 37:4–11.

Smith, A. H., K. Nichols, and J. Mclachlan. 1984. Cultivation of seamoss (*Gracilaria*) in St. Lucia, West Indies. *Hydrobiologia* 116/117:249–251.

Sokal, R. R., and F. J. Rohlf. 1969. *Biometry: The Principles and Practice of Statistics in Biological Research*. San Francisco: W. H. Freeman.

Sullivan, J. J. 1983. Seaweed aquaculture in the United States—Progress and promise. In *Proceedings of the Second North Pacific Aquaculture Symposium*, 29–46.

Taylor, W. R. 1960. *Marine Algae of the Eastern Tropical and Subtropical Coasts of the Americas*. Ann Arbor, MI: University of Michigan Press.

Thomas, T. E., P. J. Harrison, and D. H. Turpin. 1987. Adaptations of *Gracilaria pacifica* (Rhodophyta) to nitrogen procurement at different intertidal locations. *Mar. Biol.* 93:569–580.

Tseng, C.K. 1982. Commercial cultivation. In C. S. Lobban and M. J. Wynne (Eds.). *The Biology of Seaweeds*, 680–685. Oxford: Backwell Scientific Publications.

Young, D. K., M. A. Buzas, and M. W. Young. 1976. Species densities of macrobenthos associated with seagrass: A field experimental study of predation. *J. Mar. Res.* 34:577–592.

Zimmerman, R., R. Gibson, and J. Harrington. 1979. Herbivory and detritivory among Gammaridean amphipods from a Florida seagrass community. *Mar. Biol.* 54:41–47.

CHAPTER **18**

Sustainable Reef Design to Optimize Habitat Restoration

Mara G. Haseltine

CONTENTS

Introduction: Principles of Sustainable Reef Design 245
Note on Materials 246
Oyster Restoration in New York City 246
Oyster Restoration Project at College Point, Queens, New York City 250
Gill Reef 253
Floating Reef 255
Solar-Powered Reef Buoys 257
Radiolarian Reef 257
Coastline Management/Floating Island 257
Amphibious Reef 259
References 261

INTRODUCTION: PRINCIPLES OF SUSTAINABLE REEF DESIGN

Effective and sustainable habitat restoration requires innovative reef designs that optimize biological processes, alternative reef building methods that do not use plastic or concrete and that use renewable energy from solar, tidal, wind, and wave power as energy sources. Special attention must be paid to maximize the flow of water and light through the reef, providing reef inhabitants the maximum amount of oxygen, food, and nutrients while flushing away wastes. Reefs should be designed so that their shape increases the growth of structure builders such as coral and oysters, while providing habitat for other reef organisms to hide and spawn, increasing biodiversity.

This is especially important if the reef is to be used for sustainable fish and shellfish farming. Specially sized and shaped nooks and crannies can be designed into artificial reefs so that particular species will live there and can be successfully farmed without the use of nets, antibiotics, growth hormones, or external food addition. Special attention must also be paid to surrounding water and benthic habitat so there is as little impact as possible, and light as well as water can pass through the structure. Designs of structures presented in this chapter utilize the concept of "biomimicry." Biomimicry, or innovation inspired by nature, adjusts its blueprints from Mother Nature's designs which have evolved over billions of years, on a wide variety of scales from microscopic to megascopic

(Benyus, 1997). Designs based on these principles should allow restoration of degraded habitat that is much more biologically diverse and productive than conventional artificial reef structures.

Please note that these structures are designed both with functionality and beauty in mind. The concept is that they could also be learning tools and be championed by their community to become sites for education and ecotourism, creating job opportunities for local people.

NOTE ON MATERIALS

There is no need to use plastic or concrete to make any of these structures. In places where the salinity is brackish or salty, my top recommendation would be to use Biorock in a mineral accretion process whereby low-voltage DC electricity is fed into metal. If grown slowly, Biorock is self-repairing and three times the strength of concrete. Additional materials include high-fire ceramic coated with calcium carbonate and foamed glass mixed with calcium carbonate.

A recipe for oyster growth that can be used for coral as well is as follows.

The most crucial ingredient, calcium carbonate, the substance shells are made from, is the basic building block for oyster reefs. This recipe is designed for areas where reefs no longer exist, so alternative materials must be made to substitute for naturally occurring shells.

1. Submerge metal in seawater. Add low voltage direct-current electricity. Any kind of renewable power source will do. If grown slowly, the calcium carbonate is three times the strength of concrete and is naturally self-repairing. This method uses materials that boost the metabolism of the living organisms (be it oysters or coral), making them more impervious to disease and climate change.
2. Collect shell. On the East Coast of the United States, the shell must be from waters north of New York City, as southern shell is riddled with disease. Do not use restaurant shell. It is against the law to use restaurant shell in New York State. The reason for this law is that the shell from oysters served in restaurants are often imported from other states, and there is a chance that they could be contaminated with diseases from their native water bodies and could therefore spread invasive disease to local oyster population. It is of note that local water-quality groups such as the NY/NJ Baykeeper have protested this law, because there is much evidence that if the shell is aged outside in the sunlight for a few months, it poses no threat of spreading invasive disease, as all the organisms from foreign water bodies die off. In fact, this law was so strongly contested that I began to age restaurant shell in an art piece called "The New School Midden" with my students in 2009. Another advantage to aging this shell in garden situations is that the calcium carbonate or lime "sweetens" acidic soils.
3. Coat hard-fired ceramic such as porcelain with calcium carbonate or pure lime. Other forms of calcium carbonate such as aragonite, chalk, or even crushed marble may be used.
4. Mix calcium carbonate with molten glass to create textured foam. This can be cast in any shape.
5. Create netted forms with natural fibers such as hemp rope, the least water-soluble fiber. Fill the nets with young oysters attached to the substrate. The oysters will grow to fill the net with reef. The net will naturally disintegrate over time, leaving no toxic residue and a healthy reef.

Note: This recipe does not use plastic or concrete. Examples are shown in Figures 18.1 through 18.8.

OYSTER RESTORATION IN NEW YORK CITY

Henry Hudson, the arctic explorer, sailed to Sandy Hook in 1609. Hudson's crew found piles of oyster shell as high as homes of the time along the shores of Staten Island. These middens or piles of discarded oyster shells were the evidence that the Lenape, the American indigenous people, thrived on oysters. American oysters are difficult to farm. Instead of farming them, Hudson engaged in trade, exchanging tools for beans, corn, and oysters. This was the beginning of the North American oyster trade.

Figure 18.1 The ultimate recipe for a sustainable urban oyster reef.

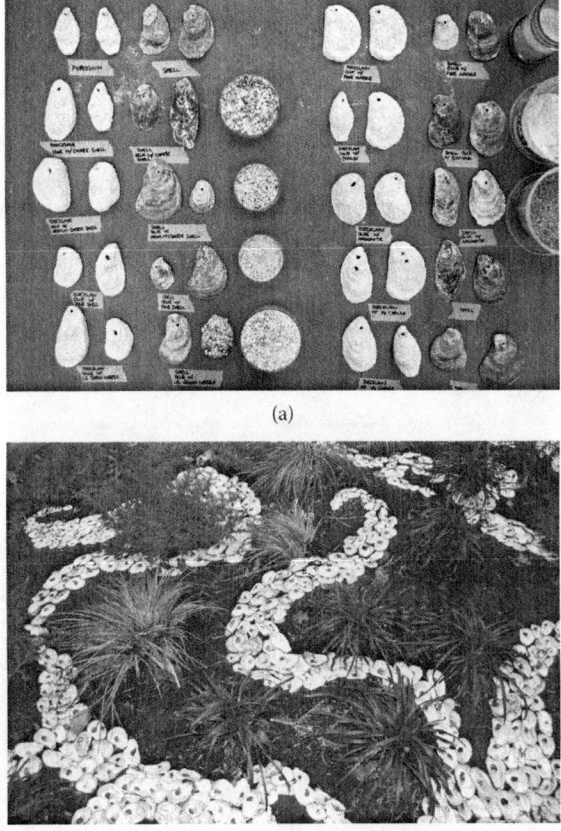

Figure 18.2 (a) Oyster substrate experiment for Cornell Marine Exchange using aged oyster shells and various forms of calcium carbonate 2010, collaboration between the New School and Cornell Marine Exchange for a class taught by Mara G. Haseltine entlited "The Art of Urban Oyster Restoration." (b) The New School Midden, an environmental artwork that ages shell for restoration purposes as collaboration between myself and the New School Students in my class "The Art of Urban Art of Oyster Restoration."

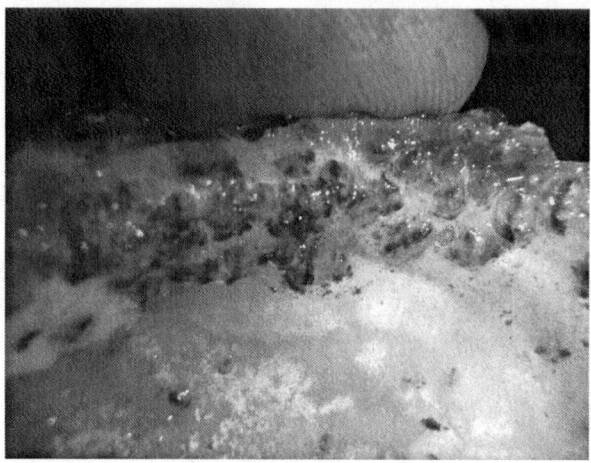

Figure 18.3 Freshly settled spat (juvenile oysters) on shell from Cornell Marine Exchange Experiment, 2010.

Figure 18.4 The most colonized substrate from Cornell Marine Exchange was hard-fired ceramic coated with aragonite, 2010.

Oysters were the foundation of the commerce-driven New York, later called New Amsterdam. Hudson, an Englishman employed by the Dutch, ignited the first international oyster trade. It had its beginning with the Dutch New Netherlands Company (1614) that eventually became the Dutch West India Company. Rich and poor alike ate oysters. They were sold off in bundles of 600,000 to one customer at a time. Renowned for the best taste among oysters worldwide, New Amsterdam commerce thrived.

With 350 square miles of oyster reef, New York earned its title as the "Oyster Capital of the World" (Kurlansky, 2006). Beginning with the American Industrial Revolution, the original oyster reefs disappeared from the Lower Hudson River Estuary (Waldman, 1999). Oyster reefs declined due to overharvesting, pollution, disease, and rising water temperature. In addition to the demise of

Figure 18.5 Detail from artwork "Oyster Island: the perfect submersible habitat for future aquatic life" made from netted hemp, glass, hard-fired ceramic, metal, crushed marble, and solar array 2010.

Figure 18.6 Detail of sand cast glass with surface texture that oysters cling to if mixed with calcium carbonate. (Courtesy of Mara G. Haseltine.)

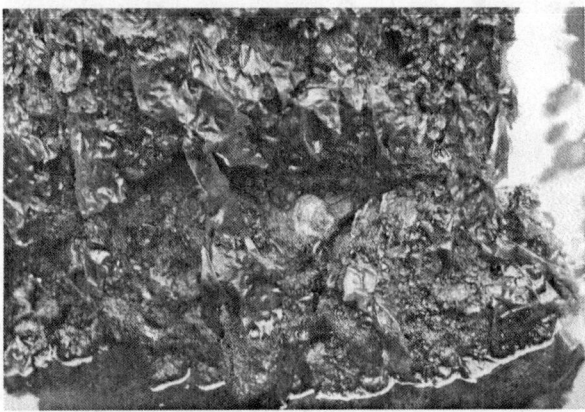

Figure 18.7 Oysters settled on foamed glass mixed with calcium carbonate. (Courtesy of NY/NJ Baykeeper and Paul Mankiewicz Ph.D of the Gaia Institute, New York, NY.)

Figure 18.8 Detail from artwork "Oyster sweet spot" is a sculptural illustration of preferred oyster substrate that does not use plastic or concrete. (Courtesy of Bobby Octavanino.)

the species, environmental factors have prevented its natural rehabilitation. Habitat-related stresses such as loss of substrate, decreases in salinity, changing currents, and a decrease in dissolved oxygen prevented oyster reef rejuvenation. Historian Mark Kurlansky, author of *The Big Oyster: History on the Half Shell*, states that perhaps "Ten million people produce far too much garbage" to be nurtured by one estuary. John Waldman, author of *Heartbeats in the Muck*, who also chronicled New York Harbor through the demise of the oyster, stated that the current health of New York Harbor is like "a healthy victim given chemotherapy and reconstructive surgery for problems she never had. Although the harbor's native biological glory will never be relived, the prognosis is good for recuperation to a satisfactory functional level."

OYSTER RESTORATION PROJECT AT COLLEGE POINT, QUEENS, NEW YORK CITY

The advantage of this structure is multifaceted. For one thing, the spiral of the helical shape takes up a minimal amount of space while providing a maximum amount of surface area for oysters to colonize. In addition, the shape has many open spaces for water to flow through, allowing for a maximum current bringing microparticles and plankton to the oysters and/or coral for nutrients or food. Because the structure contains a range of elevations spanning the height of the intertidal zone and has elements all with different orientations with regard to prevailing waves, it is ideal for experimental research on the importance of these crucial factors affecting food supplies, physical stress on growth rates, and settlement.

The vertical twisted forms are optimal for oyster beds because their vertical shape allows water to flow freely, thus allowing oysters to receive the maximum amount of nutrients and to capture the maximum amount of planktonic food, as well as keeping them above the sediment at the bottom. Vertical structures with plenty of ventilation have proven to be the best way to grow healthy oysters. This design is actually an improvement on the typical natural oyster bank, in which the oysters at the bottom are often smothered in sediment or covered by other oysters at the top of the pile, so the ones on the bottom literally starve. To offset this problem, I created a design inspired by the double helix that allows water to flow freely in order for the oysters to receive the maximum amount of food and nutrients, while also efficiently flushing away their wastes. I utilized the architecture of preexisting wooden pilings as a structural support to create vertical structures (Figures 18.9 through 18.13).

Figure 18.9 Newly installed helices, College Point, Queens: spiral shapes waiting to be accreted and settled with oysters, 2007.

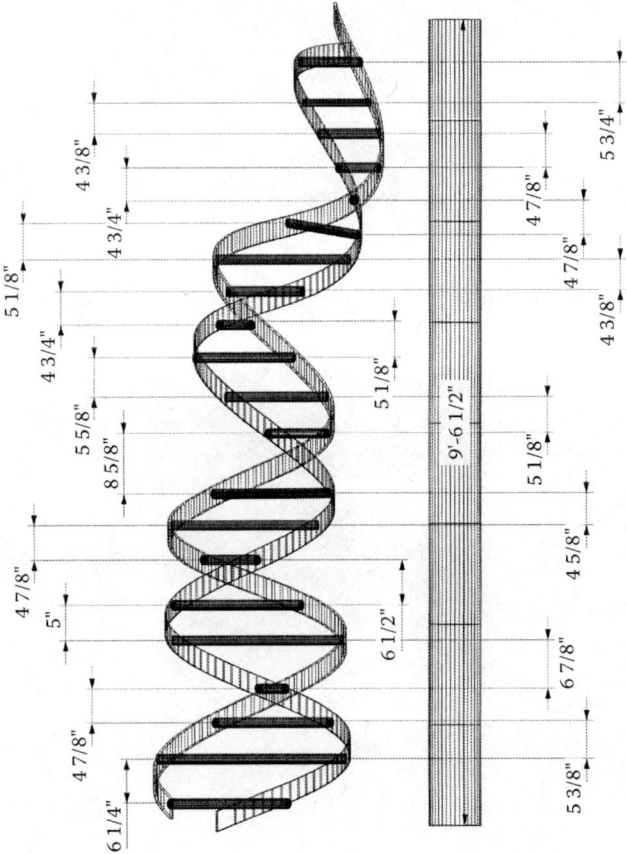

Figure 18.10 Technical drawing of double helix shape for maximum oyster settlement.

Figure 18.11 Helix detail with oysters affixed to it to measure the growth rate under solar-panel electric charge as part of a Biorock experiment.

Figure 18.12 Helix charged with electricity from a solar panel; the white areas show mineral accretion as part of an ongoing Biorock experiment.

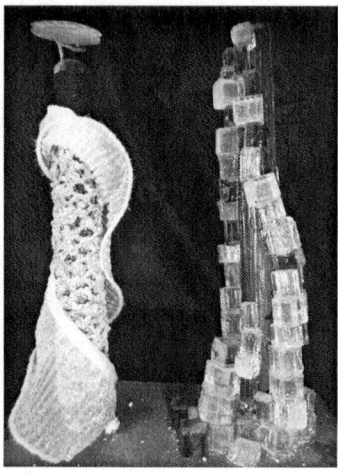

Figure 18.13 Models of two pilings, one with mineral accretion and one using foamed recycled glass; these models demonstrate that restoration can be done in a functional yet artistic manner.

This is an environmentally sound and elegant use of old wooden pier pilings, which are found in vast numbers throughout the coastline of New York City's waterways. They cannot be legally removed in New York City, because they are considered hazardous waste due to the toxic creosote coatings or copper and arsenic solutions impregnating them to prevent the action of marine boring organisms. The use of the pilings has the advantage of greater strength as well as saving space, while still enjoying the advantages of recreating a biodiverse habitat, and does not interfere with benthic (bottom-dwelling) fish.

GILL REEF

The design for this sculpture is based on the extremely efficient and elegant oyster gill (Figure 18.14). One oyster in the height of the season can filter up to 50 gallons of water per day. I looked at the figures of microscopic oyster gills and then employed the concept of biomimcry, Mother Nature's blueprints, to create the structure's design. While working on the sculpture, I incorporated feedback from my students at the time—marine biologists, marine engineers, and the New York Department of Environmental Conservation.

Every aspect of the design strives to create the ultimate urban habitat for oysters (Figures 18.14 and 18.15). The sculpture is engineered for 12–15 ft deep waters with underwater currents ranging between four and six knots. Oysters thrive in conditions where they receive the maximum amount of water flow, so the design is open and reticulated. This habitat is designed to be out of harm's way—submerged at the bottom of the river. The reef structure is lifted above the sediment, allowing every oyster access to nutritious currents. To further the likelihood of spat settlement, the reef structure is coated with lime-based substrate. An open channel in the middle and between each rib gives entry to divers who can monitor growth.

The gill can also operate as a land sculpture and teaching tool (Figures 18.16 and 18.17).

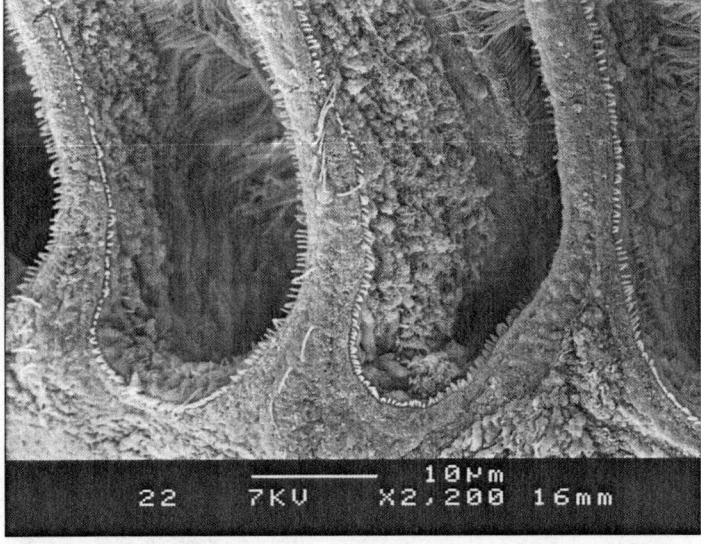

Figure 18.14 Scanning electron microscope image of an oyster gill. (Courtesy of Dr. Peter Benniger, Rozenn Cannuel Faculty of Science, University of Nantes, France.)

Figure 18.15 Rendering of oyster gill reef in intertidal zone, where oysters thrive.

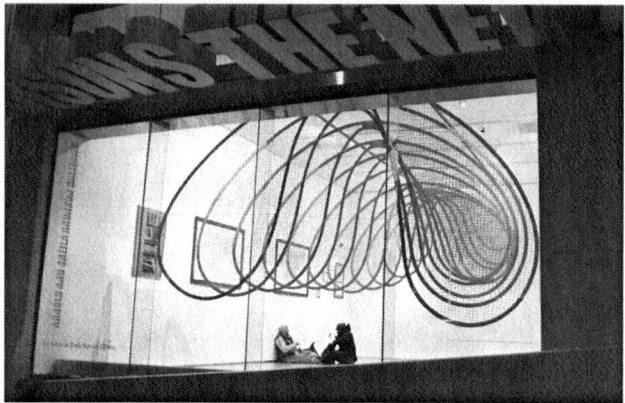

Figure 18.16 Model/sculpture of oyster gill reef to scale 22 ft in depth, 9 ft in height, and 20 ft wide at maximum width, 2009.

Figure 18.17 Rendering of the gill reef on land as a sculpture to be used as an interactive teaching tool.

FLOATING REEF

Fishermen know when they see a log floating in the water that there are fish schooling beneath it. This same principle was applied to the floating reef design; however, lightweight honeycomb structures have several advantages over a log. This floating reef design illustrates how reefs can be tethered to the ground, but can also be free-floating (Figures 18.18 through 18.22). It is excellent for sustainable fish farming, because its lightweight honeycomb structure fans out allowing light to hit it; therefore, coral can grow on it. Also, it has many nooks and crannies for fish to congregate and hide in, and the structure allows water to flow through it, supplying food and removing wastes.

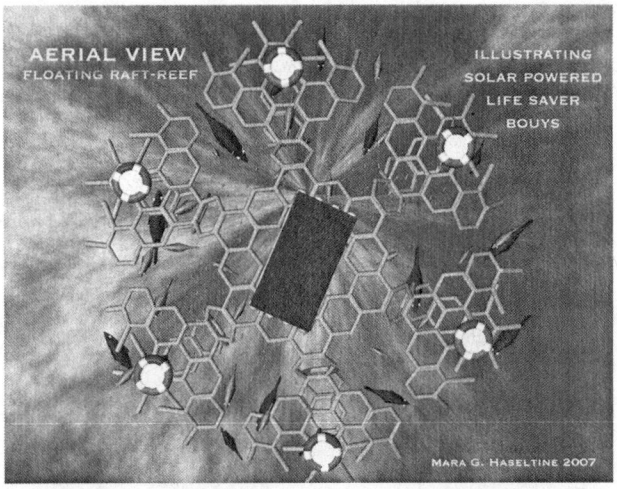

Figure 18.18 Aerial view of floating raft reef that uses solar power for mineral accretion.

Figure 18.19 Side view of raft reef depicting the different size fish that inhabit the floating reef at different depths.

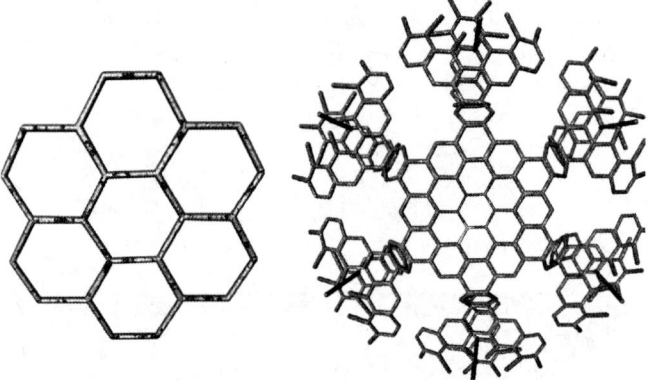

Figure 18.20 Modular welded rebar honeycomb design for floating reefs.

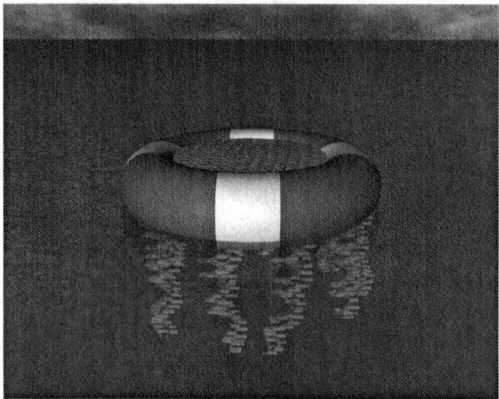

Figure 18.21 Solar-powered reef buoys have the potential to power their own reefs and create hydrogen gas for fuel.

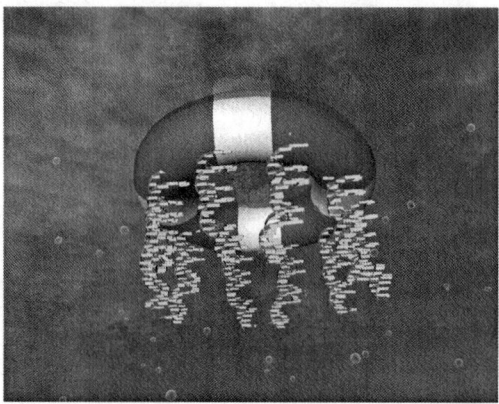

Figure 18.22 Solar-powered reef buoy as seen from below.

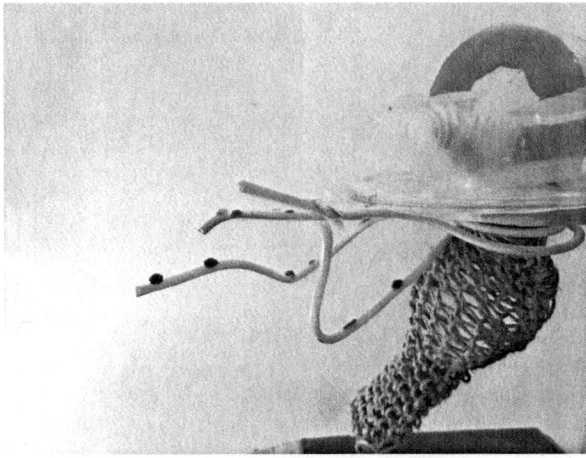

Figure 18.23 "Madam Jelly Fish" model of solar power buoy with dome to catch hydrogen gas as well as net underneath that would hold oysters and booster their immune systems and growth rates, benefiting from electricity generated by the solar panels on top.

SOLAR-POWERED REEF BUOYS

The solar-powered reef buoy is a self-contained unit that acts not only as a navigational marker, but also is its own miniature floating reef (Figure 18.23). If done properly, the solar panel could also generate hydrogen as well as power the reef's structure below it. The solar-powered reef uses minerals dissolved in seawater to grow Biorock mineral accretion. Note: It is important to gauge over time how much buoyancy the solar buoy would require, because as mineral accretion and biological life-forms grow on it, the buoy would get heavier.

RADIOLARIAN REEF

This structure's design is based on the siliceous skeleton of a radiolarian zooplankton (Figure 18.24). This model depicts the plans for a reef structure that would act as a teaching tool about the importance of these microscopic structures, as well as become a home for coral and aquatic life that depends on healthy reefs. The circular shape is great for reef-making; Wolf Hilbertz often made dome shapes as his structures (Figure 18.25), because, if they were low enough, waves just washed over them, and they do not break and also offered protection for fish habitats, flow-through, and many surface areas for coral to grow.

COASTLINE MANAGEMENT/FLOATING ISLAND

This depiction is of a floating reticulated reef designed specifically for coastline protection as well as sustainable habitat (Figures 18.26 and 18.27). As in the early drawing, the reef fits together like bones, so when a large wave comes, it absorbs the shock of the wave, protecting property and nature on its coasts. These reticulated reefs could be especially important now that sea-level rise is imminent. The later depiction is of a man-made island in which these floating reticulated reefs protect it. Note: As the mineral accretion grows on these structures, more buoyancy will have to be added.

Figure 18.24 Radiolarian reef based on the skeleton of microscopic plankton. (Courtesy of Bobby Octavanino.)

Figure 18.25 Dome reef designed by Wolf Hilbertz and made by Biorock Workshop, Gili Trawangan, Lombok, Indonesia, 2006. (Courtesy of Ari Spenhoff.)

Figure 18.26 Rendering of floating reticulated reef for coastline protection.

Figure 18.27 Rendering of human-made Spiral Floating Islands surrounded by reticulated reefs for coastline protection to colonize the sea sustainably.

AMPHIBIOUS REEF

"Enchanted Star Sand" is inspired by the unique shapes of sand found on Okinawa's beaches (Figures 18.28 through 18.30). "Star Sand" is formed by the shells of foraminiferans, which are single-celled calcareous marine organisms that live on the ocean floor. In this sculpture, these tiny star shapes cling to the delicate tentacles of a ctenophore or sea gooseberry. The ctenophore, also a planktonic marine organism, is perched on a cliff while its tentacles are submerged into the water, forming the reef below. This living sculpture would enhance the beauty of Miyako Island and protect its biodiversity, creating the perfect habitat for future aquatic and amphibious life. The concept for this project would be to create a living sculpture, which existed both below and above the intertidal zone. As transplanted coral develops on the structure, the reef will become a self-repairing breaker, protecting Miyako's shorelines. Solutions for coastal protection are necessary in an age of sea-level rise due to climate change. Enchanted Star Sand would be created in an area on Miyako Island, where coastal erosion was already taking place or would be likely to take place in the future due to climate change and water-level rise. Enchanted Star Sand embodies the concept of geotherapy (Richard Grantham, 1992a, 1922b) or restorative action to heal the planet.

In conclusion, the underlying principle behind all of my reef work is geotherapy, which is based on the concept that the earth is now sick and needs to be healed through decisive human intervention. We must become caretakers for our ailing biosphere. I think of my reef work as a series of art actions to create both awareness and a physical manifestation of possible future solutions. Now more than ever, our cultural evolution affects our biological evolution (Van Rensselaer Potter, 1971). We do not have time for the process of natural selection to occur, so we must rely on cultural adaptation to accelerate our evolution to acclimate to the threat of man-made climate change. It is my hope as an artist that combining innovative methods of sustainable reef restoration with aesthetics will inspire people to build not only a sustainable future but also a beautiful one.

260 INNOVATIVE METHODS OF MARINE ECOSYSTEM RESTORATION

Figure 18.28 Enchanted Star Sand rendering of an amphibious reef that would provide coastline protection as well as ecotourism using Biorock above and below the waterline.

Figure 18.29 Enchanted Star Sand detail 1.

Figure 18.30 Enchanted Star Sand detail 2 showing traditional style Japanese glass buoys and floating solar panels.

REFERENCES

Benyus, J. M. 1997. *Biomimicry: Innovation Inspired by Nature*. Harper, New York, NY.

Grantham, R. 1992a. Geotherapy as evolutionary choice. *Global Environmental Change* 2:2–4.

Grantham, R. 1992b. Geotherapy and global environmental ethics: Towards acceptable survival. *Global Environmental Change* 2:258–61.

Kurlansky, M. 2006. *The Big Oyster: History on the Half Shell*. Ballantine, New York, NY.

Potter, V. R. 1971. *Bioethics: Bridge to the Future*. Prentice-Hall, Englewood Cliffs, NJ.

CHAPTER 19

Marine Ecosystem Electrotherapy
Practice and Theory

Thomas J. Goreau

CONTENTS

Introduction 263
Effects on Settlement 264
Effects on Growth 269
Effects on Healing 270
Effects on Cell Growth, Division, Budding, and Branching 272
Effects on Coral Color and Fluorescence 274
Effects on Resistance to Stress 274
Electrotherapy Enhances Biodiversity, Speeds Up Growth, and
Provides Complex Sustainable Mariculture without Food 278
Cell Membrane Voltage Gradients and Energy Metabolism 282
Conclusion 287
References 287

INTRODUCTION

Electrolysis has long been studied primarily for its physical and chemical effects rather than its biological ones, even though the first book on the effects of static electrical charges on frogs' legs, *Animal Electricity* by Luigi Galvani, was published in 1791, preceding the invention of the battery by Alessandro Volta in 1800. Some time later, in 1800, batteries had already been used by William Nicholson and Anthony Carlisle to separate water into its constituent elements. Shortly afterward, Humphrey Davy and Michael Faraday applied electrolysis to practical corrosion problems, and a great deal of research on isolating elements from fused salts or acid solutions followed.

Wolf Hilbertz's pioneering studies, published in 1979, first recognized the potential of seawater electrolysis to produce calcium and magnesium minerals that had a wide range of building applications. Depending on the conditions, either very hard material, primarily limestone (calcium carbonate in the form of aragonite), with up to three times the compressive strength of ordinary concrete, could be produced in any size or shape. Under other conditions, the much softer mineral brucite (magnesium hydroxide) could be produced (Buster et al. 2006), which is readily converted to even harder cement capable of absorbing CO_2 from the atmosphere. The physical and chemical properties of these "Biorock" materials produced under various conditions; the various other potentially useful materials generated in side reactions, and their potential applications are reviewed in Chapter 3 and Goreau (2012), to which readers are referred for more information.

Hilbertz's work from 1976 until 1986 focused on the potential applications of these minerals as building materials (Chapter 1 and 2). He found that his very first project was completely overgrown by multiple layers of oysters in a few months. Starting in 1987, Hilbertz began his work with the author of this chapter on applications of electrolysis to coral reef restoration, and it was immediately found that coral growth rates were greatly accelerated to record rates on Biorock materials (Goreau and Hilbertz, Chapter 4).

It was initially thought that the key factor that stimulated coral growth was the higher pH produced on the surface of the growing Biorock structure, which would make it easier for corals to grow their skeletons. If that were the primary mechanism, then organisms like corals and oysters with limestone shells should be the major beneficiaries. Further work revealed that soft corals and organisms lacking limestone skeletons, like sponges and tunicates, also seem to settle and grow at extraordinary rates. This indicated that much broader biological benefits were involved. At first, it was also thought that only organisms directly on the Biorock structures would benefit, but it was soon noticed that there was much higher coral settlement and growth in the areas surrounding the structures, and that corals that were broken off structures and fell onto sandy or rock bottom beneath them survived and proliferated, forming dense coral masses.

This chapter focuses on reviewing the biological effects of the electrical fields, in particular to understand the causes of the remarkable effects that the other chapters in this volume have demonstrated, including greatly increased settlement, growth, survival, and resistance to stress of most marine organisms.

EFFECTS ON SETTLEMENT

The first Biorock experiments, set up in front of the lab dock at the Discovery Bay Marine Laboratory, Jamaica, in the late 1980s, showed remarkably high settlement of corals and of sand-producing calcareous algae, even though this was an intensely eutrophic habitat where corals and sand-producing algae had been smothered and killed by masses of weedy algae as the result of high nutrient levels from sewage (Goreau and Hilbertz, Chapter 4). After three years of growth, a piece of the Biorock material about 10 cm across was cut out with a hacksaw and examined under the microscope, and all juvenile corals (single polyp stages about 1–2 mm across) counted. One juvenile coral was found every 0.7 cm^2 on average. This is an extraordinary rate of recruitment even on clean reefs, much less on an algae-smothered reef site where no natural coral recruitment had taken place for years.

The site had about ten different Biorock structures of different sizes and shapes growing at different rates. On the fastest-growing structures, the Biorock material overgrew and smothered the juvenile corals before they were big enough to outgrow it, but on the slower-growing structures, juvenile coral recruits were able to rapidly grow and form new colonies, including the branching coral *Acropora palmata*, the plate coral *Agaricia agaricites*, and the head corals *Porites astreoides* and *Diploria strigosa* (Goreau and Hilbertz, Chapter 4). The most striking results were found on an old piece of heavily rusted chicken wire, retrieved out of the bushes about a year after Hurricane Gilbert. This received only a very small amount of power for no more than a few weeks before the power was cut off, and less than one millimeter of minerals grew over the steel mesh surface. One year later, the site was revisited, and hundreds of young corals of *Favia fragum* and *Agaricia agaricites*, about 2 cm across, were found to have settled and grown all over the mesh. No recruits were seen in nearby reef areas (Figure 19.1a and b).

Figure 19.1 Spontaneous settlement of (a) *Favia fragum* and (b) *Agaricia agaricites* on Biorock 1-inch spacing (2.54 cm) chicken wire mesh. Discovery Bay Jamaica, 1991. Hundreds of these settled and grew in a few square meters in less than one year. (Photograph by Wolf Hilbertz.)

Limestone settling plates were placed next to Biorock structures in Pemuteran, Bali, Indonesia, and away from them, during thesis research by Putra Nyoman Dwija at Udayana University (Dwija 2003). He found extremely high settlement of juvenile corals around Biorock substrates of 508 per square meter after only three months (169.3 per square meter per month). These rates are orders of magnitude higher than reported total numbers in the literature when their data are converted to common units (arranged here in roughly increasing order by density): 0.01–0.35 per square meter per month in Seychelles (Turner, Klaus, and Engelhardt 2000), 0.21 per square meter per month in Maldives (Loch et al. 2002), 0.21–0.46 per square meter per month on rubble and rocks in Komodo, Indonesia (Fox et al. 2001, 2003), 0.11–2.2 per square meter per month on large dumped rocks in Komodo (Fox and Pet 2001), 2.42 per square meter per month in Maldives (McClanahan 2000), 0–6.8 corals per square meter/month on artificial reef substrate in Hawaii (Fitzhardinge and Bailey-Brock 1989), 17.8 per square meter per month in Barbados (Tomascik 1991), 21.67 per square meter per month in the Great Barrier Reef (Mundy 2000), and 15.36 per square meter per month on artificial substrates versus 30.33 per square meter per month on bare reef limestone substrate at Wakatobi, Sulawesi, Indonesia (Salinas de Leon et al. 2011). However, settlement rates on coral limestone plates near Biorock structures in Pemuteran are nearly equal to rates

Figure 19.2 Dense spontaneous settlement of encrusting, plate, and branching coral juveniles on Biorock substrate in Pemuteran, Bali, Indonesia. The area shown is about 5 by 10 cm. (Close-up photograph by Thomas Sarkisian, December 2010.)

of juvenile coral recruitment found directly on Biorock substrate in Jamaica of 0.7 juveniles per square centimeter, or 7000 per square meter, measured after three years (194.4 per square meter per month) in a severely nutrient-polluted reef where no natural recruitment was taking place (Goreau and Hilbertz, Chapter 4). Therefore, settling rates on Biorock minerals appear to be in the range of one to three orders of magnitude higher than reported on other substrates (Figure 19.2).

Not only reef-building stony corals were observed to show extraordinary rates of recruitment on Biorock, the same was also noted for soft corals (spontaneously settling on and entirely covering some structures in Indonesia), oysters (as first seen by Hilbertz in 1976, see Chapter 2), mussels, sponges (Nitzsche, Chapter 8), tunicates (no work has yet been done on them, but the rates of spontaneous recruitment of tunicates, especially of the *Didemnidae* in Bali, is extraordinary, and far greater than is seen on surrounding reefs), and calcareous tube-dwelling polychaetes such as *Filograna huxleyi* on Biorock structures in the Turks and Caicos Islands. At first, the high numbers of sponges on some Biorock structures in Indonesia and the Maldives were thought to be due to higher growth rates, but the data of Nitzsche (Chapter 8) clearly shows that it must be due to much higher recruitment rates. No detailed recruitment studies have yet been done on any organisms other than hard corals (Figure 19.3a and b).

In Ko Samui, Thailand, a series of Biorock structures were grown at various levels of power. The slower the minerals grew, the greater the spontaneous recruitment, primarily *Pocillopora damicornis*, *Pocillopora verrucosa*, *Porites* spp., and *Acropora* spp. The slowest-growing structures were nearly completely covered with spontaneously settling corals within two years. It is thought that this higher coverage may be due not to higher settlement on slower-growing structures but to greater survival due to lack of overgrowth by electrochemical mineral growth (Figure 19.4a and b).

Invertebrate larvae, developing vertebrate embryos, and the larvae of marine algae have long been known to show polarity between a "head" and "tail" section even when they appear completely spherical, and this is marked by a distinct polarity in their electrical charge, with one end being positively charged and the other end negatively charged, so they orient themselves in applied electrical fields (Levin 2003; McCaig et al. 2005). It therefore appears that they use natural electrical field gradients for orientation, or "electro-tropism," to move toward potential substrate to settle on.

One can clearly see these effects in three Biorock mesh substrates of differing thickness that were laid out side-by-side in waters in the Straits of Georgia, British Columbia, Canada. The central structure, which got the most power, is completely overgrown with mussels, the slower-growing one on the left had less mussels; and the slowest-growing ones on the right had the least mussels. Although it is clear that these results have tremendous implications for mussel mariculture, unfortunately no counts of mussels, their average weight, or the thickness of mineral growth were directly measured that would have allowed quantitative data to be obtained (Figure 19.5).

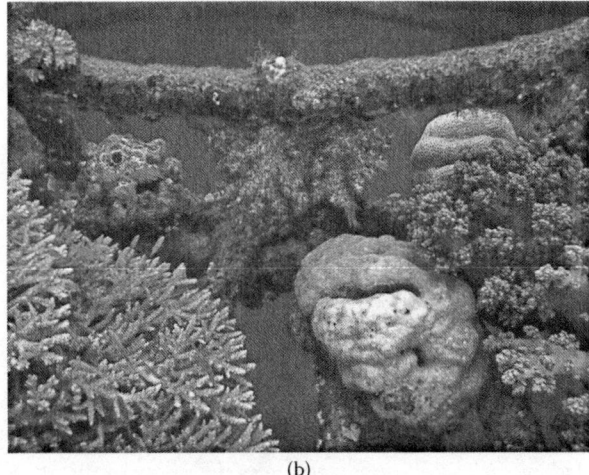

Figure 19.3 (a) Dense spontaneous settlement and growth of soft corals surround hard corals on Biorock structure in Pemuteran, Bali. (Photograph by Rani Morrow-Wuigk, 2011.) (b) Spontaneously settling soft corals, sponges, tunicates, and other organisms grow on Biorock structures. (Photograph by Rani Morrow-Wuigk, Pemuteran, Bali, 2011.)

The Biorock electrical fields apparently cause larvae of many species to move toward the negatively charged Biorock reef, greatly accelerating recruitment rates. Biorock reefs therefore quickly become oases of biodiversity that stand out from their surroundings. It is likely that the effects on attracting different species will vary with the electrical field strength, but so far no work has been done on this. Research currently under way by Solomon Viitaasari is mapping out electrical field gradients around Biorock projects, and his work will provide enormous insight into future settling experiments. We have repeatedly noticed much higher coral recruitment in the areas surrounding Biorock projects than in areas farther away, and in addition, extraordinary rates of growth and settlement around the insulated electrical cables leading from power supplies to Biorock projects. Because they are insulated, there should be no electrical field around them, but there will be a magnetic field induced by the current in surrounding areas, and it is possible that may also play an important role. Further work is needed to understand these phenomena better.

Figure 19.4 (a) Spontaneous settlement of corals and oysters nearly cover Biorock structure at Ko Samui, Thailand. (Photograph by Thomas Sarkisian, March 9, 2009.) (b) A year later, the most severe bleaching event recorded in Thailand killed more than 90% of the corals. But survival and recovery were much higher on Biorock than on the surrounding reef. Photograph during the height of coral bleaching showing surviving spontaneous recruits on Biorock structure at Ko Samui, Thailand. (Photograph by Thomas Sarkisian, June 15, 2010.)

In addition, extraordinary recruitment of fishes, especially juveniles, has been noticed on Biorock structures. When the power is turned off for some weeks, the number of juvenile fishes decreases, and rapidly increases again within days of the power being restored. It therefore seems likely that electrotropism is used by juvenile fishes in order to select habitat. It is completely unknown how they do so. The teleost, or bony fishes, clearly greatly prefer Biorock to surrounding reef (Jompa et al. Chapter 5, and Arifin et al., Chapter 6), but they are not known to have any electrical sense organs (Arvedlund and Kavanagh 2009), so the mechanism by which they identify and select Biorock habitats is a mystery that will require further research to resolve. In contrast, the elasmobranch or cartaliginous fish, like sharks and rays, are long known to have very well-developed electric sense organs that they use to identify and capture prey (Kalmijn 1971, 1982; Wueringer et al. 2012), as do mammals like the duckbill platypus and river dolphins living in turbid waters (Czech-Damal et al. 2011). Although

Figure 19.5 Spontaneous recruitment of mussels on three different Biorock substrates getting different amounts of power. The center structure had the most electrical current, the one at left an intermediate amount, and the one at right the lowest. (Photograph by Eric Vanderzee at Sechelt Inlet, Straits of Georgia, British Columbia, Canada, 2011.)

stingrays and nurse sharks have taken up residence in Biorock structures, this is infrequent, so it is hard to say that they are systematically selecting them based on the electrical field. Sharks are known to be able to sense electrical-field gradients as small as one volt in a million kilometers, which is much less than those produced by Biorock, and have been known to attack shielded marine cables, but they avoid higher electrical-field gradients (Kalmijn 1982). It is therefore possible that the Biorock electrical fields overload their sensory capabilities and are confusing and repelling them rather than attracting them. Turtles are known to sense magnetic fields for migration (Fisher and Slater 2010), but while turtles have often been filmed and photographed in the vicinity of Biorock projects, they seem to pass right by without any abnormal behavior.

EFFECTS ON GROWTH

The first Biorock projects in Louisiana showed spectacular growth of oysters, which settled spontaneously on the structures and grew to adult size in months. No measurements were made, although samples we kept showed up to three superimposed layers of large oysters (Chapter 2). The first experiments in Discovery Bay, Jamaica, transplanted small fragments of finger corals, *Porites porites*, which tripled in size in three months. Subsequent experiments with staghorn coral, *Acropora cervicornis*, showed growth rates of about 8 cm in 10 weeks, or rates around five times higher than had ever been recorded in this species (Goreau and Hilbertz, Chapter 4). Corals that are attached to Biorock structures show visible signs of growth over the substrate within a day if inspected closely, and quickly overgrow and cement themselves onto Biorock (Figures 19.6 and 19.7).

The chapters in this volume show record growth rates for many species of hard corals, (typically 2–8 times faster than genetically identical controls in the same habitat, but in some cases even higher, depending on species and conditions) (Jompa et al., Chapter 5; Arifin et al., Chapter 6), such as soft corals (Fitri and Aspari, Chapter 9), and oysters (Berger et al., Chapter 12; Karissa et al., Chapter 11; Shorr et al., Chapter 13). Extremely rapid growth has been seen for many other species of invertebrates that have not been measured, especially tunicates. This does not only affect growth of animals; marine plants also respond positively, as shown by the data in the chapters on seagrasses (Vaccarella and Goreau, Chapter 14) and salt marsh (Cervino et al., Chapter 15) responses to electrical fields. On the first Biorock structures in Jamaica, dense growths of branching calcareous algae produced large

Figure 19.6 One of the world's rarest and most unusual corals, *Acropora suharsonoi*, after one year on Biorock. The original fragment had only around six branches. (Photograph by Emma Woollacott, Gili Trawangan, Indonesia, 2006.)

amounts of sand around them when they died or were dislodged by storm waves (Goreau and Hilbertz, Chapter 4). A fine algal layer is frequently noticed on Biorock structures, where it is avidly grazed by herbivorous fish that school around them. It therefore seems that the property of electrical field growth enhancement is a completely general property that affects essentially all forms of marine life.

This effect is not confined only to the structures themselves. The rate of coral settlement and growth for meters around the structures is visibly higher than farther away, as shown by Nitzsche (Chapter 8). Coral fragments broken off Biorock structures by waves continue to grow after falling onto the bottom. Seagrass growth is also seen to increase around Biorock structures. Therefore, the effects are due to the electrical field and not the reactions taking place on the surface of the Biorock structures themselves. Mapping out the fields will provide more insight into how far the benefits extend.

EFFECTS ON HEALING

Broken and damaged corals rolling around on hard bottom or half buried in sand and mud are observed to show extraordinarily rapid recovery, changing from pale and unhealthy colors to vibrant colors and full polyp expansion within a day when placed loose in the electrical field. Broken corals can often be seen by eye to visibly start overgrowing and attaching to Biorock in less than a day (Figure 19.8a and b).

Freshly broken branching coral tips that are transplanted onto Biorock have been observed to heal over very rapidly and release no mucus, which is the typical general sign of coral stress. In contrast, identical controls transplanted to non-Biorock substrates at the same time continued to release mucus for two weeks afterward (Dwija 2003).

There is a vast literature, going back to the 1800s, on the role of direct-current electrical fields greatly speeding up wound healing and regeneration of limbs in frogs and salamanders (Becker and Selden 1985). Direct-current fields have long been used to speed up bone healing and repair following fractures (Bassett 1993; Oschman 2000, 2003) and are widely used in sports medicine.

Figure 19.7 (a) Corals on Biorock have dense budding. (Photograph by Thomas Sarkisian, Pemuteran, Bali, 2010). (b) Biorock corals are very brightly pigmented. This close-up shows intense purple coral pigments. High density and division rates of *Symbiodinium* are seen in brown lines. (Photograph by Thomas Sarkisian, Pemuteran, Bali, 2010.)

When a skin cut has a DC field of the right orientation applied across it, the wound rapidly closes. If the field is reversed, it opens up again, and if reversed yet again it closes up (Alvarez et al. 1983; Song et al. 2002; McCaig et al. 2005; Zhao et al. 2006). This strongly indicates that electrical fields play a fundamental role in stimulating membrane repair, cell proliferation, and tissue healing, and would explain the results that we see in the field.

Figure 19.8 (a) Broken and injured corals rescued from reef and placed loose on Biorock mesh are bright and healthy one day later. (Photograph by Leong Sze Wong, Gili Trawangan, Lombok, Indonesia, December 2006.) (b) Broken and injured soft coral fragment rescued from reef and placed loose on Biorock mesh without being attached is bright and fully expanded one day later. (Photograph by Leong Sze Wong, Gili Trawangan, Lombok, Indonesia, December 2006.)

EFFECTS ON CELL GROWTH, DIVISION, BUDDING, AND BRANCHING

Falugi, Grattarola, and Prestipino (1987) exposed sea urchin larvae to low-intensity direct electrical current fields, comparable in intensity and gradients to those used in Biorock and to those used in medical applications for healing fractured bones. They found that this greatly accelerated division of the early egg; for example, eggs in electrical fields had 18.08 times faster division to the third cell cleavage stage. They found that embryos exposed to electrical fields matured much faster, showing advanced cell structures and incipient skeleton formation at an earlier stage, without any cellular or developmental abnormalities.

Biorock corals show exceptionally dense and perfect branching (for photographs, please see the presentation in the CD in the back of this book on the Karang Lestari project prepared for receipt of the 2012 Equator Award from the United Nations Development Programme, the top award for Community-Based Development), and they bud and branch more densely than genetically identical clones in the same habitat but away from Biorock. This difference is analogous to the difference in growth of genetically identical seeds grown in rich soil versus poor soil: the former show much more branches, leaves, growth rates, and fruit bearing. Zamani et al. (Chapter 7) show clear evidence that even small electrical fields result in elevated coral budding and branching, and Stromberg et al. (Chapter 10) show increased budding of deep-sea cold-water corals on Biorock compared to controls. These results suggest that cell division and tissue growth are accelerated even more than skeletal growth, and imply that the primary mechanism for increased growth rates is increased cell and tissue growth rather than increased calcification caused by higher pH. That is consistent with the higher growth rates noticed for non-calcareous organism like tunicates (Figure 19.9a and b).

Figure 19.9 (a) Density of polyps is unusually high on Biorock. (Photograph by Thomas Sarkisian, Pemueran, Bali, 2010.) (b) Terminal polyps are elongated, and new polyps and branches rapidly grow from broken tips. (Photograph by Thomas Sarkisian, Pemuteran, Bali, 2010.)

Direct evidence for increased cell division also comes from measurements of coral symbiotic algae (*Symbiodinium* spp.) on Biorock compared to genetically and environmentally identical colonies of six different major reef-building coral genera off Biorock. Biorock corals had an average of 24.7% higher *Symbiodinium* densities than controls, and they had an average of 74.3% higher division rates (mitotic index) (Goreau, Cervino, and Pollina 2004). The much greater increase in symbiotic alga cell division than in densities implies that they were growing faster than is beneficial for the cora and that the surplus was being expelled. This predicted higher rate of algal expulsion can be directly tested. Curiously, the control corals had 44.7% higher chlorophyll per *Symbiodinium* cell than the Biorock corals. This is analogous to the lower chlorophyll content per algal cell in corals in high sunlight compared to those in shade (Wethey and Porter 1976; Porter et al. 1984; Hennige et al. 2009), which has been interpreted as evidence of a control mechanism to reduce excessive productivity of the symbiont, again suggestive that symbionts in electrical fields are more productive than the coral's needs.

It is not yet known if higher cell division rates also translates into greater reproduction, at least in corals, but we have noticed in Indonesia that pearl oysters achieve reproductive maturity and gonad formation at a much younger age on Biorock than controls do.

EFFECTS ON CORAL COLOR AND FLUORESCENCE

In general the brightness of coral tissue fluorescence appears to be an excellent measure of coral health, and in particular of the abundance of zooplankton food supplies. Corals show enormous variability in tissue fluorescence from place to place and seasonally. In many reefs, the corals are never brightly colored; in others, they are positively glowing with fluorescence. You can see this; for example, comparing the same species in lagoonal reefs versus outer slope reefs on atolls in the Marshall Islands, in reefs along atoll inflow-outflow passages compared to those further away, and in the Bahamas between reefs in eastern Abaco exposed to open Atlantic waters versus those on shallow banks. Indonesian corals have astonishing fluorescence, probably due to high currents.

Biorock electrical stimulation also greatly increases fluorescence. Corals growing on Biorock are visibly much more fluorescent than genetically identical mother colonies growing in the same habitat from which they have been transplanted. If the power is turned off for a few weeks, they become much paler, like the surrounding corals, and when the power is restored one can see the increase in fluorescence brightness within hours. It is astonishing how fast they respond. Much more work is needed on changes in coral fluorescent pigments in response to electrical fields (Figures 19.10 and 19.11).

A similar astonishing increase in brightness and fluorescence in electrical fields is seen with giant clams, which also harbor symbiotic *Symbiodinium* dinoflagellate algae. A Biorock project in the Marshall Islands was off power for a year or more because the cable linking it to solar panels was cut by a storm. While the power was off, a number of giant clams, including *Tridacna maxima*, *Tridacna squamosa*, and *Hippopus hippopus* were moved under the structure. They were pale and not very noticeable when the power cable was repaired. Three weeks after the power was restored, they were deeply colored and glowing with fluorescence (Figure 19.12a through f).

EFFECTS ON RESISTANCE TO STRESS

Corals growing on Biorock in the Maldives showed 16–50 times higher survival (NB, that is times, not percent!) after severe high-temperature bleaching events in 1998 (Goreau, Hilbertz, and Hakeem 2000; Goreau and Hilbertz 2005). During severe bleaching events in the Gulf of Thailand in 2010 Biorock corals had markedly lower bleaching, faster recovery, and higher survival than

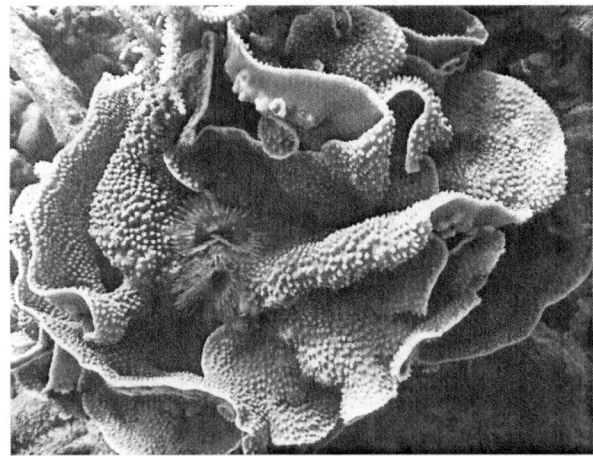

Figure 19.10 Corals on Biorock show much more intense fluorescence. This coral was grown from a fragment about the size of the small circular growth at top center, and is much brighter yellow and faster-growing than the mother colony 10 m away from which it was transplanted. (Photograph by Rani Morrow-Wuigk, Pemuteran, Bali 2011.)

Figure 19.11 Intense pigmentation of a rare purple *Porites divaricata* on Biorock. When the power is off they get paler, but they darken quickly when the power is restored. (Photograph from Antigua by Martha Watkins Gilkes, 2012.)

corals on surrounding reefs (Goreau and Sarkisian 2010). The same has been repeatedly noticed during mild bleaching events in Bali and Lombok, Indonesia (Figure 19.13a and b).

In Ko Tao, Thailand, corals on one Biorock project under low power had markedly less bleaching and higher survival than surrounding reefs (C. Scott, personal communication). In contrast, another Biorock reef, which had achieved spectacular coral growth, especially of table corals, unfortunately had the power supply damaged by electrical power surges, and it was not replaced. In the 2010 bleaching event, all the corals died on the Biorock structure without power (C. Scott, personal communication). This indicates that the higher survival is the result of ongoing electrical fields at the time of stress, and that there is no residual benefit, so, if the power is cut off, the coral soon becomes equally vulnerable to stress as do surrounding corals.

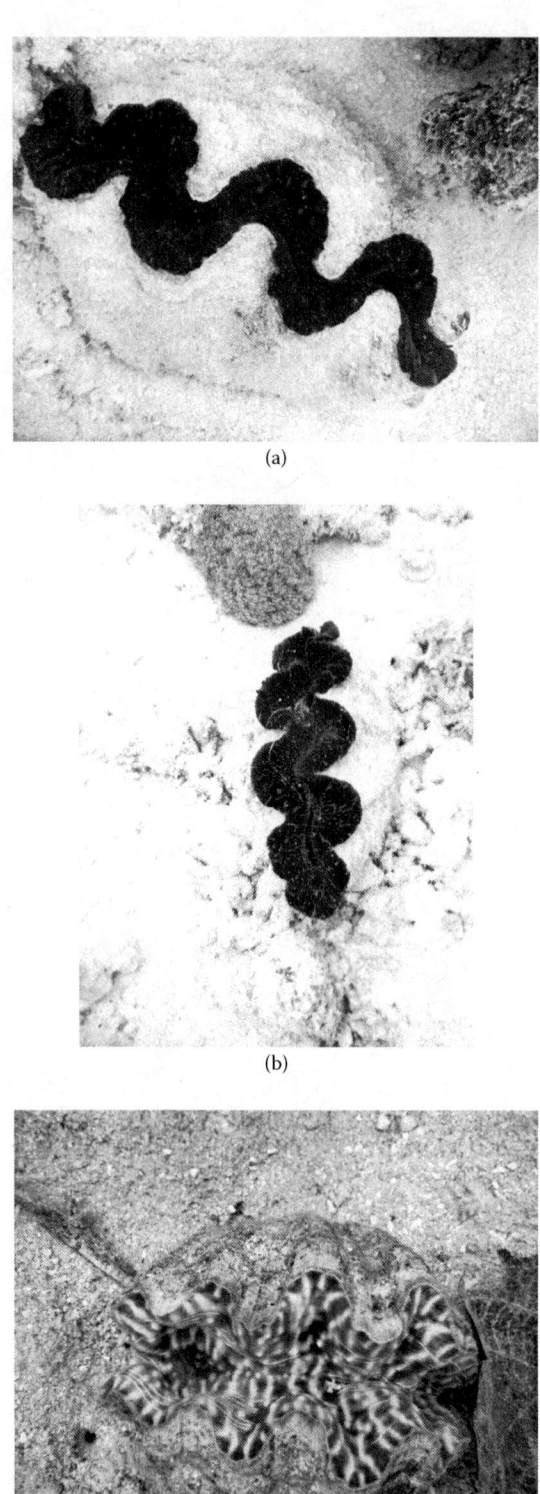

Figure 19.12 (a) through (f) Giant clams quickly become much brighter in color in Biorock projects, each with a unique color pattern. (Photograph at Majuro, Marshall Islands, 2010, by Thomas J. Goreau.)

MARINE ECOSYSTEM ELECTROTHERAPY 277

Figure 19.12 *(Continued)*

On Biorock projects in extremely muddy locations in Panama, Dominican Republic, and Thailand, we have been able to grow coral species that could not tolerate such conditions.

Further examples of greatly enhanced stress resistance are shown in the much higher overwinter survival of Biorock oysters than controls (Shorr et al., Chapter 13), in the ability of seagrasses to colonize rocky wave-swept bottoms they normally could not attach to (Vaccarella and Goreau, Chapter 14),

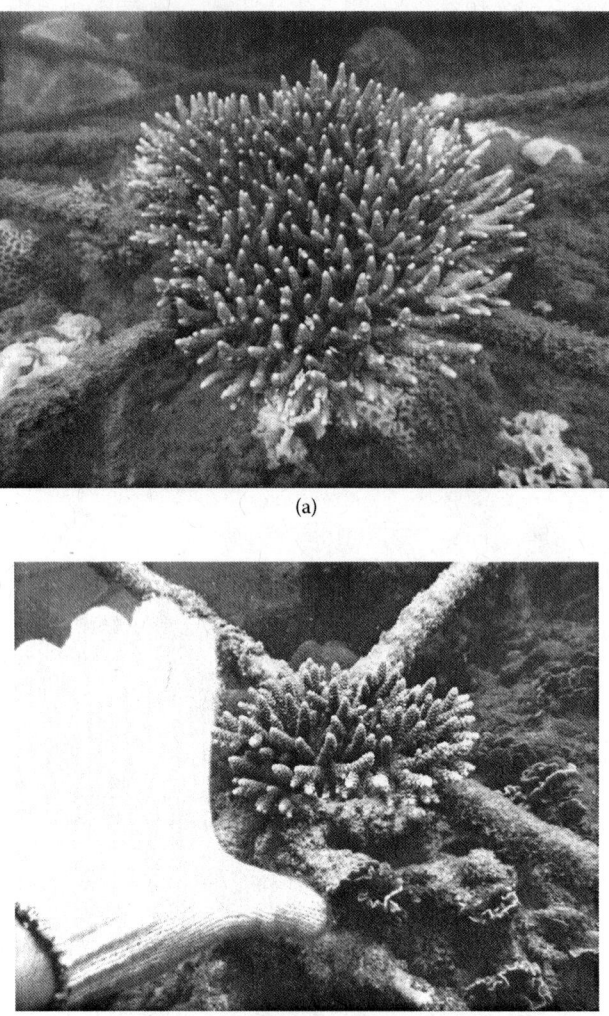

Figure 19.13 (a) Acropora completely unbleached on Biorock structure in Ko Samui, Thailand, during the height of the severe bleaching event that killed almost all Acropora in Thailand. Biorock reef corals showed less bleaching, faster recovery, and higher survival than surrounding reefs. Photograph by Thomas Sarkisian, June 15, 2010. (b) The same coral a year earlier, showing dramatic growth. (Photograph by Thomas Sarkisian, March 9, 2009.)

and in the much higher overwinter survival of Biorock saltmarsh grass than controls and their ability to grow deeper in the intertidal than they normally could (Cervino et al., Chapter 15), allowing salt marshes to be extended seaward of their normal limit.

ELECTROTHERAPY ENHANCES BIODIVERSITY, SPEEDS UP GROWTH, AND PROVIDES COMPLEX SUSTAINABLE MARICULTURE WITHOUT FOOD

The massive destruction, damage, and degradation of coral reefs worldwide has severely impacted tropical coastal fisheries on a scale that conservation of good areas is incapable of addressing without large-scale active restoration of degraded coral-reef habitats. There have been many efforts to transplant corals using cements and glues, and these work well as long as the water quality is good, but the corals almost all die whenever it becomes too hot or polluted, because conventional methods do not increase coral settlement, growth rates, survival, or resistance to environmental stress like the Biorock process does. Here, we will not review conventional methods, or their failure to keep up with the pace of destruction and degradation, because those points are comprehensively reviewed by Rinkevich (2005, 2008) and Shaish et al. (2010), to whom readers are referred.

Normally on Biorock projects, we transplant only naturally broken, badly injured, and seriously damaged corals in order to nurse them back to health. But we quickly notice that all reef organisms quickly settle on them or migrate to them. The result is extraordinarily diverse ecosystems, the closest to real natural coral reefs that have ever been produced through human effort. Our goal is to maintain complete reef ecosystems and maintain them as Coral Arks to preserve biodiversity through the mass extinction crisis that global warming is now causing to the world's coral reefs (Hayes and Goreau 1991; Goreau et al. 2000). We are currently growing about 80% of all the genera of reef-building corals in the world, and around half of all the species. Our goal is to grow them all, along with all the other organisms that are part of healthy reef ecosystems.

Naturally, all organisms do not respond equally to the improved condition that Biorock technology provides them. Some organisms are observed to grow much faster than others and to become weeds, overgrowing each other. The effect is much like spreading a bag of fertilizer all over an empty lot and walking away. All species of plants benefit, but some weeds are much more efficient at taking advantage of high nutrient levels and overgrow and kill those that grow more slowly. Some will release toxins (allellochemicals) to attack, impair, and even kill their neighbors. Pests may sense the increased growth and flock to eat the plants. If we want roses and not pests, we will have to pull up the weeds, or, as Votaire's Candide says, "we must tend our gardens." Although many Biorock projects thrive for years with no management at all, we find that we can get much superior results by encouraging the beneficial species and removing the pests, so Biorock farming is much more like horticulture than like forestry.

Biorock mariculture therefore allows the creation of extraordinarily biodiverse ecosystems even in formerly barren habitat, allowing ecosystems to be kept alive under conditions that would otherwise kill them, and growing back complex habitats in a few years in places where coals had died and failed to recover (Goreau and Hilbertz 2007). Most of our projects have been built on barren sands or rocks and quickly created vibrant ecosystems full of corals, oysters, and fishes, which repopulate fisheries stocks in surrounding areas and are award-winning ecotourism attractions (Goreau and Hilbertz 2008).

This is a paradigm diametrically opposed to conventional mariculture, which generally grows a single species, usually a single clone, at high density. The lack of genetic diversity pollutes the genetic capacity for adaptation of surrounding wild populations when the mariculture stock escapes, as some inevitably do. The dense populations incubate disease epidemics, and when one dies of disease, they all die, because of crowding and lack of genetic diversity and resistance. They promote dense populations of parasites, which then infest surrounding wild populations. They are fundamentally feedlot operations, dependent on high-energy inputs and dense supplies of artificial foods, often provided from the destruction of remote ecosystems. The dense populations and food requirements cause intense pollution from excrement and rotting food, causing severe eutrophication of surrounding ecosystems. Such unsustainable mariculture produces expensive foods for export that local people cannot afford to eat, usually for the profit of local elites or rich foreign investors, while local subsistence fishermen see their fisheries stocks, and their food and income, destroyed by parasites, disease, pollution, and exclusion from their ancestral fishing grounds.

In contrast, Biorock mariculture is sustainable, extremely biodiverse, promotes species and genetic diversity, enriches all marine populations, grows its own food through complex food chains with no external food imported, and can be powered purely by nonpolluting energy provided by the sun, winds, waves, and ocean currents (Goreau 2010). Fishermen in areas surrounding Biorock projects are at first wary of them because they are afraid that areas will be withdrawn from fisheries, but they quickly become very strong supporters when they see for themselves the dramatic increase in fish populations on the projects spilling over into the surrounding fisheries, providing them with increased fish stocks and fish diversity. Fishing communities, once they see the results, soon come to us and ask us for many more projects.

Unfortunately, at present, there is no serious funding from any government, international agency, or big international NGO (BINGO) for serious habitat restoration. The overwhelming paradigm is marine-protected areas (MPAs), excluding fishermen from designated regions, with the claim that coral reef and fisheries will bounce back in a "resilient" way all by themselves. Yet, MPAs are full of dead and dying coral being killed by global warming that no MPA can possibly protect them from (Hayes and Goreau 1991). And if there is no habitat for fish populations, the fisheries will not recover no matter how many fishermen and their families starve. Without restoration of degraded habitats, fisheries cannot possibly recover (Goreau 2010).

At present, no government seriously invests in training and loans for subsistence coastal fishermen. They know that they are destroying their children's future by overharvesting, but their families need to eat today. They would gladly learn to be more productive and less destructive, but need training in new techniques and capital to invest in the materials needed. Sadly, governments and international agencies regard poor coastal fishermen as expendable, and ignore their needs while they use taxpayers' money for billions of dollars in perverse subsidies to open ocean fishing fleets that are stripping the oceans bare with fine drift nets, long lines, bottom trawlers, and sophisticated sonar methods and real-time satellite mapping that are finding, and killing, the last survivors. These perverse subsidies should be removed from destructive offshore fisheries and instead invested in helping coastal fishermen restore their fisheries habitat for sustainable management (Goreau 2010).

Biorock reefs show dramatic increases in fish populations that are immediately apparent to any observer. The data in Jompa et al., Chapter 5 and Arifin et al., Chapter 6 are the first data on this increase. Astonishingly, Biorock methods are capable of producing much greater fisheries stocks and production than even the richest natural reefs, and doing so in areas that are completely barren. As Biorock can be built in any size or shape, it is possible to produce many layers of habitat at one place. A natural reef usually has one layer of holes, and the populations of fishes, shellfish, lobsters, crabs, octopus, etc. are limited by the number of shelter holes of the right size and shape that they can find and defend. With Biorock, there is no limit to the number of layers and holes and shapes that can be produced. Every species has a different preference, so we get very different results with different shapes. We are learning what different species prefer when they show us what they want by moving in at high density. We have unintentionally produced extraordinary densities of spiny lobsters by accidentally building structures in the shapes that they prefer.

In addition, we can create habitat specifically for juvenile fishes and restock them with post-larval juvenile fish captured from the open ocean immediately after they metamorphose from larvae, which turns something like about 95% mortality to predators into something like 95% survival (about 19 times greater!) by combining Biorock juvenile fish habitat with Ecoean methods of postlarval capture and culture (Lecaillon, Chapter 16). Doing so will be the fastest possible way of restoring coastal fisheries, as it eliminates the three major barriers to fisheries restoration simultaneously: namely lack of recruitment, lack of habitat and hiding places, and lack of food. Furthermore, such structures are readily powered by untapped clean local energy sources, such as solar, wind, waves, and currents.

Besides creating complex fisheries habitat, Biorock can also be used for low-cost stabilization of barren rubble substrates, for example after hurricanes, ship groundings, or bomb fishing. Figure 19.14

shows such a reef grown on barren rubble and sand in a few years by use of fencing material. Some corals were transplanted onto it, but most spontaneously settled. An even cheaper method is to wire reefs with thin steel binding wire, and we have done so very quickly at low cost with very impressive results in Indonesia (Figure 19.15). The main disadvantage of such methods is that while it is cheaper, less fish habitat is produced per unit area, and corals are vulnerable to bottom-crawling predators like Crown of Thorns starfish (*Acanthaster—planci*), which climbs on corals, extrudes its stomach over the coral, and digests it completely (Goreau 1962; Goreau et al. 1972), coral-eating snails *Drupella* spp., (Turner 1994) and *Coralliophila* spp., and the coral-eating polychaete worm *Hermodice carunculata* (Ott and Lewis 1972), which should be removed as needed.

These methods are not limited to coastal fisheries but are easily extended to the open ocean by growing floating Biorock reefs to grow reef fish out in blue open water and greatly increase populations of pelagic fish, such as tunas and mahi-mahi, that are attracted to such habitat.

Figure 19.14 (a) Biorock meshes cover large areas cheaply and stabilize loose coral rubble, such as areas impacted by bomb blast and ship-grounding damage. Small loose live corals quickly grow through and over them and rapidly turn dead rubble fields back into reef. (Photograph by Thomas Sarkisian, Pemuteran, Bali, 2010.) (b) This Biorock mesh, placed on a barren rubble area, has been extensively covered with corals a few years later. (Photograph by Thomas Sarkisian, Pemuteran, Bali, 2010.)

Figure 19.15 (a) Reefs can be most cheaply and quickly restored by wiring them with cheap construction wire. This wire placed on dead rubble has had corals spontaneously settle and grow rapidly on it. (Photograph by Thomas Sarkisian, Pemuteran, Bali, 2010.) (b) Biorock wires only a millimeter or two in diameter have been buried under half a meter of rapidly growing coral in a couple of years. (Photograph by Thomas Sarkisian, Pemuteran, Bali, 2010.)

CELL MEMBRANE VOLTAGE GRADIENTS AND ENERGY METABOLISM

The original results with corals were thought to be primarily due to high pH generated on the cathode by the electrolysis of water (Hilbertz 1979; Goreau 2012). Reef-building corals and calcareous algae create high pH conditions within their tissues by the removal of CO_2 in photosynthesis and need to use metabolic energy to pump calcium ions to the site of calcification (Goreau and Bowen 1955; Goreau and Goreau 1959; Goreau 1961, 1963; Hayes and Goreau 1977) so it was assumed that by providing them with higher pH through the Biorock process, metabolic energy would be freed up for growth, and the dependence of skeleton formation on photosynthesis would be decoupled to some degree. If this were the primary mechanism, then only organisms with calcareous shells would benefit, but it was quickly noticed that non-calcareous organisms, such as tunicates, also showed extraordinary growth rates, so the mechanism must be more general.

The extraordinary biological benefits documented above for Biorock systems imply that very fundamental biophysical and biochemical mechanisms are being induced, leading to vastly improved metabolic and physiological health. The role of electrical and magnetic fields in biology has been riddled with charlatans, impostors, and self-deluded people, giving the entire field a bad name, and our work is constantly ridiculed as "electrocution," "electrical shock therapy," and "Frankenstein" technology. This is done by people confusing our extremely mild, low-voltage, low-amperage, direct-current "electro-tickle" trickle charging with those who have in the past promoted high voltages, high current densities, and alternating current, whose negative effects are all too well known, symbolized by the people who dry their hair standing in the bathtub, and die when the hair drier falls in the water and electrocutes them! But we get a charge out of our electrifying results because the conditions are so very different. We can hold onto the anode and the cathode simultaneously with our bare hands, shorting out the system, and feel nothing at all, because the current is passing through the much more conductive seawater rather than through our bodies.

We observe no organisms to be repelled by the electrical fields, and all of them seem to be attracted, as if they can sense the exceptional life-enhancing health benefits of Biorock and actively seek it out. Nor is there any surprise why this should be so. Low-voltage electrical fields provide the very foundational mechanism that all forms of life use to make biochemical energy to power their growth, reproduction, healing, and resistance to environmental stress, and we are greatly enhancing their ability to do so with life-enhancing Biorock technology.

All cells maintain an electrical voltage gradient of about one-tenth of a volt across the cell membrane that separates the inside of their cell and the outside world. The interior of the cell is more negative than the outside medium, so the gradient organisms are exposed to on Biorock structures are of the correct sign to enhance natural gradients. This voltage gradient causes negatively charged electrons and positively charged protons to flow across the gradient in opposite directions. All forms of life use a common set of enzymes that evolved billions of years ago when life first evolved, which are able to tap this biophysical voltage gradient to make biochemical energy metabolites that are the "energy currency" for all forms of life. The higher the concentrations of these metabolites, the healthier they are. However, organisms must spend a large part of that energy maintaining that voltage gradient across their membrane, which they do by using energy-consuming enzymes to pump protons and charged ions across the membrane in order to maintain those gradients.

The Biorock process provides organisms with this voltage gradient for free, at much less metabolic cost to the organism, leaving much more energy for growth. Of course, this gradient must be in the right range to be most beneficial. Consequently, the Biorock method provides the most profound and natural way of enhancing metabolic energy and health, and is so basic that it is not likely to be improved. Of course, every organism may respond a little differently because of variations in the efficiency with which they take advantage of these benefits, so the optimal conditions are likely to vary somewhat for different organisms. Needless to say, different organisms should have the same fundamental effects, but vary somewhat in their optimum ranges.

Organisms grown in electrical gradients have been shown to produce more ATP and NADP, the fundamental biochemical energy currency of all life, as well as higher protein synthesis (Cheng et al. 1982). Cheng's data, measured on rat skin, are plotted in the graphs (Figures 19.16 and 19.17). Strong biochemical benefits are found in a clear, broad, optimum range of electrical currents.

There are three further very important points, namely that (1) the results are due to the electrical field itself, not electrolysis per se, (2) that the results are instantaneous with no residual effect, and (3) that magnetic fields may also play a role.

Point (1) is most clearly shown by the work of Nitzsche (Chapter 8), who found that corals on Biorock structures—underneath them but not on the structures—and three meters away all grew at similar rates, and that these rates were higher than corals growing 10, 30, and 100 meters away. Nitzsche, in fact, found that corals near, but not on Biorock structures grew slightly faster than those on the structures themselves. That could be explained if there is a small inhibitory effect on growth from hydrogen bubbling, which can act to strip oxygen out of the water to some degree. The work currently under way by Viitaasari is mapping out these field effects in much more detail.

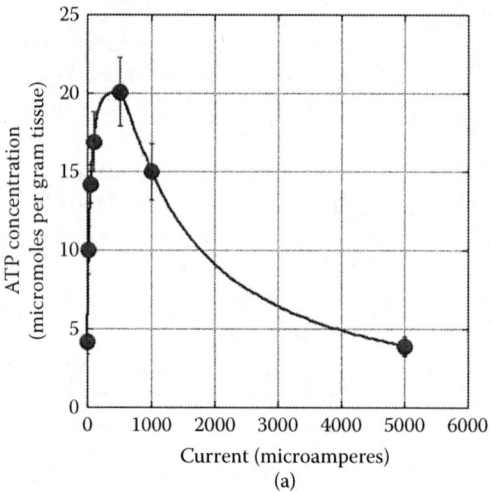

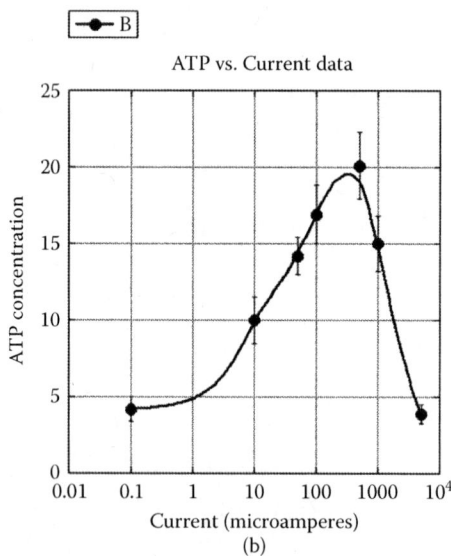

Figure 19.16 (a) Adenosine triphosphate (ATP), the biochemical that is the universal currency of energy in all forms of life, is increased up to five times by electrical currents in the right range. (b) When plotted on a semilogarithmic graph, ATP concentrations are proportional to the log of the current over a very wide range (from nearly one to a thousand microamperes), reach a peak, and then decline at high-current densities. (Graphs by T. Goreau; data from Cheng et al., *Clin. Orthop. Relat. Res.*, 171, 264, 1982.)

Point 2 is clearly demonstrated by growth-weight measurements of the hydrozoan coral *Millepora alcicornis*, grown in an extremely hot, polluted, and stagnant area (Beddoe 2007; Beddoe, Agard, and Phillip 2008, Beddoe et al. 2010). So severe were the conditions that the control corals steadily lost weight, something we had never seen before. In contrast, Biorock corals grew very rapidly, increasing in weight more than four times under the same conditions over 16 weeks, until the cable was cut. At that point, they immediately began losing weight parallel to the controls for 16 weeks; but when the power cable was repaired they immediately began growing rapidly again for 16 weeks. Unfortunately, the data analysis by Beddoe, Agard, and Phillip (2008) fitted a single curve to the entire data set, and failed to distinguish the importance of the separate period with the power off. We have fitted each segment separately (Figure 19.18). No residual effect

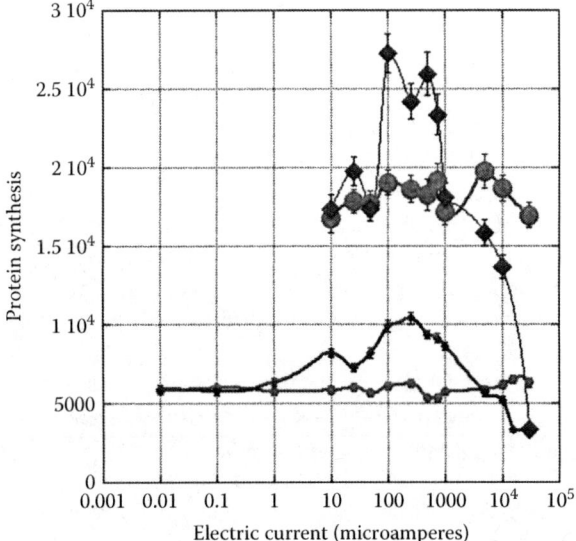

Figure 19.17 Protein synthesis rises in the same current range due to higher ATP energy availability. The lower pair of lines show glycine amino-acid uptake into protein synthesis, and the upper pair show aminoisobutyric acid uptake. In each pair, the relatively flat line is the control treatment with no electrical current, and the curve that reaches a peak around 100–1000 µA is the electrical treatment. The electrically treated tissue makes more protein over a broad optimal range, but at excessive currents the effects on protein synthesis turn negative. (Graph by T. Goreau; data from Cheng et al., *Clin. Orthop. Relat. Res.*, 171, 264, 1982.)

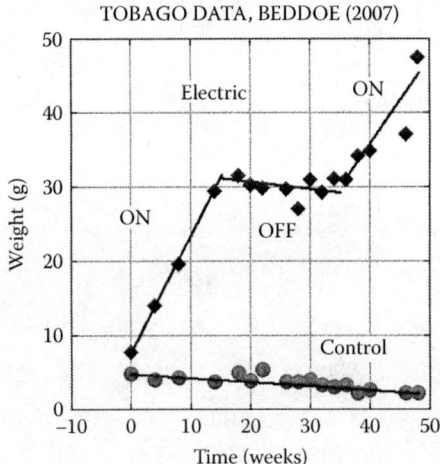

Figure 19.18 Coral was grown in an extremely poor-water-quality site. The controls with no electricity steadily lost weight due to adverse conditions. The Biorock corals increased in weight around four times in 16 weeks, then, when the power was turned off for 16 weeks, they lost weight like the controls and resumed rapid growth when the power was turned back on for 16 weeks. This shows that the effects of the current are instantaneous and not residual. (Graph by T. Goreau; data from Beddoe, *Investigating Mineral Accretion for Coral Reef Restoration*, MSc Thesis, University of the West Indies at St. Augustine, Trinidad, 2007.)

is seen in the data (average weights of 40 corals in each treatment). So, it is clear that the benefits are instantaneous and do not continue after the current is turned off. We have also seen this at projects that were cut off power before severe bleaching. Those projects off power lost all of their corals to bleaching like surrounding reef areas, but those under power had survival many times greater than surrounding reefs.

Point 3 is illustrated by the fact that just as greatly elevated settlement and growth is noticed in the surroundings of Biorock structures, a similar effect is also noticed around the insulated electrical cables delivering DC current to the structures. These cables, being insulated, have no electrical field around them, but a magnetic field is set up in the surrounding medium whenever an electrical current flows through the cable, according to Maxwell's equations. These results are profound, but no detailed studies have yet been done, although they will clearly be very rewarding. Figure 19.19a and b show the same area where cables were laid down on nearly barren bottom, and tremendous spontaneous settlement and growth that have taken place around them in ten years.

Certainly, the use of the wrong conditions also will not work as well, and training and the right materials, designs, and operating conditions are critical to success. We have taught hundreds of people around the world how to do this in Biorock training courses and projects. Hands-on experience of the craft and maintenance are essential for the best results. Many people incorrectly think that they can achieve the same results by copying our work without training. Improper materials, design, and incorrect operating conditions have been invariably used by untrained imitators of

(a)

(b)

Figure 19.19 In 2001, insulated electrical cables carrying low-voltage direct current were installed to power Biorock projects at Pemuteran, Bali. The area was essentially barren of live corals. (Photograph by Rani Morrow-Wuigk.) (b) The same area in 2011 shows prolific coral settlement and growth in the entire area around the cables. (Photograph by Rani Morrow-Wuigk.)

Biorock methods, and have inevitably led to inferior results. These people, who were not trained in Biorock methods, used conditions that were far suboptimal, and they blame their failures on the technology, not on their lack of mastery of it. Thanks to the academic "publish or perish" mentality, they publish their poor or failed results anyway (Schumacher and Schillak 1994; Van Treeck and Schumacher 1997, 1998, 1999; Schumacher et al. 2000; Schumacher 2002; Sabater and Yap 2002, 2004; Eggeling 2006; Borell 2008, Borell, Romatzki, and Ferse 2009). Almost all of them grossly overcharged in order to get fast results. In one case, in which the authors actually claimed that electrical fields reduced the coral growth rates (Borell 2008; Borell, Romatzki, and Ferse 2009) in total contradiction to all other results, the authors deliberately concealed what they knew, namely that the apparent lower growth rate of one species of coral was due to the fact that every single growing tip had been systematically bitten off as a breeding territory marker by a territorial terminal male parrotfish! We set up the experiment for them and personally saw the freshly broken fragments lying all over the bottom around it. One imitator, http://coralreefcreator.com/, is notorious for breaking off corals off living reefs and killing them through use of incorrect conditions.

CONCLUSION

Much more work is needed on the fundamental biophysics and biochemistry to determine the optimal electrical and magnetic field conditions for each species. When that is done, there is little doubt that a fundamental revolution in restoring health of organisms and entire ecosystems will result, with profound results for all areas of marine coastal management, and also for agriculture and medicine.

REFERENCES

Alvarez, O. M., P. M. Mertz, R. V. Smerbeck, and W. H. Eaglstein. 1983. The healing of superficial skin wounds is stimulated by external electrical current. *Journal of Investigative Dermatology* 81:144–148.

Arvedlund, M., and K. Kavanagh. 2009. The senses and environmental cues used by marine larvae of fish and decapod crustaceans to find tropical coastal ecosystems. 135–184 in I. Nagelkerken (Ed.). *Ecological Connectivity Among Tropical Coastal Ecosystems Part 2*. Springer Verlag, New York, doi 10.1007/978-90-481-2406-0_5.

Bassett, C. A. L. 1993. Beneficial effects of electromagnetic fields. *Journal of Cellular Biochemistry* 51:397–393.

Becker, R. O., and G. Selden. 1985. *The Body Electric: Electromagnetism and the Foundation of Life*. William Morrow, New York.

Beddoe, L. A. 2007. *Investigating Mineral Accretion for Coral Reef Restoration*. MSc Thesis. University of the West Indies, St. Augustine, Trinidad.

Beddoe, L., J. Agard, and D. T. Phillip. 2008. Electrical enhancement of coral growth in Tobago. Proceedings of the 11th International Coral Reef Symposium. Fort Lauderdale FL.

Beddoe, L., T. J. Goreau, J. B. R. Agard, M. George, and D. A. T. Phillip. 2010. Electrical enhancement of coral growth: A pilot study. 116–122 in A. Lawrence and H. P. Nelson (Eds.). Proceedings of the 1st Research Symposium on Biodiversity in Trinidad and Tobago, University of the West Indies.

Borell, E. M. 2008. Coral photophysiology in response to thermal stress, nutritional status, and seawater electrolysis. Thesis. Centre for Tropical Marine Ecology, University of Bremen, Bremen.

Borell, E. M., S. B. C. Romatzki, and C. A. Ferse. 2009. Differential physiological responses of two congeneric scleractinian corals to mineral accretion and an electrical field. *Coral Reefs*, doi 10.1007/s00338-009-0564-y.

Buster, N. A., C. W. Holmes, T. J. Goreau, and W. Hilbertz. 2006. Crystal habits of the magnesium hydroxide mineral brucite within coral skeletons. Proceeding of the American Geophysical Union Annual Meetings, Abstract and Poster.

Cheng, N., H. Van Hoof, E. Bockx, M. J. Hoogmartens, J. C. Mulier, F. J. De Ducker, W. M. Sansen, and W. De Loecker. 1982. The effects of electric currents on ATP generation, protein synthesis, and membrane transport in rat skin. *Clinical Orthopaedics and Related Research* 171:264–272.

Czech-Damal, N. U., A. Liebschner, L. Miersch, G. Klauer, F. D. Hanke, C. Marshall, G. Dehnhardt, and W. Hanke. 2011. Electroreception in the Guiana dolphin (*Sotalia guianensis*). Proceedings of the Royal Society of London B, doi 10.1098/rspb.2011.1127.

Dwija, P. N. 2003. Investigation of coral growth and recruitment diversity at Pemuteran Village, Gerokgak District, Buleleng Province. Thesis, Biology Faculty, Udayana University, Bali, Indonesia.

Eggeling, D. 2006. Electro-mineral accretion assisted coral growth: An aquarium environment. Report, Queensland Aquarium Townsville.

Falugi, C., M. Grattarola, and G. Prestipino. 1987. Effects of low-intensity pulsed electromagnetic fields on the early development of sea urchins. *Biophysical Journal* 51:999–1003.

Fisher, C., and M. Slater. 2010. 1–26 in Effects of electromagnetic fields on marine species: A literature review. Oregon Wave Energy Trust.

Fitzhardinge, R. C., and J. H. Bailey-Brock. 1989. Colonization of artificial reef materials by corals and other sessile organisms. *Bulletin of Marine Science* 44:567–579.

Fox, H. E., and J. Pet. 2001. Pilot study suggests viable options for reef restoration in Komodo National Park. *Coral Reefs* 20:219–220.

Fox, H. E., J. S. Pet, R. Dahuri, and R. L. Caldwell. 2001. Coral reef restoration after blast fishing in Indonesia. Proceedings of the 9th International Coral Reef Symposium. Bali, Indonesia. 2:969–976.

Fox, H. E., J. S. Pet, R. Dahuri, and R. L. Caldwell. 2003. Recovery in rubble fields: Long term impacts of blast fishing. *Marine Pollution Bulletin* 46:1024–1031.

Goreau, T. F. 1961. Problems of growth and calcium deposition in reef corals. *Endeavour* 20:32–39.

Goreau, T. F. 1962. On the predation of coral by the spiny starfish *Acanthaster planci* (L.) in the southern Red Sea. *Bulletin of the Sea Fisheries Research Station, Haifa* 35:1–5.

Goreau, T. F. 1963. Calcium carbonate deposition by coralline algae and corals in relation to their roles as reef builders. *Annals of the New York Academy of Sciences* 109:127–167.

Goreau, T. J. 2010. Coral reef and fisheries habitat restoration in the Coral Triangle: the key to sustainable reef management. 244–253. In J. Jompa, R. Basuki, Suraji, M. Tesoro, and E. T. Lestari (Eds.). Proceedings of the COREMAP Symposium on Coral Reef Management in the Coral Triangle. Invited Keynote Talk, World Ocean Congress, Manado, Sulawesi, Indonesia.

Goreau, T. J. 2012 (in press). Marine electrolysis for building materials and environmental restoration. In V. LInkov (Ed.). *Electrolysis*, InTech Publishing. Rijeka, Croatia.

Goreau, T. F., and V. T. Bowen. 1955. Calcium uptake by corals. *Science* 122:1188–1189.

Goreau, T. F., and N. I. Goreau. 1959. The physiology of skeleton formation in corals. II. Calcium deposition by hermatypic corals under various conditions in the reef. *Biological Bulletin* 117:239–250.

Goreau, T. J., and W. Hilbertz. 2005. Marine ecosystem restoration: Costs and benefits for coral reefs. *World Resource Review* 17:375–409.

Goreau, T. J., and W. Hilbertz. 2007. Reef restoration as a fisheries management tool. In P. Safran (Ed.). *Fisheries and Aquaculture*. Encyclopedia of Life Support Systems (EOLSS), UNESCO Eolss Publishers, Oxford. http://www.eolss.net.

Goreau, T. J., and W. Hilbertz. 2008. Bottom-up community-based coral reef and fisheries restoration in Indonesia, Panama, and Palau. 143–159, In R. France (Ed.). *Handbook of Regenerative Landscape Design*. CRC Press, Boca Raton.

Goreau, T. J., J. M. Cervino, and R. Pollina. 2004. Increased zooxanthellae numbers and mitotic index in electrically stimulated corals. *Symbiosis* 237:107–120.

Goreau, T. F., J. C. Lang, E. A. Graham, and P. D. Goreau. 1972. Structure and ecology of the Saipan reefs in relation to predation by *Acanthaster planci* (Linnaeus). *Bulletin of Marine Science* 22:113–152.

Goreau, T. J., W. Hilbertz, A. Azeez, and A. Hakeem. 2000. Increased coral and fish survival on mineral accretion reef structures in the Maldives after the 1998 bleaching event. Abstracts 9th International Coral Reef Symposium. Bali. 263.

Goreau, T. J., T. McClanahan, R. Hayes, and A. Strong. 2000. Conservation of coral reefs after the 1998 global bleaching event. *Conservation Biology* 14:5–15.

Goreau, T. J., and T. Sarkisian. 2010. Biorock® Corals bleach less, recover faster, and have higher survival after heat shock: Biorock Ark electrotherapy to save coral reefs from global warming. Global Coral Reef Alliance, http://www.globalcoral.org/BIOROCK%20CORALS%20BLEACH%20LESS,%20RECOVER%20FASTER,%20AND%20HAVE%20HIGHER%20SURVIVAL%20AFTER%20HEAT%20SHOCK.htm, accessed August 28, 2012.

Hayes, R. L., and N. I. Goreau. 1977. Intracellular crystal-bearing vesicles in the epidermis of scleractinian corals, *Astrangia danae* (Agassiz) and *Porites porites* (Pallas). *Biological Bulletin* 152:26–40.

Hayes, R. L., and T. J. Goreau. 1991. The tropical coral reef ecosystem as a harbinger of global warming. Proceedings of the 2nd International Conference on Global Warming. *World Resources Review* 3:306–322.

Hennige, S. J., D. J. Suggett, M. E. Warner, K. E. McDougall, and D. J. Smith. 2009. Photobiology of *Symbiodinium* revisited: Bio-physical and bio-optical signatures. *Coral Reefs* 28:179–195.

Hilbertz, W. H. 1979. Electrodeposition of minerals in sea water: Experiments and applications. *IEEE Journal on Oceanic Engineering* 4:1–19.

Kalmijn, A. J. 1971. The electric sense of sharks and rays. *Journal of Experimental Biology* 55:371–383.

Kalmijn, A. J. 1982. Electric and magnetic field detection in elasmobranch fishes. *Science* 218:916–918.

Levin, M. 2003. Bioelectromagnetics in morphogenesis. *Bioelectromagnetics* 24:295–315.

Loch, K., W. Loch, H. Schuhmacher, and W. R. See. 2002. Coral recruitment and regeneration on a Maldivian reef after the coral bleaching event of 1998. *Marine Ecology* 23:219–236.

McCaig, C. D., A. M. Rajnicek, B. Song, and M. Zhao. 2005. Controlling cell behavior electrically: Current views and future potential. *Physiological Reviews* 85:943–978.

McClanahan, T. R. 2000. Bleaching damage and recovery potential of Maldivian coral reefs. *Marine Pollution Bulletin* 40:587–597.

Mundy, C. N. 2000. An appraisal of methods used in coral recruitment studies. *Coral Reefs* 19:124–131.

Oschman, J. L. 2000. *Energy Medicine: The Scientific Basis*. Churchill Livingstone, Edinburgh.

Oschman, J. L. 2003. *Energy Medicine in Therapeutics and Human Performance*. Butterworth Heinemann, Oxford.

Ott, B., and J. B. Lewis. 1972. The importance of the gastropod *Coralliophila abbreviata* (Lamarck) and the polychaete *Hermodice carunculata* (Pallas) as coral reef predators. *Canadian Journal of Zoology* 50:1651–1656.

Porter, J. W., L. Muscatine, Z. Dubinsky, and P. G. Falkowski. 1984. Primary production and photoadaptation in light- and shade-adapted colonies of the symbiotic coral *Stylophora pistillata*. *Proceedings of the Royal Society of London B* 222:161–180.

Rinkevich, B. 2005. Conservation of coral reefs through active restoration measures: Recent approaches and last decade progress. *Environmental Science and Technology* 39:4333–4342.

Rinkevich, B. 2008. Management of coral reefs: We have gone wrong when neglecting active reef restoration. *Marine Pollution Bulletin* 56:1821–1824.

Sabater, M. G., and H. T. Yap. 2002. Growth and survival of coral transplants with and without electrochemical deposition of CaCO3. *Journal of Experimental Marine Biology and Ecology* 272:131–146.

Sabater, M. G., and H. T. Yap. 2004. Long term effects of mineral accretion on growth, survival, and corallite properties *Porites cylindrica* Dana. *Journal of Experimental Marine Biology and Ecology* 311:355–374.

Salinas de Leon, P., A. Costales-Carrera, S. Zeljkovic, D. J. Smith, and J. J. Bell. 2011. Scleractinian settlement patterns to natural cleared reef substrata and artificial settlement panels on an Indonesian coral reef. *Estuarine, Coastal, and Shelf Science* doi: 10.1016/j.ecss.2011.02.016.

Schumacher, H. 2002. Use of artificial reefs with special reference to the rehabilitation of coral reefs. *Bonner Zoologische Monographien* 50:81–108.

Schumacher, H., and L. Schillak. 1994. Integrated electrochemical and biogenic deposition of hard material—a nature-like colonization substrate. *Bulletin of Marine Science* 55:672–679.

Schumacher, H., P. Van Treeck, M. Eisinger, and M. Paster. 2000. Transplantation of coral fragments from ship groundings on electro-chemically formed reef structures. Proceedings of the 9th International Coral Reef Symposium. Bali, Indonesia. 2:23–27.

Shaish, L., G. Levy, G. Katzir, and B. Rinkevich. 2010. Coral reef restoration (Bolinao, Philippines) in the face of frequent natural catastrophes. *Restoration Ecology* 18:285–299.

Song, B., M. Zhao, J. V. Forrester, and C. D. McCaig. 2002. Electrical cues regulate the orientation and frequency of cell division and the rate of wound healing in vivo. *Proceedings of the National Academy of Sciences* 99:13577–13582.

Tomascik, T. 1991. Settlement patterns of Caribbean scleractinian corals on artificial substrata along a eutrophication gradient in Barbados, West Indies. *Marine Ecology Progress Series* 77:261–269.

Turner, S. J. 1994. The biology and population outbreaks of the corallivorous gastropod *Drupella* on Indo-Pacific reefs. *Oceanography and Marine Biology Annual Reviews* 32:461–530.

Turner, J., R. Klaus, and U. Engelhardt. 2000. The Reefs of the Seychelles granitic islands, Coral Reef Degradation in the Indian Ocean. http://www.oceandocs.org/bitstream/1834/466/1/CORDIO8.pdf

Van Treeck, P., and H. Schumacher. 1997. Initial survival of coral nubbins transplanted by a new coral transplantation technology—options for reef rehabilitation. *Marine Ecology Progress Series* 150:287–292.

Van Treeck, P., and H. Schumacher. 1998. Mass diving tourism—a new dimension calls for new management approaches. *Marine Pollution Bulletin* 37:499–504.

Van Treeck, P., and H. Schumacher. 1999. Artificial reefs created by electrolysis and coral transplantation: An approach ensuring the compatibility of environmental protection and diving tourism. *Estuarine, Coastal, and Shelf Science* 49:75–81.

Wethey, D. S., and J. W. Porter. 1976. Sun and shade differences in productivity of reef corals. *Nature* 262:281–282.

Wueringer, B. E., L. Squire, S. M. Kajiura, N. S. Hart, and S. P. Collin. 2012. The function of the sawfish's saw. *Current Biology* 22:R150–R151.

Zhao, M., B. Song, J. Pu, T. Wada, B. Reid, G. Tai, F. Wang, et al. 2006. Electrical signals control wound healing through phosphatidylinositol-3-OH kinase-γ and PTEN. *Nature* 442:457–460.

Index

A

Acropora cervicornis, 39, 41, 269
Acropora cytherea, 85
Acropora formosa, 56
    average growth rate of, 71–72
    growth on biorock substrate, 70
    growth rate of coral, 63–65
Acropora microphthalma, 94
    growth rate of, 97, 98
Acropora palmata, 41
Acropora pulchra, 126, 127
Acropora suharsonoi, 270
Acropora tenuis, 85
Acropora, unbleached on biorock structure, 278
Acropora valenciennesi, 56
Acropora yongei, 126, 127
Acroporid corals, 48
Active restoration, 6, 10
Adenosine triphosphate (ATP), 125–126, 285
    concentrations, 284
    in skin-tissue samples, 114
Agar
    contents of *Gracilaria,* 210–211
    extractions and determinations, 201–202
    production, 190–192
Agaricia agaricites, spontaneous settlement of, 264–265
Age treatments, of pearl oyster, 132, 135
Algae, 42
    sand-producing, 264
American oyster, 141, 151, 152
Ammonium uptake, 228, 230, 237–238
Amphibious reef, 259–260
Amphipods, grazing by, 203, 234
Analysis of variance (ANOVA), 118, 202, 221
    Gracilaria crassissima growth rates, 223
    Gracilaria domingensis growth rates, 219–220
    one-way, 120
    of survival rate, 135
Animals maintenance, 106–107
Anode materials, 37
ANOVA, *see* Analysis of variance
Antierosion reefs, 63
Aquaculture, 184
Aquaria cultures, 225–226
Aquarium fish farming, 183–184
Aquatic environment parameters, 135
Aragonite substrate (AS) treatment, coral response to, 120–124
Artemia salina, 116
Artificial oyster nursery, benefits of, 159
Artificial reefs, 35, 36, 82, 153, 154
    adhering oysters to, 143
    method, 132
    species on, 42–43
    structure, 38
    vs. natural reefs, 41
ATP, *see* Adenosine triphosphate

B

Back-reef (BR1)
    Gracilaria crassissima cultures, 223
    Gracilaria domingensis cultures, 219
    Gracilaria terete cultures, 203, 213, 218–219
Barrang Lompo Island
    coral reefs in, 49
    experiment
        coral growth rates, 52–53
        coral survival rates, 52
Barrier reefs, crest nets setting up on, 181
BAW, *see* Biorock Anti-Wave
Big international nongovernmental organizations (BINGOs), 6
BioRestore©, 185
BioRestore®, 185–186
Biorock
    design and setting, 106
    electrical field, 267, 269
    electricity with and without, 133
    experiment effects
        on cell growth, division, budding, and branching, 272–274
        on growth, 269–270
        on healing, 270–272
        on resistance to stress, 274–278
        on settlement, 264–269
    gorgonian mortality, 108
    low-voltage method, 137
    mesh, 164, 165
    method, system and principle of, 61–63
    oyster restoration, 153–154
    superior marine construction material, 22–25
    vicinity of transplants, 85
Biorock Anti-Wave (BAW), 27, 29, 30
Biorock corals, 3, 84, 271, 272
    bleaching, 274–275
    weight loss, 284, 285
Biorock reefs, 3, 25, 33, 43, 280–281
    biorock electrical fields, 267
    restoration, 60–63
Biorock shore protection
    in Indonesia, 27–33
    in Maldives, 25–27
Biorock structures, 62, 65, 66
    growth of *Clathria* sponges on, 75–76
    growth rate of corals on, 48
Biorock technology, 8, 9, 62
    mechanism of, 143
    oyster growth study using, 142
    oyster reef formation, 143

BR1, *see* Back-reef
Brucite, 263

C

Cage cultures, 197
Calcareous organisms, 43
CARE, *see* Collect by Artificial Reef Eco-Friendly
Caribbean reefs, 233
Carlo Erba 1106 Elemental Analyzer, 202
Cathode materials, 37
Cell division, effects on, 272–274
Cell growth, effects on, 272–274
Cell membrane, voltage gradients, 282–287
Ceramium, 203, 206, 209, 215
    biomass of, 233
    fouling of cultures by, 220–221
Chi-square (χ^2) tests, 118, 119, 120
Clathria sponge
    growth on biorock structure, 75–76
    survival rate of, 66–68
Coastal fish, 180
Coastal management in Louisiana, 9
Coastal wetlands, 177
Coastline management, 257–259
Cold-water coral reefs, 113–114
    experimental design, 115
    feeding and monitoring, 116–117
    growth measurements and statistics, 118
    mineral accretions, 119–120
    response to treatments, 120–124
    tanks and substrates, 116
Collect by Artificial Reef Eco-Friendly (CARE), 182–183
Conchiolin, 132
Control corals, 84
Controlled grazing experiments, 198–199
Conventional methods, 279
Coral Arks, 279
Coral-attaching devices, 95
Coral branch, in environmental parameters, 83, 84–85
Coral exoskeleton, 132
Coral fishes
    density of, 69–70
    observation of, 68–69
    of *Pomacentridae* and *Labridae*, 78–79
Coral fragments, 83
    growth rate and branch number of, 85–86
Coral growth, 96, 99–100
    in environmental parameters, 83, 84–85
    limiting factors of, 65–66
    rates, 52–53
Coral limestone plates, settlement rates on, 265–266
Coral-reef fishery techniques, 179
Coral-reef restoration projects, 92, 142
Coral reefs, 12, 20, 21, 35
    on barren rubble and sand, 281
    biorock wires, 282
    cell growth, division, budding, and branching, 272–274

    cold-water, *see* Cold-water coral reefs
    color and fluorescence, 274
    destruction, damage, and degradation of, 278–279
    deterioration in Indonesia, 81–82
    effects on healing, 270–272
    and fishery, 280
    growth rates of, 269–270
    in Indonesia, 48
    resistance to stress, 274–278
    settlement of, 264–269
Coral survival rates, 52
Coral transplanting, 52
Crassotrea virginica, 155
    growth rate and survival of, 158
Crest nets setting up on barrier reefs, 181
Crown of Thorns starfish, 281
Ctenophore, 259
C treatment, coral response to, 121, 123
Culture sites
    growth, epiphyte, and grazing differences
        Gracilaria crassissima, 216–217
        Gracilaria domingensis, 214–216
        Gracilaria terete, 213–214
    salinity/nutrient relationships of, 211–213
Cybertecture, 1

D

Density stratification, 211
Depletion curves for cultures, 227, 229
Depth of culture, *Gracilaria*, 234–235
Diatom fouling of *Gracilaria domingensis*, 215–216
Diploria strigosa, 40
Direct-current electrical fields, 270–271, 272
Discovery Bay, 231
Dome reef, 257, 258

E

East Discovery Bay, 225–226
Eastern oyster, *see* American oyster
East River, 154
EFH, *see* Essential Fish Habitat
Eggs of marine fish, 180
Electrical field gradient
    biorock, 269, 283
    marine algae larvae using, 266
Electrical stimulation
    2010 growth rate with and without, 155
    2011 growth rate with and without, 156
Electrical treatment, pearl oyster, 132, 135
Electricity on sponges, negative impact of, 100
Electrified corals, 53–55
Electrifying structures, sponge extension increase on, 100–102
Electrodeposition, 114
    of minerals in seawater, 153
Electrolysis, 263
    corals transplantation using, 52
    of seawater, 36, 57

INDEX

Electrotherapy, 278–282
Electro-tropism
    marine algae larvae using, 266
    used by juvenile fishes, 268
Enchanted Star Sand, 259–260
Energy metabolism, 282–287
Enteromorpha, 209, 215
Environmental quality parameters, 137
Epiphytes
    accumulation, 213
    fouling of *Gracilaria* cultures in, 233
    observations and quantification, 199
Essential Fish Habitat (EFH), 8, 9
Estuarine aquarium system, 142
Eutrophication, causes of, 152

F

Favia fragum, spontaneous settlement of, 264–265
Fishery
    coral reef and, 280
    management, 153
Fishes in seagrass growth, 166
Fishing techniques, 179
Fish Structure 3, 94–97
Floating island, 257–259
Floating reef, 255–256
Flow-through culture
    Gracilaria domingensis, 227–231
    physical parameters and nutrient levels, 200

G

Galvanic elements (GEs) treatment, coral response to, 120–124
Galveston, 13
    shore erosion, 14–20
Genetic diversity, 185
Gili Trawangan
    coral reefs in, 49
    experiments
        coral growth in, 55
        fish populations in, 55–57
    laboratory for biorock reef restoration, 60–61
    methods of biorock in, 63
Gill reef, oyster, 253–254
Giovinazzo, seagrasses restoration in, 162, 163–165
Global sea-level rise, 12–13
Global shore-erosion crisis, 11–12
Global warming, 12–13
Gorgonian soft corals, 105, 106
    growth and survival of, 108–109
    physical parameters, 107
Gracilaria
    agar
        contents of, 210–211
        production, 190–192
    collection sites, 195
    cultures in back-reef habitats, 232–235
    field sampling of, 195
    growth and productivity of, 231
    in Jamaican nitrate-enriched back-reef habitats, 193–194
    natural populations and culture sites, 196
    net culture, 197
    physical and chemical parameters, 201
    potential in tropical countries, 192–193
    rope culture, 196–197
    in situ culture, 195–196
    statistical analysis, 202
Gracilaria blodgettii, weight increase and growth rates of, 207–208
Gracilaria cervicornis, weight increase and growth rates of, 208, 209
Gracilaria crassissima
    growth, epiphyte, and grazing differences, 216–217
    growth rates of, 208–209
    mean growth rates of back reef (BR1), 223
    nutrient enrichment experiments, 222–224
    phosphate uptake rates, 227
    rope cultures in Discovery Bay, 231
Gracilaria damaecornis, initial growth rate of, 209–210
Gracilaria domingensis
    batch cultures of, 232
    cultured and natural populations of, 224–225
    diatom fouling of, 215–216
    flow-through culture, 227–231
    growth, epiphyte and grazing differences, 214–216
    initial density and corresponding growth rates of, 207–208
    mean growth rates of, 215
    natural populations, 235–239
    nutrient enrichment experiments
        flow-through culture, 222
        growth rates of back-reef (BR1), 219–220
        growth rates of lagoon (L2), 221–222
    phosphate uptake
        kinetics, 238
        rate of, 227
    weight increase and growth rate of, 206–207
Gracilaria foliifera, 210, 232
Gracilaria mammillaris, 210
Gracilaria sjoestedtii, 210
Gracilaria terete
    batch cultures of, 232
    depth of culture, 206
    growth, epiphyte and grazing differences, 213–214
    mean growth rates of, 213–214
    nutrient enrichment experiments, 217
    phosphate uptake rate, 227
    rope culture
        in bamboo raft, 197
        in Discovery Bay, 231
    strains, 203
    weight increase and growth rates of, 202–203
Gracilaria tikvahiae
    in Florida Keys, 231
    phosphate uptake rates of, 238–239

Grazing
    by amphipods, 203
    observations and quantification of, 199
Groundwater-enriched seawater culture, 225

H

Habitat replacement strategy, 153
Hard corals
    growth rates of, 269
    soft corals surround, 266, 267
Hard shore protection, 13
Harlem River, 154
Helix shape for maximum oyster settlement, 251
Herbivory, *Gracilaria* cultures, 233–234
Hippuristanol, 105
Hoa nets, setting up between small islands on reef ridges, 181–182
Horizontal rope cultures, 208, 209, 213
    Gracilaria terete, 203
Hudson-Raritan Estuary (HRE), 141
Hudson River, 154
Hypnea macroalgal community, 234

I

Ihuru island, 26, 27
Index of species dominance, 70
Indonesia
    biorock shore protection in, 27–33
    coral reefs deterioration in, 81–82
In situ biological observations, 83, 84
In situ culture, 226
    BR1, 224–225
    categories of epiphytes, 199
    characteristic growth rates of, 204–205
    growth rates of, 202
    irradiance of flow-through and, 201
    methods, 196
    pulse treatments, *Gracilaria* species
        nitrate uptake rates, 226
        phosphate uptake rates, 227
    sites, 195–196
In situ Gracilaria cultures, nutrient limitation of
    growth rates, 235–236
    nitrate and phosphate uptake, 237–239
    nutrient status, 236–237
Intense diatom fouling, 233
Intergovernmental Panel on Climate Change, 152
Intertidal marine plants, 171
Invertebrates, 42
Isis hippuris, 105–106, 108–110

J

Juvenile corals, settlement of, 264–266
Juvenile fishes
    electro-tropism used by, 268
    habitat for, 280–281
Juvenile oysters, 152, 153, 248

K

Karang Lestari Coral Reef Restoration project, 93
Keystone species, 141, 151

L

Labridae coral fishes, density of, 78–79
Large-scale restoration, 5–7
Life cycle
    of coastal marine species, 187
    of postlarvae collection, 180–181
Light trap, 182
LIII treatment, coral response to, 120–124
LII treatment, coral response to, 120–124
Limestone, 39, 44, 61
    groundwater, 193
Linear regression analysis, nitrate uptake, 227, 228
LI treatment, coral response to, 120–124
Lophelia pertusa, 113, 114
Louisiana, ecosystem restoration, 7–9
Low-voltage electrical fields, biorock, 283, 286
Low-voltage electrical treatment, *Pinctada maxima,* 132, 133, 137, 138

M

Macroalgae, 232
    nitrate uptake by, 237–238
    species of, 57
Madam Jelly Fish model, 257
Magnesium hydroxide deposits, organisms on, 166
Maldives, biorock shore protection in, 25–27
Man and Biosphere program (MAB), 186
Marine fishery, 179
Marine intertidal wetlands, 170
Marine-protected areas (MPAs), 5, 6, 184, 185, 280
McNeill Park, 170
Mechanical harvesting, destroyed oyster beds, 152
MET, *see* Microcurrent electrical therapy
Michaelis–Menten equations, 237
Microcurrent electrical therapy (MET), 114
Millepora alcicornis, 284
Mineral accretion, 3, 44, 132, 252, 257
    cementation by, 37
    method, 143
        to biorock structure, 148–149
        on cold-water coral, 114, 119–120, 124
    process, 246
    solar power for, 255
    structure, 41, 43
Mineral deposition, 85
    formation of, 61
Minerals, growth rates of, 52, 163
Miracle Gro® fertilizer, 173
Modified persulfate digestion method, 202
Montipora digitata, growth rate of, 63–65, 72–74
MPAs, *see* Marine-protected areas
Mussels, spontaneous recruitment of, 266, 269

INDEX

N

Natural reefs, artificial reefs *vs.*, 41
New York City's Green Infrastructure Program, 151
Nitrate concentration of *Gracilaria* cultures, 232
Nitrate-enriched back-reef habitats, *Gracilaria* in, 193–194, 231
Nitrate–salinity relationships, 211
Nitrate uptake
    of *Gracilaria domingensis*, 228, 230
    of *Gracilaria* species, 226
    linear regression analysis, 227, 228
    by macroalgae, 237–238
Nutrient depletion measurements during pulse treatments, 227
Nutrient enrichment experiments, 198–199
    Gracilaria domingensis
        flow-through culture, 222
        growth rates of lagoon (L2), 221–222
    weather observations, 199
Nutrient uptake experiments, 200–201

O

Ocean acidification, 142, 152
Oceans crisis, reversed through large-scale restoration, 5–7
Open-ocean fishing techniques, 179
Otranto electroduct, 163, 165–166
Overfishing, oyster decline, 152
Oyster gill reef, 253–254
Oyster Growth Study, 142, 143
Oyster larvae, 151–152
Oyster reefs, 7, 8, 20, 22, 141
    accretion, 148–149
    biology, 151–152
    decline, causes and consequences of, 152
    formation, 142, 143
    growth comparison, 146–147
    growth rate, 269
        in 2007 and 2008, 145
        with and without electrical stimulation, 155, 156
    growth study using biorock technology, 142
    layout and numbering system, 144
    materials and methods, 154–155
    mortality, 147–148
    restoration
        in New York city, 246–250
        project, 250–253
    restoration strategies, 153
    spontaneous settlement of, 268
    winter 2010–2011 survival data, 156–158

P

Pale pink sponge, 100, 101
    extension of, 93
PCC, *see* Postlarval capture and culture
Pearl oyster, 132
    Pinctada maxima, *see Pinctada maxima*
Pearl oyster juveniles of *Pinctada maxima*, 132
Pearson Correlations testing, 118, 120, 124
Pemuteran experiments, coral growth rates, 54–55
Pemuteran village, coral reefs in, 49
Phosphate concentration of *Gracilaria* cultures, 232–233
Phosphate-enriched cultures, 220–221
Phosphate–salinity relationships, 211
Phosphate uptake
    depletion during groundwater enrichment, 227, 228
    of *Gracilaria* species, 227, 237–239
Pinctada maxima, 131
    electricity with and without accretion, 135–137, 138
    experimental design, 132, 135
    juvenile stocking, 133
    low-voltage electrical treatment, *see* Low-voltage electrical treatment
    shell length and weight measurements, 134
Pocillopora damicornis, 127
Pollution effects in *Spartina* salt marshes, 171–173
Polyps on biorock, 273
Pomacentridae coral fishes, density of, 78–79
Porites astreoides, 40
Porites cylindrica, 125
Porites divaricata, 41, 275
Porites porites, 269
    fragments of, 39
Posidonia oceanica beds, 162
Postlarvae collection, 180
    life cycle and nonimpact of, 180–181
Postlarvae fishing techniques
    CARE, 182–183
    crest nets set up on barrier reefs, 181
    market opportunities, 183
Postlarval capture and culture (PCC), 180, 187
    as provider of eco-jobs, 186
    technology, 183
Power in Torre Guaceto, 165
Pulsed groundwater enrichment, 228, 232

R

Radiolarian reef, 257, 258
Raft reef, floating, 255
RCRA, *see* Resource Conservation and Recovery Act
Reef ridges, *hoa* nets setting up between small islands on, 181–182
Repletion strategy, oyster shells, 153
Resource Conservation and Recovery Act (RCRA), 170
Restocking, 184–185
    as mitigation for maritime impact, 185–186
Restoration of seagrasses, *see* Seagrass restoration
Rope culture, 196–197

S

Salinity for *Gracilaria* culture, 232
Salt marshes, 7, 8, 169
    site history of, 170
    Spartina, *see Spartina* salt marshes

Samalona Island
    coral reefs in, 49
    experiments, coral growth rate, 53–54
Seagrass growth rates, 270
Seagrass restoration, 161
    in Giovinazzo, 163–165
    materials and methods, 162–163
    in Otranto Electroduct, 165–166
    in Torre Guaceto, 165
Seascaping, 3
Seribu Islands, 82
Shallower station meshes, 163
Simpson index, *see* Index of species dominance
Small islands on reef ridges, *hoa* nets setting up between, 181–182
Soft corals
    broken and injured, 272
    gorgonian, 105, 106, 109–110
        growth and survival of, 108–109
        physical parameters, 107
    surround hard corals, 266, 267
Soft shore protection, 20–22
Solar photovoltaic panels, 154
Solar-powered reef, 256, 257
Solar radiation
    intensity of, 66
    for photosynthesis, 71
Solid breakwaters, 13
Spartina salt marshes
    ecology and physiology, 170–171
    experiment methods, 173–177
    pollution effects in, 171–173
Spat, *see* Juvenile oysters
Species association with biorock at Torre Guaceto, 166
Sponge *Clathria*
    growth on biorock structure, 75–76
    survival rate of, 66–68
Sponge extension
    and distance to electrical source, 97–98
    first and second measurement period, 99
    reasons for, 100–102
Sponge-feeding fish, 100
Sponge-infested sample corals, 96–97
Sponges
    on biorock structures, 266
    impact of electricity on, 100
    survival rate of transplanting, 77–78
Spring phytoplankton bloom, 152
Star Sand, 259
Statistical significance, 84
Steel frame for coral, 49
Structure 22, 94, 95, 98

Superfund toxic waste site, 154
Survival rates, 83–84
    of corals, 86–88
    of *Pinctada maxima,* 135, 136
Sustainable reef design, principles of, 245–246
Symbiodinium, 105, 110, 271, 274

T

Temperature for *Gracilaria* culture, 232
The River Project (TRP), 142–143
Tidal Wetlands Act, 170
Tissue analysis for C, N and P contents, 202
Torre Guaceto, 162, 163
    power, 165
    species associated with biorock at, 166
Transplanting corals, 39, 41, 52
    survival rate of, 66
Transplanting fragments, 86, 88
Transplanting sponge fragments, basal growth rate of, 77
Tunicates, growth rates of, 264, 269, 273, 282

U

Unenriched seawater culture, 225
Unfertilized *in situ Gracilaria* cultures, 204–205
US-based National Fish and Wildlife Foundation, 184

V

Vertical rope cultures, 198, 208
    Gracilaria blodgettii growth rates, 208
    Gracilaria domingensis, 206
    Gracilaria terete growth rates, 203
    moderate growth rates, 207
Vexar cage culture, 197, 208, 210
    moderate growth rates, 207
Visual census technique, 69

W

Water quality, 151, 153
Weather conditions variability of *Gracilaria* growth rate, 235
Wetland restoration experimental site, 171

X

X-ray diffractogram petrographic analysis, 164

Z

Zero day oyster measurements, 156

DATE DUE